Fundamentals of Wavelets

Fundamentals of Wavelets

Theory, Algorithms, and Applications

JAIDEVA C. GOSWAMI
ANDREW K. CHAN
Texas A&M University

A WILEY-INTERSCIENCE PUBLICATION
JOHN WILEY & SONS, INC.
New York / Chichester / Weinheim / Brisbane / Singapore / Toronto

This book is printed on acid-free paper. ♾

Published simultaneously in Canada.

Library of Congress Cataloging-in-Publication Data:

Goswami, Jaideva C.
Fundamentals of wavelets : theory, algorithms, and applications / Jaideva C. Goswami, Andrew K. Chan.
p. cm. — (Wiley series in microwave and optical engineering)
Includes index.
ISBN 0-471-19748-3 (cloth)
1. Signal processing—Mathematics. 2. Wavelets (Mathematics) 3. Image processing—Mathematics. 4. Electromagnetic waves—Scattering—Mathematical models. 5. Boundary value problems. I. Chan, Andrew K. II. Title. III. Series.
TK5102.9.G67 1999
621.3′01′5152433—dc21 98-26348

Printed in the United States of America

10 9 8 7 6 5 4 3 2 1

To

Shrimati Sati and Shri Chandra Nath Goswami
—Jaideva C. Goswami

My Lord Jesus Christ from whom I received wisdom and knowledge
and
my wife Sophia, for her support and encouragement
—Andrew K. Chan

Contents

Preface xv

1 What This Book Is About 1

2 Mathematical Preliminaries 5

2.1 Linear Spaces 5

2.2 Vectors and Vector Spaces 7

2.3 Basis Functions 9

2.3.1 Orthogonality and Biorthogonality 10

2.4 Local Basis and Riesz Basis 13

2.5 Discrete Linear Normed Space 15

2.6 Approximation by Orthogonal Projection 16

2.7 Matrix Algebra and Linear Transformation 18

2.7.1 Elements of Matrix Algebra 18

2.7.2 Eigenmatrix 19

2.7.3 Linear Transformation 20

2.7.4 Change of Basis 21

2.7.5 Hermitian Matrix, Unitary Matrix, and Orthogonal Transformation 22

2.8 Digital Signals 23

2.8.1 Sampling of Signal 23

2.8.2 Linear Shift-Invariant Systems 24

2.8.3 Convolution 24

2.8.4 z-Transform 25
2.8.5 Region of Convergence 26
2.8.6 Inverse z-Transform 28

2.9 Exercises 29

References 30

3 Fourier Analysis **31**

3.1 Fourier Series 31

3.2 Examples 32
3.2.1 Rectified Sine Wave 32
3.2.2 Comb Function and the Fourier Series Kernel $K_N(t)$ 33

3.3 Fourier Transform 35

3.4 Properties of the Fourier Transform 37
3.4.1 Linearity 37
3.4.2 Time Shifting and Time Scaling 38
3.4.3 Frequency Shifting and Frequency Scaling 38
3.4.4 Moments 38
3.4.5 Convolution 39
3.4.6 Parseval's Theorem 40

3.5 Examples of the Fourier Transform 40
3.5.1 Rectangular Pulse 41
3.5.2 Triangular Pulse 41
3.5.3 Gaussian Function 42

3.6 Poisson's Sum 43
3.6.1 Partition of Unity 45

3.7 Sampling Theorem 46

3.8 Partial Sum and the Gibbs Phenomenon 49

3.9 Fourier Analysis of Discrete-Time Signals 50
3.9.1 Discrete Fourier Basis and Discrete Fourier Series 50
3.9.2 Discrete-Time Fourier Transform 52

3.10 Discrete Fourier Transform 54

3.11 Exercises 55

References 56

4 Time–Frequency Analysis **57**

4.1 Window Function 58

4.2 Short-Time Fourier Transform 60

4.2.1 Inversion Formula 61

4.2.2 Gabor Transform 61

4.2.3 Time–Frequency Window 62

4.2.4 Properties of STFT 62

4.3 Discrete Short-Time Fourier Transform 64

4.3.1 Examples of STFT 64

4.4 Discrete Gabor Representation 65

4.5 Continuous Wavelet Transform 67

4.5.1 Inverse Wavelet Transform 69

4.5.2 Time–Frequency Window 70

4.6 Discrete Wavelet Transform 72

4.7 Wavelet Series 73

4.8 Interpretations of the Time–Frequency Plot 74

4.9 Wigner–Ville Distribution 76

4.10 Properties of the Wigner–Ville Distribution 80

4.10.1 A Real Quantity 80

4.10.2 Marginal Properties 80

4.10.3 Correlation Function 81

4.11 Quadratic Superposition Principle 81

4.12 Ambiguity Function 83

4.13 Exercises 84

4.14 Computer Programs 85

4.14.1 Short-Time Fourier Transform 85

4.14.2 Wigner–Ville Distribution 86

References 88

5 Multiresolution Analysis **89**

5.1 Multiresolution Spaces 89

5.2 Orthogonal, Biorthogonal, and Semiorthogonal Decomposition 92

5.3 Two-Scale Relations 96

5.4 Decomposition Relation 97

5.5 Spline Functions 98
5.5.1 Properties of Splines 102
5.6 Mapping a Function into MRA Space 103
5.7 Exercises 104
5.8 Computer Programs 106
5.8.1 B-Splines 106
References 107

6 Construction of Wavelets 108

6.1 Necessary Ingredients for Wavelet Construction 109
6.1.1 Relationship Between Two-Scale Sequences 109
6.1.2 Relationship Between Reconstruction and Decomposition Sequences 110
6.2 Construction of Semiorthogonal Spline Wavelets 112
6.2.1 Expression for $\{g_0[k]\}$ 113
6.3 Construction of Orthonormal Wavelets 114
6.4 Orthonormal Scaling Functions 118
6.4.1 Shannon Scaling Function 118
6.4.2 Meyer Scaling Function 119
6.4.3 Battle–Lemarié Scaling Function 123
6.4.4 Daubechies Scaling Function 125
6.5 Construction of Biorthogonal Wavelets 129
6.6 Graphical Display of Wavelets 132
6.6.1 Iteration Method 132
6.6.2 Spectral Method 132
6.6.3 Eigenvalue Method 134
6.7 Exercises 134
6.8 Computer Programs 138
6.8.1 Daubechies Wavelet 138
6.8.2 Iteration Method 139
References 139

7 Discrete Wavelet Transform and Filter Bank Algorithms 141

7.1 Decimation and Interpolation 141
7.1.1 Decimation 142

7.1.2 Interpolation 144
7.1.3 Convolution Followed by Decimation 147
7.1.4 Interpolation Followed by Convolution 147
7.2 Signal Representation in the Approximation Subspace 148
7.3 Wavelet Decomposition Algorithm 149
7.4 Reconstruction Algorithm 153
7.5 Change of Bases 154
7.6 Signal Reconstruction in Semiorthogonal Subspaces 156
7.6.1 Change of Basis for Spline Functions 157
7.6.2 Change of Basis for Spline Wavelets 160
7.7 Examples 163
7.8 Two-Channel Perfect Reconstruction Filter Bank 165
7.8.1 Spectral-Domain Analysis of a Two-Channel PR Filter Bank 168
7.8.2 Time-Domain Analysis 176
7.9 Polyphase Representation for Filter Banks 180
7.9.1 Signal Representation in the Polyphase Domain 180
7.9.2 Filter Bank in the Polyphase Domain 181
7.10 Comments on DWT and PR Filter Banks 182
7.11 Exercises 183
7.12 Computer Programs 184
7.12.1 Algorithms 184
References 186

8 Fast Integral Transform and Applications 187

8.1 Finer Time Resolution 188
8.2 Finer Scale Resolution 190
8.3 Function Mapping into the Interoctave Approximation Subspaces 194
8.4 Examples 196
8.4.1 IWT of a Linear Function 197
8.4.2 Crack Detection 202
8.4.3 Decomposition of Signals with Nonoctave Frequency Components 203
8.4.4 Perturbed Sinusoidal Signal 203
8.4.5 Chirp Signal 204

8.4.6 Music Signal with Noise 204
8.4.7 Dispersive Nature of the Waveguide Mode 205
References 209

9 **Digital Signal Processing Applications** **210**

9.1 Wavelet Packets 211
9.2 Wavelet Packet Algorithms 212
9.3 Thresholding 214
9.3.1 Hard Thresholding 216
9.3.2 Soft Thresholding 217
9.3.3 Percentage Thresholding 218
9.3.4 Implementation 218
9.4 Interference Suppression 219
9.5 Faulty Bearing Signature Identification 221
9.5.1 Pattern Recognition of Acoustic Signals 221
9.5.2 Wavelets, Wavelet Packets, and FFT Features 226
9.6 Two-Dimensional Wavelets and Wavelet Packets 228
9.6.1 Two-Dimensional Wavelets 228
9.6.2 Two-Dimensional Wavelet Packets 231
9.7 Wavelet and Wavelet Packet Algorithms for Two-Dimensional Signals 233
9.7.1 Two-Dimensional Wavelet Algorithm 233
9.7.2 Wavelet Packet Algorithm 234
9.8 Image Compression 235
9.8.1 Image Coding 235
9.8.2 Wavelet Tree Coder 236
9.8.3 EZW Code 238
9.8.4 EZW Example 239
9.8.5 Spatial-Oriented Tree 242
9.8.6 Generalized Self-Similarity Tree 244
9.9 Microcalcification Cluster Detection 244
9.9.1 CAD Algorithm Structure 244
9.9.2 Partitioning of Image and Nonlinear Contrast Enhancement 245
9.9.3 Wavelet Decomposition of the Subimages 245
9.9.4 Wavelet Coefficient Domain Processing 246

9.9.5 Histogram Thresholding and Dark Pixel Removal 248
9.9.6 Parametric ART2 Clustering 248
9.9.7 Results 249

9.10 Multicarrier Communication Systems 249
9.10.1 OFDM Multicarrier Communication Systems 250
9.10.2 Wavelet Packet–Based MCCS 252

9.11 Three-Dimensional Medical Image Visualization 252
9.11.1 Three-Dimensional Wavelets and Algorithms 255
9.11.2 Rendering Techniques 256
9.11.3 Region of Interest 258
9.11.4 Summary 258

9.12 Computer Programs 258
9.12.1 Two-Dimensional Wavelet Algorithms 258
9.12.2 Wavelet Packets Algorithms 263

References 265

10 Wavelets in Boundary Value Problems 267

10.1 Integral Equations 268

10.2 Method of Moments 272

10.3 Wavelet Techniques 273
10.3.1 Use of Fast Wavelet Algorithm 273
10.3.2 Direct Application of Wavelets 274
10.3.3 Wavelets in Spectral Domain 275
10.3.4 Wavelet Packets 280

10.4 Wavelets on the Bounded Interval 280

10.5 Sparsity and Error Considerations 282

10.6 Numerical Examples 285

10.7 Semiorthogonal Versus Orthogonal Wavelets 291

10.8 Differential Equations 294

10.9 Expressions for Splines and Wavelets 295

References 297

Index 301

Preface

This textbook on wavelets evolves from teaching undergraduate and postgraduate courses in the Department of Electrical Engineering at Texas A&M University, and teaching several short courses at Texas A&M University as well as in conferences such as the Progress in Electromagnetic Research Symposium (PIERS), IEEE Antenna and Propagation (IEEE-AP) Symposium, IEEE Microwave Theory and Technique (IEEE-MTT) Conference, and those of the Association for Computational Electromagnetic Society (ACES). The participants at the short courses came from industries as well as universities and had backgrounds mainly in electrical engineering, physics, and mathematics with little or no prior understanding of wavelets. In preparing material for the lectures, we referred to many books on this subject, some catering to the needs of mathematicians and physicists, others to engineers with a signal processing background. We felt the need for a book that would combine the theory, algorithms, and applications of wavelets and present them in such a way that the reader can easily learn the subject and be able to apply them to solve practical problems. That being our motivation, we have tried to keep a balance between mathematical rigor and practical applications of wavelet theory. Many mathematical concepts have been elucidated through figures.

The book is organized as follows. The first chapter gives an overview of the book. The rest of the book is divided into four parts. In Chapters 2 and 3 we review some basic concepts of linear algebra, Fourier analysis, and discrete signal analysis. The next three chapters are devoted to discussing theoretical aspects of time–frequency analysis, multiresolution analysis, and construction of various types of wavelets, and in Chapters 7 and 8 we give several algorithms to compute wavelet transforms and implement them through filter bank approach. In Chapter 8 and in Chapters 9 and 10 we present many interesting applications of wavelets to signal processing and boundary value problems.

In preparing this book we have benefited from the assistance of a number of people. We learned a lot about wavelets through our association with Professor Charles Chui, to whom we are very grateful. We would like to thank Professors Raj

Mittra, Linda Katehi, and Hao Ling for inviting us to speak at short courses at IEEE-AP and MTT conferences. Thanks are also due to Professor L. Tsang for inviting us to organize the short course at PIERS. Parts of Chapters 4 and 9 come from our collaboration with graduate students at Texas A&M University, notable among them being Minsen Wang, Howard Choe, Nai-wen Lin, Tsai-fa Yu, and Zhiwha Xu. We thank all of them for their contributions. We wish to express our deep sense of appreciation to Michelle Rubin, who typed and proofread most of this book. We thank Professor Kai Chang for his encouragement to write this book. Thanks are also due to Mr. George J. Telecki, Executive Editor, and Ms. Angioline Loredo, Associate Managing Editor, for their efficient management and supervision during the production of this book. Last but not the least, we thank Mousumi Goswami and Sophia Chan for their encouragement and support during the preparation of this book.

JAIDEVA C. GOSWAMI
ANDREW K. CHAN

Fundamentals of Wavelets

CHAPTER ONE

What This Book Is About

The concept of wavelet analysis has been in place in one form or another since the beginning of the twentieth century. The Littlewood–Paley technique and Calderón–Zygmund theory in harmonic analysis and digital filter bank theory in signal processing can be considered as forerunners of wavelet analysis. However, in its present form, wavelet theory attracted attention in the 1980s through the work of several researchers from various disciplines—Strömberg, Morlet, Grossmann, Meyer, Battle, Lemarié, Coifman, Daubechies, Mallat, Chui—to name a few. Many other researchers have also made significant contributions.

In applications to discrete data sets, wavelets may be considered as basis functions generated by dilations and translations of a single function. Analogous to Fourier analysis, there are wavelet series (WS) and integral wavelet transforms (IWTs). In wavelet analysis, WS and IWTs are intimately related. The IWT of a finite-energy function on the real line evaluated at certain points in the time–scale domain gives the coefficients for its wavelet series representation. No such relation exists between Fourier series and Fourier transform, which are applied to different classes of functions; the former is applied to finite-energy periodic functions, whereas the latter is applied to functions that have finite energy over the real line. Furthermore, Fourier analysis is global in the sense that each frequency (time) component of the function is influenced by all the time (frequency) components of the function. On the other hand, wavelet analysis is a local analysis. This local nature of wavelet analysis makes it suitable for time–frequency analysis of signals.

Wavelet techniques enable us to divide a complicated function into several simpler ones and study them separately. This property, along with fast wavelet algorithms, which are comparable in efficiency to fast Fourier transform algorithms, makes these techniques very attractive in analysis and synthesis problems. Different types of wavelets have been used as tools to solve problems in signal analysis, image analysis, medical diagnostics, boundary value problems, geophysical signal processing, statistical analysis, pattern recognition, and many others. While wavelets have gained popularity in these areas, new applications are continually being investigated.

A reason for the popularity of wavelet is its effectiveness in representation of nonstationary (transient) signals. Since most of natural and human-made signals are transient in nature, different wavelets have been used to represent this much larger class of signals than Fourier representation of stationary signals. Unlike Fourier-based analyses that use global (nonlocal) sine and cosine functions as bases, wavelet analysis uses bases that are localized in time and frequency to represent nonstationary signals more effectively. As a result, a wavelet representation is much more compact and easier to implement. Using the powerful multiresolution analysis, one can represent a signal by a finite sum of components at different resolutions so that each component can be processed adaptively based on the objectives of the application. This capability to represent signals compactly and in several levels of resolution is the major strength of wavelet analysis. In the case of solving partial differential equations by numerical methods, the unknown solution can be represented by wavelets of different resolution, resulting in a multigrid representation. The dense matrix resulting from an integral operator can be made less dense using wavelet-based thresholding techniques to attain an arbitrary degree of solution accuracy.

There have been many research monographs on wavelet analysis as well as textbooks for specific application areas. However, there seems to be no textbook that provides a systematic introduction to the subject of wavelets and its wide areas of applications. This was the motivating factor for us to write this introductory text. Our aim is (1) to present this mathematically elegant analysis in a formal yet readable fashion, (2) to introduce to readers many possible areas of applications in both signal processing and boundary value problems, and (3) to provide several algorithms and computer codes for basic hands-on practices. The level of writing will be suitable for college seniors and first-year graduate students. However, sufficient details will be given so that practicing engineers without a background in signal analysis will find it useful.

The book is organized logically to develop the concept of wavelets. It is divided into four major parts. Rather than vigorously proving theorems and developing algorithms, the subject matter is developed systematically from the very basics of signal representation using basis functions. Wavelet analysis is explained via a parallel with Fourier analysis and the short-time Fourier transform. Multiresolution analysis is developed to demonstrate the decomposition and reconstruction algorithms. Filter bank theory is incorporated so that readers may draw a parallel between the filter bank algorithm and the wavelet algorithm. Specific applications in signal processing, image processing, electromagnetic wave scattering, boundary value problems, wavelet imaging systems, and interference suppression are included. A detailed chapter-by-chapter outline of the book follows.

Chapters 2 and 3 are devoted to reviewing some basic mathematical concepts and techniques and to setting the tone for time–frequency and time–scale analysis. To have a better understanding of wavelet theory, it is necessary to review the basics of linear functional space. Concepts in Euclidean vectors are extended to spaces in higher dimensions. Vector projection, basis functions, local and Reisz bases, orthogonality, and biorthogonality are discussed in Chapter 2. In addition, least-squares approximation of functions as well as such mathematical tools as matrix algebra and z-transform are discussed. Chapter 3 provides a brief review of Fourier analysis to set the foundation for the development of continuous wavelet transform and discrete wavelet series. The main objective of this chapter is not to redevelop Fourier theory but to remind readers of some important issues and relations in Fourier analysis that are relevant to later development. The principal properties of Fourier series and Fourier transform are discussed. Lesser known theorems, including Poisson's sum formulas, partition of unity, sampling theorem, and the Dirichlet kernel for partial sum are developed in this chapter. Discrete-time Fourier transform and discrete Fourier transform are also mentioned briefly for the purpose of comparing them with continuous and discrete wavelet transforms. Some advantages and drawbacks of Fourier analysis in terms of signal representation are presented.

Development of time–frequency and time–scale analysis forms the core of the second major section of this book. Chapter 4 is devoted to a discussion of the short-time Fourier transform (time–frequency analysis) and the continuous wavelet transform (time–scale analysis). The similarities and differences between these two transforms are pointed out. In addition, window widths as measures of localization of a time function and its spectrum are introduced. This chapter also contains the major properties of the transform, such as perfect reconstruction and uniqueness of inverse. Discussions of the Gabor transform and Wigner–Ville distribution complete this chapter on time–frequency analysis. Chapter 5 contains an introduction to and discussion of multiresolution analysis. The relationships between nested approximation spaces and wavelet spaces are developed via a derivation of two-scale relations and decomposition relations. Orthogonality and biorthogonality between spaces and between basis functions and their integer translates are also discussed. This chapter also contains a discussion of the semiorthogonal B-spline function as well as techniques for mapping the function onto multiresolution spaces. In Chapter 6, methods and requirements for wavelet construction are developed in detail. Orthogonal, semiorthogonal, and biorthogonal wavelets are constructed via examples to elucidate the procedure. Biorthogonal wavelet subspaces and their orthogonal properties are also discussed in this chapter. A derivation of the formulas used in methods to compute and display the wavelet is presented at the end of this chapter.

The algorithm development for wavelet analysis is contained in Chapters 7 and 8. Chapter 7 provides the construction and implementation of decomposition and reconstruction algorithms. The basic building blocks for these algorithms are discussed at the beginning of the chapter. Formulas for decimation, interpolation, discrete convolution, and their interconnections are derived. Although these algorithms are general for various types of wavelets, special attention is given to the compactly supported semiorthogonal B-spline wavelets. Mapping formulas between the spline spaces and the dual spline spaces are derived. The algorithms for a perfect reconstruction filter bank in digital signal processing are developed via z-transform in this chapter. The time-domain and polyphase-domain equivalents of the algorithms are discussed. Examples of biorthogonal wavelet construction are given at the end of the chapter. In Chapter 7, limitations to the discrete wavelet algorithms, including the time-variant property of the discrete wavelet transform and the sparsity of data distribution are pointed out. To circumvent the difficulties, the fast integral wavelet transform (FIWT) algorithm is developed in Chapter 8 for the semiorthogonal spline wavelet. Starting with an increase in time resolution and ending with an increase in scale resolution, a step-by-step development of the algorithm is presented in this chapter. A number of applications using FIWT are included to illustrate its importance.

The final section of this book is on the application of wavelets to engineering problems. Chapter 9 includes their applications to signal and image processing, and in Chapter 10 we discuss the use of wavelets in solving boundary value problems. In Chapter 9, the concept of wavelet packets is discussed first as an extension of wavelet analysis to improve the spectral domain performance of the wavelet. Wavelet packet representation of signal is seen as a refinement of the wavelet in the spectral domain by further subdividing the wavelet spectrum into subspectra. This is seen to be useful in subsequent discussions of radar interference suppression. Three types of amplitude thresholding are discussed in this chapter and are used in subsequent applications to image compression. Signature recognition on faulty bearing completes the one-dimensional wavelet signal processing. The wavelet algorithms in Chapter 7 are extended to two dimensions for processing of images. Major wavelet image-processing applications included in this chapter are image compression and target detection and recognition. Details of tree-type image coding are not included because of limited space. However, the detection, recognition, and clustering of microcalcifications in mammograms are discussed in moderate detail. The application of wavelet packets to multicarrier communication systems and the application of wavelet analysis to three-dimensional medical image visualization are also included. Chapter 10 concerns wavelets in boundary value problems. The traditional method of moments and the wavelet-based method of moments are developed in parallel. Different ways to use wavelets in the method of moments are discussed. In particular, wavelets on a bounded interval as applied to solving integral equations arising from electromagnetic scattering problems are presented in some detail. These boundary wavelets are also suitable to avoid edge effects in image processing. Finally, an application of wavelets in the spectral domain is illustrated by applying them to solve a transmission-line discontinuity problem.

Most of the material is derived from lecture notes prepared for undergraduate and graduate courses in the Department of Electrical Engineering at Texas A&M University as well as for short courses taught at several conferences. The material in this book can be covered in one semester. Topics can also be selectively amplified to complement other signal processing courses in an existing curriculum. Exercises are included in some chapters for the purpose of practice. A number of figures have been included in each chapter to expound mathematical concepts. Suggestions on computer code generation are also included at the end of some chapters.

CHAPTER TWO

Mathematical Preliminaries

The purpose of this chapter is to familiarize the reader with some of the mathematical notations and tools that are useful in an understanding of wavelet theory. Since wavelets are continuous functions that satisfy certain admissibility conditions, it is prudent to discuss some definitions and properties of functional spaces. For a more detailed discussion of functional spaces, the reader is referred to standard texts on real analysis. The wavelet algorithms discussed in later chapters involve digital processing of coefficient sequences. A fundamental understanding of topics in digital signal processing, such as sampling, the z-transform, linear shift-invariant systems, and discrete convolution, is necessary for a good grasp of wavelet theory. In addition, a brief discussion of linear algebra and matrix manipulations is included that is very useful in discrete-time domain analysis of filter banks. Readers already familiar with its contents may skip this chapter.

2.1 LINEAR SPACES

In the broadest sense, a functional space is simply a collection of functions that satisfy a certain mathematical structural pattern. For example, the finite-energy space $L^2(-\infty, \infty)$ is a collection of functions that are square integrable; that is,

$$\int_{-\infty}^{\infty} |f(x)|^2 \, dx < \infty. \tag{2.1}$$

Some of the requirements and operational rules on linear spaces are stated as follows:

1. The space S must not be empty.
2. If $x \in S$ and $y \in S$, then $x + y = y + x$.
3. If $z \in S$, then $(x + y) + z = x + (y + z)$.

4. There exists in S a unique element $\mathbf{0}$, such that $x + \mathbf{0} = x$.
5. There exists in S another unique element $-x$ such that $x + (-x) = \mathbf{0}$.

Besides these simple yet important rules, we also define *scalar multiplication* $y = cx$ such that if $x \in S$, then $y \in S$, for every *scalar* c. We have the following additional rules for the space S:

1. $c(x + y) = cx + cy$.
2. $(c + d)x = cx + dx$ with scalar c and d.
3. $(cd)x = c(dx)$.
4. $1 \cdot x = x$.

Spaces that satisfy these additional rules are called *linear spaces*. However, up to now, we have not defined a measure to gauge the size of an element in a linear space. If we assign a number $\|x\|$, called the *norm* of x, to each function in S, this space becomes a *normed linear space* (i.e., a space with a measure associated with it). The norm of a space must also satisfy certain mathematical properties:

1. $\|x\| \geq 0$ and $\|x\| = 0 \Longleftrightarrow x = 0$.
2. $\|x + y\| \leq \|x\| + \|y\|$.
3. $\|ax\| = a\,\|x\|$, where a is scalar.

The norm of a function is simply a measure of the distance of the function to the origin (i.e., 0). In other words, we can use the norm

$$\|x - y\| \tag{2.2}$$

to measure the difference (or distance) between two functions, x and y.

There are a variety of norms that one may choose as a measure for a particular linear space. For example, the finite-energy space $L^2(-\infty, \infty)$ uses the norm

$$\|x\| = \left[\int_{-\infty}^{\infty} |f(x)|^2\,dx\right]^{1/2} < \infty, \tag{2.3}$$

which we shall call the L^2*-norm*. This norm has also been used to measure the overall difference (or error) between two finite-energy functions. This measure is called the root-mean-square error (RMSE) defined by

$$\text{RMSE} = \left[\lim_{T\to\infty} \frac{1}{T} \int_{-T/2}^{T/2} |f(x) - f_a(x)|^2\,dx\right]^{1/2}, \tag{2.4}$$

where $f_a(x)$ is an approximating function to $f(x)$. The expression in (2.4) is the approximation error in the L^2 sense.

2.2 VECTORS AND VECTOR SPACES

Based on the basic concepts of functional spaces discussed in Section 2.1, we now present some fundamentals of vector spaces. We begin with a brief review of geometric vector analysis.

A vector $\mathbf{V}$ in a three-dimensional Euclidean vector space is defined by three complex numbers $\{v_1, v_2, v_3\}$ associated with three orthogonal unit vectors $\{\mathbf{a}_1, \mathbf{a}_2, \mathbf{a}_3\}$. The ordered set $\{v_j\}_{j=1}^{3}$ represents the scalar components of the vector $\mathbf{V}$, where the unit vector set $\{\mathbf{a}_j\}_{j=1}^{3}$ spans the three-dimensional Euclidean vector space. Any vector $\mathbf{U}$ in this space can be decomposed into three vector components $\{u_j\mathbf{a}_j\}_{j=1}^{3}$ (see Figure 2.1d).

The addition and scalar multiplication of vectors in this space are defined by:

1. $\mathbf{U}+\mathbf{V} = \{u_1 + v_1, u_2 + v_2, u_3 + v_3\}$.
2. $k\mathbf{V} = \{kv_1, kv_2, kv_3\}$.

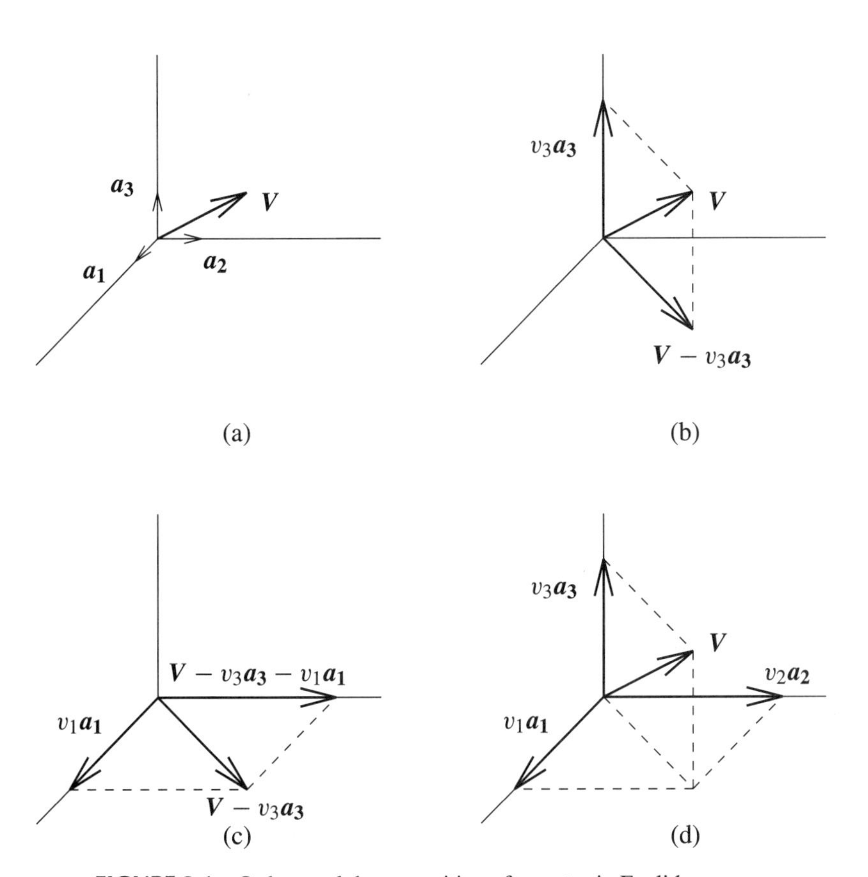

FIGURE 2.1 Orthogonal decomposition of a vector in Euclidean space.

In addition to these operations, vectors in a three-dimensional Euclidean space also obey the commutative and associative laws:

1. $\mathbf{U} + \mathbf{V} = \mathbf{V} + \mathbf{U}$.
2. $(\mathbf{U} + \mathbf{V}) + \mathbf{W} = \mathbf{U} + (\mathbf{V} + \mathbf{W})$.

We may represent a vector by a column matrix,

$$\mathbf{V} = \begin{bmatrix} v_1 \\ v_2 \\ v_3 \end{bmatrix}, \tag{2.5}$$

since all of the mathematical rules above apply to column matrices. We define the length of a vector similar to the definition of the norm of a function, by

$$|\mathbf{V}| = \sqrt{v_1^2 + v_2^2 + v_3^2}. \tag{2.6}$$

The scalar product (inner product) of two vectors is a very important operation in vector algebra that we should consider here. It is defined by

$$\begin{aligned} \mathbf{U} \cdot \mathbf{V} &:= |\mathbf{U}|\,|\mathbf{V}| \cos \angle \mathbf{U}, \mathbf{V} \\ &= u_1 v_1 + u_2 v_2 + u_3 v_3 \\ &= \begin{bmatrix} u_1 & u_2 & u_3 \end{bmatrix} \begin{bmatrix} v_1 \\ v_2 \\ v_3 \end{bmatrix} \\ &= \begin{bmatrix} u_1 \\ u_2 \\ u_3 \end{bmatrix}^t \begin{bmatrix} v_1 \\ v_2 \\ v_3 \end{bmatrix}, \end{aligned} \tag{2.7}$$

where the superscript t indicates matrix transposition and $:=$ is the symbol for definition. It is known that the scalar product obeys the commutative law: $\mathbf{U} \cdot \mathbf{V} = \mathbf{V} \cdot \mathbf{U}$. Two vectors $\mathbf{U}$ and $\mathbf{V}$ are *orthogonal* to each other if $\mathbf{U} \cdot \mathbf{V} = 0$. We define the *projection* of a vector onto another vector by

$$\begin{aligned} \frac{\mathbf{U} \cdot \mathbf{V}}{|\mathbf{V}|} &= \mathbf{U} \cdot \mathbf{a}_v \\ &= \text{projection of } \mathbf{U} \text{ in the direction of } \mathbf{a}_v \\ &= \text{a component of } \mathbf{U} \text{ in the direction of } \mathbf{a}_v. \end{aligned} \tag{2.8}$$

Projection is an important concept which will be used often in later discussions. If one needs to find the component of a vector in a given direction, simply project the vector in that direction by taking the scalar product of the vector with the unit vector of the direction desired.

We may now extend this concept of basis and projection from the three-dimensional Euclidean space to an N-dimensional vector space. The components of a vector in this space form an $N \times 1$ column matrix, while the basis vectors $\{\mathbf{a}_j\}_{j=1}^{N}$ form an orthogonal set such that

$$\mathbf{a}_k \cdot \mathbf{a}_\ell = \delta_{k,\ell} \qquad \forall\, k, \ell \in \mathbb{Z}, \tag{2.9}$$

where $\delta_{k,\ell}$ is the Kronecker delta, defined as

$$\delta_{k,\ell} = \begin{cases} 1, & k = \ell \\ 0, & k \neq \ell. \end{cases} \tag{2.10}$$

One can obtain a specific component v_j of a vector $\mathbf{V}$ (or the projection of $\mathbf{V}$ in the direction of $\mathbf{a}_j$) using the inner product

$$v_j = \mathbf{V} \cdot \mathbf{a}_j, \tag{2.11}$$

and the vector $\mathbf{V}$ is expressed as a linear combination of its vector components,

$$\mathbf{V} = \sum_{k=1}^{N} v_k \mathbf{a}_k. \tag{2.12}$$

It is well known that a vector can be decomposed into elementary vectors along the direction of the basis vectors by finding its components one at a time. Figure 2.1 illustrates this procedure. The vector $\mathbf{V}$ in Figure 2.1a is decomposed into $\mathbf{V}_p = \mathbf{V} - v_3\mathbf{a}_3$ and its orthogonal complementary vector $v_3\mathbf{a}_3$. The vector $\mathbf{V}_p$ is further decomposed into $v_1\mathbf{a}_1$ and $v_2\mathbf{a}_2$. Figure 2.1d represents the reconstruction of the vector $\mathbf{V}$ from its components.

The example shown in Figure 2.1, although elementary, is analogous to the wavelet decomposition and reconstruction algorithm. There the orthogonal components are wavelet functions at different resolutions.

2.3 BASIS FUNCTIONS

We extend the concept of Euclidean geometric vector space to normed linear spaces. That is, instead of thinking about a collection of geometric vectors, we think about a collection of functions. Instead of basis vectors, we have basis functions to represent arbitrary functions in that space. These basis functions are basic building blocks for functions in that space. We will use the Fourier series as an example. The topic of Fourier series is considered in more detail in Chapter 3.

Example: Let us recall that a periodic function $p_T(t)$ can be expanded into a series

$$p_T(t) = \sum_{k=-\infty}^{\infty} c_k e^{jk\omega_0 t}, \tag{2.13}$$

where T is the periodicity of the function, $\omega_0 = 2\pi/T = 2\pi f$ is the fundamental frequency, and $e^{jn\omega_0 t}$ is the nth harmonic of the fundamental frequency. Equation (2.13) is identical to (2.12) if we make the equivalence between $p_T(t)$ with **V**, c_k with v_k, and $e^{jk\omega_0 t}$ with $\mathbf{a}_k$. Therefore, the function set $\left\{e^{jk\omega_0 t}\right\}_{k\in\mathbb{Z}}$ forms the basis set for the Fourier space of discrete frequency. Here $\mathbb{Z}$ is the set of all integers, $\{\ldots,-1,0,1,\ldots\}$. The coefficient set $\{c_k\}_{k\in\mathbb{Z}}$ is often referred to as the *discrete spectrum*. It is well known that the discrete Fourier basis is an orthogonal basis. Using the inner product notation for functions

$$\langle g, h\rangle = \int_{\Omega} g(t)\overline{h(t)}\,dt, \tag{2.14}$$

where the overbar indicates complex conjugation, we express the orthogonality by

$$\frac{1}{T}\int_{-T/2}^{T/2} e^{jk\omega_0 t}e^{-j\ell\omega_0 t}\,dt = \delta_{k,\ell} \qquad \forall\, k,\ell \in \mathbb{Z}. \tag{2.15}$$

We may normalize the basis functions (with respect to unit energy) by dividing them with $\sqrt{T}$. Hence $\left\{e^{jk\omega_0 t}/\sqrt{T}\right\}_{k\in\mathbb{Z}}$ forms the orthonormal basis of the discrete Fourier space.

2.3.1 Orthogonality and Biorthogonality

Orthogonal expansion of a function is an important tool for signal analysis. The coefficients of expansion represent the magnitudes of the signal components. In the example above, the Fourier coefficients represent the amplitudes of the harmonic frequency of the signal. If for some particular signal processing objective we decide to minimize (or make zero) certain harmonic frequencies (such as 60-Hz noise), we simply design a filter at that frequency to reject the noise. It is therefore meaningful to decompose a signal into components for observation before processing the signal.

Orthogonal decomposition of a signal is straightforward, and the computation of the coefficients is simple and efficient. If a function $f(t) \in L^2$ is expanded in terms of a certain orthonormal set $\{\phi_k(t)\}_{k\in\mathbb{Z}} \in L^2$, we may write

$$f(t) = \sum_{k=-\infty}^{\infty} c_k\phi_k(t). \tag{2.16}$$

We compute the coefficients by taking the inner product of the function with the basis to yield

$$\begin{aligned}\langle f, \phi_k\rangle &= \int_{-\infty}^{\infty} f(t)\overline{\phi_k(t)}\,dt \\ &= \int_{-\infty}^{\infty}\sum_{\ell=-\infty}^{\infty} c_\ell\phi_\ell(t)\overline{\phi_k(t)}\,dt\end{aligned}$$

$$= \sum_{\ell=-\infty}^{\infty} c_\ell \delta_{\ell,k}$$

$$= c_k. \tag{2.17}$$

Computation of an inner product such as the one in (2.17) requires knowledge of the function $f(t)$ for all t and is not real-time computable.

We have seen that an orthonormal basis is an efficient and straightforward way to represent a signal. In some applications, however, the orthonormal basis function may lack certain desirable signal processing properties, causing inconvenience in processing. Biorthogonal representation is a possible alternative to overcoming the constraint in orthogonality and producing a good approximation to a given function. Let $\{\phi_k(t)\}_{k\in\mathbb{Z}} \in L^2$ be a biorthogonal basis function set. If there exists another basis function set $\{\widetilde{\phi}_k(t)\}_{k\in\mathbb{Z}} \in L^2$ such that

$$\langle \phi_k, \widetilde{\phi}_\ell \rangle = \int_{-\infty}^{\infty} \phi_k(t)\overline{\widetilde{\phi}_\ell(t)}\, dt = \delta_{k,\ell}, \tag{2.18}$$

the set $\{\widetilde{\phi}_k(t)\}_{k\in\mathbb{Z}}$ is called the *dual basis* of $\{\phi_k(t)\}_{k\in\mathbb{Z}}$. We may expand a function $g(t)$ in terms of the biorthogonal basis

$$g(t) = \sum_{k=0}^{\infty} d_k \phi_k(t),$$

and obtain the coefficients by

$$d_n = \langle g, \widetilde{\phi}_n \rangle \tag{2.19}$$

$$= \int_{-\infty}^{\infty} g(t)\overline{\widetilde{\phi}_n(t)}\, dt. \tag{2.20}$$

On the other hand, we may expand the function $g(t)$ in terms of the dual basis

$$g(t) = \sum_{k=0}^{\infty} \widetilde{d}_k \widetilde{\phi}_k(t), \tag{2.21}$$

and obtain the dual coefficients $\widetilde{d}_k$ by

$$\widetilde{d}_\ell = \langle g, \phi_\ell \rangle \tag{2.22}$$

$$= \int_{-\infty}^{\infty} g(t)\overline{\phi_\ell(t)}\, dt. \tag{2.23}$$

We recall that in an orthogonal basis, all basis functions belong to the same space. In a biorthogonal basis, however, the dual basis does not have to be in the original space. If the biorthogonal basis and its dual belong to the same space, these bases are called *semiorthogonal*. Spline functions of an arbitrary order belong to the semiorthogonal class. More details about spline functions are considered in later chapters.

We use geometric vectors in a two-dimensional vector space to illustrate the biorthogonality. Let the vectors

$$\mathbf{b}_1 = \begin{bmatrix} 1 \\ 0 \end{bmatrix}, \qquad \mathbf{b}_2 = \begin{bmatrix} \dfrac{1}{2} \\ \dfrac{\sqrt{3}}{2} \end{bmatrix}$$

form a biorthogonal basis in the two-dimensional Euclidean space. The dual of this basis is

$$\widetilde{\mathbf{b}}_1 = \begin{bmatrix} 1 \\ -\dfrac{1}{\sqrt{3}} \end{bmatrix}, \qquad \widetilde{\mathbf{b}}_2 = \begin{bmatrix} 0 \\ \dfrac{1}{\sqrt{3}} \end{bmatrix}.$$

The bases are displayed graphically in Figure 2.2. We can compute the dual basis in this two-dimensional Euclidean space simply by solving a set of simultaneous equations. Let the biorthogonal basis be

$$\mathbf{b}_1 = \begin{bmatrix} x_1 \\ x_2 \end{bmatrix}, \qquad \mathbf{b}_2 = \begin{bmatrix} y_1 \\ y_2 \end{bmatrix}$$

and the dual basis be

$$\widetilde{\mathbf{b}}_1 = \begin{bmatrix} u_1 \\ u_2 \end{bmatrix}, \qquad \widetilde{\mathbf{b}}_2 = \begin{bmatrix} v_1 \\ v_2 \end{bmatrix}.$$

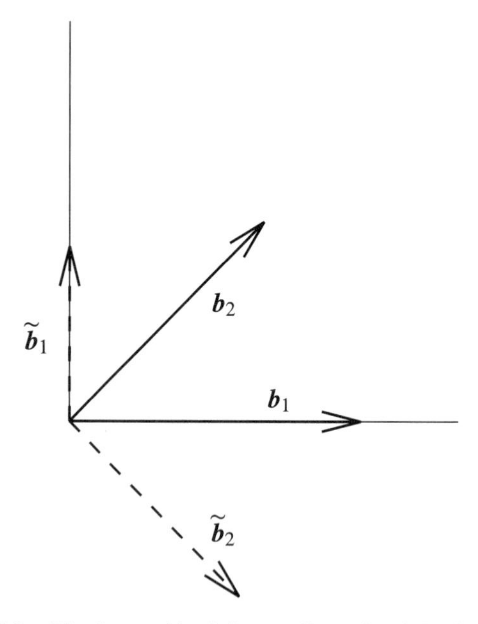

FIGURE 2.2 Biorthogonal basis in two-dimensional Euclidean space.

The set of simultaneous equations that solves for $\widetilde{\mathbf{b}}_1$ and $\widetilde{\mathbf{b}}_2$ is

$$\begin{aligned}\left\langle \mathbf{b}_1, \widetilde{\mathbf{b}_1} \right\rangle &= 1 \\ \left\langle \mathbf{b}_2, \widetilde{\mathbf{b}_1} \right\rangle &= 0 \\ \left\langle \mathbf{b}_1, \widetilde{\mathbf{b}_2} \right\rangle &= 0 \\ \left\langle \mathbf{b}_2, \widetilde{\mathbf{b}_2} \right\rangle &= 1.\end{aligned} \tag{2.24}$$

This process can be generalized into a finite-dimensional space where the basis vectors form an oblique (nonorthogonal) coordinate system. It requires linear matrix transformations to compute the dual basis. This process will not be elaborated upon here. The interested reader may refer to [1].

2.4 LOCAL BASIS AND RIESZ BASIS

We have considered orthogonal bases of a *global* nature in previous sections $[\phi(t) : t \in (-\infty, \infty)]$. Observe that *sine* and *cosine* basis functions for Fourier series are defined on the entire real line $(-\infty, \infty)$ and therefore are called *global bases*. Many bases that exist in a finite interval of the real line $[\phi(t) : t \in (a, b)]$ satisfy the orthogonality or biorthogonality requirements. We call these *local bases*. The Haar basis is the simplest example of a local basis.

Example 1: The *Haar basis* is described by $\phi_{H,k}(t) = \chi_{[0,1)}(t-k), k \in \mathbb{Z}$, where

$$\chi_{[0,1)}(t) = \begin{cases} 1, & 0 \le t < 1 \\ 0, & \text{otherwise} \end{cases} \tag{2.25}$$

is the characteristic function. The Haar basis clearly satisfies the orthogonality condition

$$\begin{aligned}\left\langle \phi_{H,j}(t), \phi_{H,k}(t) \right\rangle &= \int_{-\infty}^{\infty} \chi_{[0,1)}(t-j)\overline{\chi_{[0,1)}(t-k)}\, dt. \\ &= \delta_{j,k}.\end{aligned} \tag{2.26}$$

To represent a global function $f(t), t \in (-\infty, \infty)$ with a local basis $\phi(t), t \in (a, b)$, functions that exist outside the finite interval must be represented by integer shifts (delays) of the basis function along the real line. Integer shifts of global functions can also form bases for linear spaces. The Shannon function $\phi_{\mathrm{SH}}(t)$ is an example of such a basis.

Example 2: The Shannon function is defined by

$$\phi_{\mathrm{SH}}(t) = \frac{\sin \pi t}{\pi t}, \tag{2.27}$$

and the basis formed by

$$\phi_{\mathrm{SH},k}(t) = \frac{\sin \pi(t-k)}{\pi(t-k)}, \qquad k \in \mathbb{Z} \tag{2.28}$$

is an orthonormal basis and is global. The proof of the orthonormality is best shown in the spectral domain.

Let $g(t) \in L^2$ be expanded into a series with basis functions $\{\phi_k(t)\}_{k\in\mathbb{Z}}$:

$$g(t) = \sum_k c_k \phi_k(t). \tag{2.29}$$

The basis set $\{\phi_k(t)\}_{k\in\mathbb{Z}}$ is called a *Riesz basis* if it satisfies the following inequality:

$$R_1 \|c_k\|_{\ell^2}^2 \le \|g(t)\|^2 \le R_2 \|c_k\|_{\ell^2}^2 \tag{2.30}$$

$$R_1 \|c_k\|_{\ell^2}^2 \le \left\| \sum_k c_k \phi_k(t) \right\|^2 \le R_2 \|c_k\|_{\ell^2}^2, \tag{2.31}$$

where $0 < R_1 \le R_2 < \infty$ are called Riesz bounds. The space ℓ^2 is the counterpart of L^2 for discrete sequences with the norm defined as

$$\|c_k\|_{\ell^2} = \left(\sum_k |c_k|^2 \right)^{1/2} < \infty. \tag{2.32}$$

An equivalent expression for (2.30) in the spectral domain is

$$0 < R_1 \le \sum_k \left|\widehat{\phi}(\omega + 2\pi k)\right|^2 \le R_2 < \infty. \tag{2.33}$$

The derivation of (2.33) is left as an exercise. A hat over a function represents its Fourier transform, a topic discussed in Chapter 3.

If $R_1 = R_2 = 1$, the basis is orthonormal. The Shannon basis is an example of a Reisz basis that is orthonormal, since the spectrum of the Shannon function is 1 in the interval $[-\pi, \pi]$. Hence

$$\sum_k \left|\widehat{\phi}_{\mathrm{SH}}(\omega + 2\pi k)\right|^2 \equiv 1. \tag{2.34}$$

If the basis functions $\{\phi(t-k) : k \in \mathbb{Z}\}$ are not orthonormal, we can obtain an orthonormal basis set $\{\phi^{\perp}(t-k) : k \in \mathbb{Z}\}$ by the relation

$$\widehat{\phi^{\perp}}(\omega) = \frac{\widehat{\phi}(\omega)}{\left\{ \sum_k \left|\widehat{\phi}(\omega + 2\pi k)\right|^2 \right\}^{1/2}}. \tag{2.35}$$

The Reisz basis is also called a *stable basis* in the sense that if

$$g_1(t) = \sum_j a_j^{(1)} \phi_j(t)$$

$$g_2(t) = \sum_j a_j^{(2)} \phi_j(t),$$

a small difference in the functions results in a small difference in the coefficients, and vice versa. In other words, stability implies that

$$\text{small } \|g_1(t) - g_2(t)\|^2 \Longleftrightarrow \text{small } \left\| a_j^{(1)} - a_j^{(2)} \right\|_{\ell^2}^2 . \tag{2.36}$$

2.5 DISCRETE LINEAR NORMED SPACE

A *discrete linear normed space* is a collection of elements that are discrete sequences of real or complex numbers with a given norm. For a discrete normed linear space, the operation rules in Section 2.1 are applicable to discrete linear normed space as well. An element in an N-dimensional linear space is represented by the sequence

$$x(n) = \{x(0), x(1), \ldots, x(N-1)\}, \tag{2.37}$$

and we represent a sum of two elements as

$$w(n) = x(n) + y(n) = \{x(0) + y(0), x(1) + y(1), \ldots, x(N-1) + y(N-1)\}. \tag{2.38}$$

The inner product and the norm in discrete linear space are defined separately as

$$\langle x(n), y(n) \rangle = \sum_n x(n)\overline{y}(n) \tag{2.39}$$

$$\|x\| = \langle x, x \rangle^{1/2} = \sqrt{\sum_n |x(n)|^2}. \tag{2.40}$$

Orthogonality and biorthogonality as defined previously apply to discrete bases as well. The biorthogonal discrete basis satisfies the condition

$$\left\langle \phi_i(n), \widetilde{\phi}_j(n) \right\rangle = \sum_n \phi_i(n)\overline{\widetilde{\phi}_j(n)} = \delta_{i,j}. \tag{2.41}$$

For an orthonormal basis, the spaces are self-dual; that is,

$$\phi_j = \widetilde{\phi}_j. \tag{2.42}$$

Example 1: The discrete Haar basis, defined as

$$H_0(n) = \begin{cases} \dfrac{1}{\sqrt{2}} & \text{for } n = 0, 1 \\ 0, & \text{otherwise,} \end{cases} \tag{2.43}$$

is an orthonormal basis formed by the even translates of $H_0(n)$. The Haar basis, however, is not complete. That is, there exist certain sequences that cannot be represented by an expansion from this basis. It requires a complementary space to make it complete. The complementary space of Haar is

$$H_1(n) = \begin{cases} \dfrac{1}{\sqrt{2}} & \text{for } n = 0 \\ \dfrac{-1}{\sqrt{2}} & \text{for } n = 1 \\ 0, & \text{otherwise.} \end{cases} \tag{2.44}$$

The odd translates of $H_1(n)$ form the complementary space, so any real sequence can be represented by the Haar basis.

Example 2: The sequence

$$D_2(n) = \left\{ \frac{1+\sqrt{3}}{4\sqrt{2}}, \frac{3+\sqrt{3}}{4\sqrt{2}}, \frac{3-\sqrt{3}}{4\sqrt{2}}, \frac{1-\sqrt{3}}{4\sqrt{2}} \right\} \tag{2.45}$$

is a finite sequence with four members whose integer translates form an orthonormal basis. The proof of orthonormality is left as an exercise.

2.6 APPROXIMATION BY ORTHOGONAL PROJECTION

Assuming that a vector $u(n)$ is not a member of the linear vector space V spanned by $\{\phi_k\}$, we wish to find an approximation $u_p \in V$. We use the orthogonal projection of u onto the space V as the approximation. The projection is defined by

$$u_p = \sum_k \langle u, \phi_k \rangle \, \phi_k. \tag{2.46}$$

We remark here that the approximation error $u - u_p$ is orthogonal to the space V. That is,

$$\langle u - u_p, \phi_k \rangle = 0 \qquad \forall\, k.$$

Furthermore, the mean square error (MSE) of such an approximation is minimum.

To prove the minimality of the MSE for any orthogonal projection, consider a function $g \in L^2[a, b]$ that is approximated by using a set of orthonormal basis functions $\{\phi_k : k = 0, \ldots, N-1\}$ such that

$$g(t) \approx g_c(t) = \sum_{j=0}^{N-1} c_j \phi_j(t), \tag{2.47}$$

with

$$c_j = \langle g, \phi_j \rangle. \tag{2.48}$$

The pointwise error $\epsilon_c(t)$ in the approximation of the function $g(t)$ is

$$\epsilon_c(t) = g(t) - g_c(t) = g(t) - \sum_{j=0}^{N-1} c_j \phi_j(t). \tag{2.49}$$

We wish to show that when the coefficient sequence $\{c_j\}$ is obtained by orthogonal projection given by (2.48), the MSE $\|\epsilon_c(t)\|^2$ is minimum. To show this let us assume that there is another sequence $\{d_j : j = 0, \ldots, N-1\}$ which is obtained other than by orthogonal projection, which also minimizes the error. Then we will show that $c_j = d_j;\ j = 0, \ldots, N-1$, thus completing our proof.

With the sequence $\{d_j\}$ we have

$$g(t) \approx g_d(t) = \sum_{j=0}^{N-1} d_j \phi_j(t) \tag{2.50}$$

and

$$\begin{aligned}
\|\epsilon_d(t)\|^2 &= \left\| g(t) - \sum_{j=0}^{N-1} d_j \phi_j(t) \right\|^2 = \left\langle g(t) - \sum_{j=0}^{N-1} d_j \phi_j(t),\ g(t) - \sum_{j=0}^{N-1} d_j \phi_j(t) \right\rangle \\
&= \langle g, g \rangle - \sum_{j=0}^{N-1} d_j \left\langle \phi_j(t), g \right\rangle - \sum_{j=0}^{N-1} \overline{d_j} \left\langle g, \phi_j(t) \right\rangle + \sum_{j=0}^{N-1} \left| d_j \right|^2 \\
&= \langle g, g \rangle - \sum_{j=0}^{N-1} d_j \overline{c_j} - \sum_{j=0}^{N-1} \overline{d_j} c_j + \sum_{j=0}^{N-1} \left| d_j \right|^2.
\end{aligned} \tag{2.51}$$

To complete the square of the last three terms in (2.51), we add and subtract $\sum_{j=0}^{N-1} \left| c_j \right|^2$ to yield

$$\begin{aligned}
\|\epsilon_d(x)\|^2 &= \left\| g(x) - \sum_{j=0}^{N-1} d_j \phi_j(t) \right\|^2 \\
&= \left\| g(x) - \sum_{j=0}^{N-1} c_j g_j(x) \right\|^2 + \sum_{j=0}^{N-1} \left| d_j - c_j \right|^2
\end{aligned} \tag{2.52}$$

$$= \|\epsilon_c(x)\|^2 + \sum_{j=0}^{N-1} \left|d_j - c_j\right|^2. \tag{2.53}$$

It is clear that to have a minimum MSE, we must have $d_j = c_j;\ j = 0, \ldots, N-1$, and hence the proof.

2.7 MATRIX ALGEBRA AND LINEAR TRANSFORMATION

We have already used column matrices to represent vectors in finite-dimensional Euclidean spaces. Matrices are operators in these spaces. We give a brief review of matrix algebra in this section and discuss several types of special matrices that will be useful in the understanding of time-domain analysis of wavelets and filter banks. For details, readers are referred to [2].

2.7.1 Elements of Matrix Algebra

1. *Definition*. A *matrix* $\mathbf{A} = \left[A_{ij}\right]$ is a rectangular array of elements. The elements may be real numbers, complex numbers, or polynomials. The first integer index, i, is the row indicator, and the second integer index, j, is the column indicator. A matrix is infinite if $i, j \to \infty$. An $m \times n$ matrix is displayed as

$$\mathbf{A} = \begin{bmatrix} A_{11} & A_{12} & A_{13} & \\ A_{21} & A_{22} & & \\ A_{31} & & & \\ & & & A_{mn} \end{bmatrix}. \tag{2.54}$$

   If $m = n$, $\mathbf{A}$ is a square matrix. An $N \times 1$ column matrix (only one column) represents an N-dimensional vector.

2. *Transposition*. The transpose of $\mathbf{A}$ is $\mathbf{A}^t$, whose element is A_{ji}. If the dimension of $\mathbf{A}$ is $m \times n$, the dimension of $\mathbf{A}^t$ is $n \times m$. The transposition of a column $(N \times 1)$ matrix is a row $(1 \times N)$ matrix.

3. *Matrix sum and difference*. Two matrices may be summed if they have the same dimensions:

$$\mathbf{C} = \mathbf{A} \pm \mathbf{B} \Longrightarrow C_{ij} = A_{ij} \pm B_{ij}.$$

4. *Matrix product*. Multiplication of two matrices is meaningful only if their dimensions are compatible. *Compatibility* means that the number of columns in the first matrix must be the same as the number of rows in the second matrix. If the dimensions of $\mathbf{A}$ and $\mathbf{B}$ are $m \times p$ and $p \times q$, respectively, the dimension of $\mathbf{C} = \mathbf{AB}$ is $m \times q$. The element C_{ij} is given by

$$\mathbf{C}_{ij} = \sum_{\ell=1}^{p} \mathbf{A}_{i\ell}\mathbf{B}_{\ell j}.$$

The matrix product is not commutative since $p \times q$ is not compatible with $m \times p$. In general, $\mathbf{AB} \neq \mathbf{BA}$.

5. *Identity matrix.* An identity matrix is a square matrix whose major diagonal elements are ones and whose off-diagonal elements are zeros.

$$\mathbf{I} = \begin{bmatrix} 1 & 0 & 0 & 0 & 0 & 0 \\ 0 & 1 & 0 & 0 & 0 & 0 \\ 0 & 0 & 1 & 0 & 0 & 0 \\ 0 & 0 & 0 & 1 & 0 & 0 \\ 0 & 0 & 0 & 0 & 1 & 0 \\ 0 & 0 & 0 & 0 & 0 & 1 \end{bmatrix}.$$

6. *Matrix minor.* A minor $\mathbf{S}_{ij}$ of matrix $\mathbf{A}$ is a submatrix of $\mathbf{A}$ created by deleting the ith row and jth column of $\mathbf{A}$. The dimension of $\mathbf{S}_{ij}$ is $(m-1) \times (n-1)$ if the dimension of $\mathbf{A}$ is $m \times n$.

7. *Determinant.* The determinant of a square matrix $\mathbf{A}$ is a value computed successively using the definition of minor. We compute the determinant of a square $(m \times m)$ matrix by

$$\det(A) = \sum_{i=1}^{m} (-1)^{i+j} A_{ij} \det(\mathbf{S}_{ij}).$$

The index j can be any integer between $[1, m]$.

8. *Inverse matrix.* $\mathbf{A}^{-1}$ is the inverse of a square matrix $\mathbf{A}$ such that $\mathbf{A}^{-1}\mathbf{A} = \mathbf{I} = \mathbf{A}\mathbf{A}^{-1}$. We compute the inverse by

$$A_{ij}^{-1} = \frac{1}{\det(\mathbf{A})} (-1)^{j+i} \det(\mathbf{S}_{ji}).$$

If $\det(\mathbf{A}) = 0$, the matrix is *singular*, and $\mathbf{A}^{-1}$ does not exist.

2.7.2 Eigenmatrix

A linear transformation is a mapping such that when a vector $\mathbf{x} \in V$ (a vector space) is transformed, the result of the transformation is another vector, $\mathbf{y} = \mathbf{Ax} \in V$. The vector $\mathbf{y}$, in general, is a scaled, rotated, and translated version of $\mathbf{x}$. In particular, if the output vector $\mathbf{y}$ is the only scalar multiple of the input vector, we call this scalar the *eigenvalue* and the system an *eigensystem*. Mathematically, we write

$$\mathbf{y} = \mathbf{Ax} = \mu \mathbf{x}, \tag{2.55}$$

where $\mathbf{A}$ is an $N \times N$ matrix, $\mathbf{x}$ is an N vector, and μ is a scalar eigenvalue. We determine the eigenvalues from the solution of the characteristic equation

$$\det(\mathbf{A} - \mu \mathbf{I}) = 0. \tag{2.56}$$

If $\mathbf{x}$ is an $N \times 1$ column matrix, there are N eigenvalues in this system. These eigenvalues may or may not all be distinct. Associated with each eigenvalue is an eigenvector. Interpretations of the eigenvectors and eigenvalues depend on the nature of the problem at hand. We substitute each eigenvalue μ_j into (2.56) to solve for the eigenvector $\mathbf{x}_j$. We use the following example to illustrate this procedure. Let

$$\mathbf{A} = \begin{bmatrix} 3 & -1 & 0 \\ -1 & 2 & -1 \\ 0 & -1 & 3 \end{bmatrix}$$

be the transformation matrix. The characteristic equation from (2.56) is

$$\begin{aligned} \det(\mathbf{A} - \mu\mathbf{I}) &= \det \begin{bmatrix} 3-\mu & -1 & 0 \\ -1 & 2-\mu & -1 \\ 0 & -1 & 3-\mu \end{bmatrix} \\ &= (3-\mu)\left[(2-\mu)(3-\mu) - 1\right] - (3-\mu) \\ &= (3-\mu)(\mu-1)(\mu-4) = 0 \end{aligned}$$

and the eigenvalues are $\mu = 4$, 1, and 3. We substitute $\mu = 4$ into (2.55),

$$\begin{bmatrix} 3 & -1 & 0 \\ -1 & 2 & -1 \\ 0 & -1 & 3 \end{bmatrix} \begin{bmatrix} x_1 \\ x_2 \\ x_3 \end{bmatrix} = 4 \begin{bmatrix} x_1 \\ x_2 \\ x_3 \end{bmatrix}$$

and obtain the following set of linear equations

$$\begin{aligned} -x_1 - x_2 + 0 &= 0 \\ -x_1 - 2x_2 - x_3 &= 0 \\ 0 - x_2 - x_3 &= 0. \end{aligned} \tag{2.57}$$

This is a linearly dependent set of algebraic equations. We assume that $x_1 = \alpha$ and obtain the eigenvector $\mathbf{e}_3$ corresponding to $\mu = 4$ as

$$\alpha \begin{bmatrix} 1 \\ -1 \\ 1 \end{bmatrix}, \qquad \alpha \neq 0. \tag{2.58}$$

The reader can compute the other two eigenvectors as an exercise.

2.7.3 Linear Transformation

Using the example of the eigensystem in Section 2.7.2, we have

$$\mathbf{A} = \begin{bmatrix} 3 & -1 & 0 \\ -1 & 2 & -1 \\ 0 & -1 & 3 \end{bmatrix}$$

$$\mu_j = 1, 3, 4 \qquad \text{for } j = 1, 2, 3,$$

and the eigenvectors corresponding to the eigenvalues are

$$\mathbf{e}_1 = \begin{bmatrix} 1 \\ 2 \\ 1 \end{bmatrix}, \qquad \mathbf{e}_2 = \begin{bmatrix} 1 \\ 0 \\ -1 \end{bmatrix}, \qquad \mathbf{e}_3 = \begin{bmatrix} 1 \\ -1 \\ 1 \end{bmatrix}. \tag{2.59}$$

From the definitions of eigenvalue and eigenfunction, we have

$$\mathbf{A}\mathbf{e}_j = \mu_j \mathbf{e}_j \qquad \text{for } j = 1, 2, 3. \tag{2.60}$$

We may rearrange this equation as

$$\mathbf{A}\begin{bmatrix}\mathbf{e}_1 & \mathbf{e}_2 & \mathbf{e}_3\end{bmatrix} = \begin{bmatrix}\mu_1\mathbf{e}_1 & \mu_2\mathbf{e}_2 & \mu_3\mathbf{e}_3\end{bmatrix}. \tag{2.61}$$

To be more concise, we put (2.61) into a more compact matrix form,

$$\begin{aligned} \mathbf{AE} &= \mathbf{E}\begin{bmatrix} \mu_1 & 0 & 0 \\ 0 & \mu_2 & 0 \\ 0 & 0 & \mu_3 \end{bmatrix} \\ &= \mathbf{E}\mu, \end{aligned} \tag{2.62}$$

where μ is a diagonal matrix and $\mathbf{E}$ is the eigenmatrix. If the matrix $\mathbf{E}$ is nonsingular, we diagonalize the matrix $\mathbf{A}$ by premultiplying (2.62) by $\mathbf{E}^{-1}$:

$$\mathbf{E}^{-1}\mathbf{AE} = \boldsymbol{\mu}. \tag{2.63}$$

Therefore, we have used the eigenmatrix $\mathbf{E}$ in a linear transformation to diagonalize the matrix $\mathbf{A}$.

2.7.4 Change of Basis

One may view the matrix $\mathbf{A}$ in the preceding example as a matrix that defines a linear system

$$\mathbf{y} = \begin{bmatrix} y_1 \\ y_2 \\ y_3 \end{bmatrix} = \mathbf{Ax} = \mathbf{A}\begin{bmatrix} x_1 \\ x_2 \\ x_3 \end{bmatrix}. \tag{2.64}$$

The matrix $\mathbf{A}$ is a transformation that maps $\mathbf{x} \in \mathbb{R}^3$ to $\mathbf{y} \in \mathbb{R}^3$. The components of $\mathbf{y}$ are related to that of $\mathbf{x}$ via the linear transformation defined by (2.64). Since $\mathbf{e}_1$, $\mathbf{e}_2$, and $\mathbf{e}_3$ are linearly independent vectors, they may be used as a basis for $\mathbb{R}^3$. Therefore, we may expand the vector $\mathbf{x}$ on this basis:

$$\begin{aligned} \mathbf{x} &= x_1'\mathbf{e}_1 + x_2'\,\mathbf{e}_2 + x_3'\mathbf{e}_3 \\ &= \mathbf{Ex}', \end{aligned} \tag{2.65}$$

and the coefficient vector $\mathbf{x}'$ is computed by

$$\mathbf{x}' = \mathbf{E}^{-1}\mathbf{x}. \tag{2.66}$$

The new coordinates for the vector $\mathbf{y}$ with respect to this new basis become

$$\begin{aligned} \mathbf{y}' &= \mathbf{E}^{-1}\mathbf{y} \\ &= \mathbf{E}^{-1}\mathbf{A}\mathbf{x} \\ &= \mathbf{E}^{-1}\mathbf{A}\mathbf{E}\mathbf{x}' \\ &= \boldsymbol{\mu}\mathbf{x}'. \end{aligned} \tag{2.67}$$

Equation (2.67) states that we have modified the linear system $\mathbf{y} = \mathbf{A}\mathbf{x}$ by a change of basis to another system, $\mathbf{y}' = \boldsymbol{\mu}\mathbf{x}'$, in which the matrix $\boldsymbol{\mu}$ is a diagonal matrix. We call this linear transformation via the eigenmatrix the *similarity transformation*.

2.7.5 Hermitian Matrix, Unitary Matrix, and Orthogonal Transformation

Given a complex-valued matrix $\mathbf{H}$, we can obtain its Hermitian, $\mathbf{H}^h$, by taking the conjugate of the transpose of $\mathbf{H}$, namely

$$\mathbf{H}^h := \overline{\mathbf{H}^t}. \tag{2.68}$$

The two identities

$$\begin{aligned} \left(\mathbf{H}^h\right)^h &= \mathbf{H} \\ (\mathbf{G}\mathbf{H})^h &= \mathbf{H}^h\mathbf{G}^h \end{aligned}$$

obviously follow from the definition.

Let the basis vectors of an N-dimensional vector space be $\mathbf{b}_i, i = 1, \ldots, N$, where $\mathbf{b}_i$ is itself a vector of length N. An *orthogonal basis* means that the inner product of any two basis vectors vanishes:

$$\langle \mathbf{b}_j, \mathbf{b}_i \rangle = \left[\mathbf{b}_j\right]^t [\mathbf{b}_i] = \delta_{i,j} \qquad \forall\, i, j \in \mathbb{Z}. \tag{2.69}$$

For complex-valued basis vectors, the inner product is expressed by

$$\langle \mathbf{b}_j, \mathbf{b}_i \rangle = \left[\mathbf{b}_j\right]^h [\mathbf{b}_i].$$

If the norm of $\mathbf{b}_i$ is 1, this basis is called an *orthonormal basis*. We form an $N \times N$ matrix of transformation $\mathbf{P}$ by putting the orthonormal vectors in a row as

$$\mathbf{P} = [\mathbf{b}_1, \mathbf{b}_2, \ldots, \mathbf{b}_N]. \tag{2.70}$$

Since

$$\left[\mathbf{b}_j\right]^h [\mathbf{b}_i] = \delta_{i,j}, \tag{2.71}$$

it follows that

$$\mathbf{P}^h \mathbf{P} = \mathbf{I} \tag{2.72}$$

and

$$\mathbf{P}^h = \mathbf{P}^{-1}. \tag{2.73}$$

In addition to the column-wise orthonormality, if $\mathbf{P}$ also satisfies the row-wise orthonormality, $\mathbf{P}\mathbf{P}^h = \mathbf{I}$, matrix $\mathbf{P}$ is called a *unitary* (or *orthonormal*) matrix.

2.8 DIGITAL SIGNALS

In this section we provide some basic notations and operations pertinent to signal processing techniques. Details may be found in [3].

2.8.1 Sampling of Signal

Let $x(t)$ be an energy-limited continuous-time (analog) signal. If we measure the signal amplitude and record the result at a regular interval h, we have a discrete-time signal

$$x(n) := x(t_n), \qquad n = 0, 1, \ldots, N-1, \tag{2.74}$$

where

$$t_n = nh.$$

For simplicity in writing and convenience of computation, we use $x(n)$ with the sampling period h understood. These discretized sample values constitute a signal, called a *digital signal*.

In order to have a good approximation to a continuous bandlimited function $x(t)$ from its samples $\{x(n)\}$, the sampling interval h must be chosen such that

$$h \le \frac{\pi}{\Omega},$$

where 2Ω is the bandwidth of the function $x(t)$ [i.e., $\widehat{x}(\omega) = 0$ for all $|\omega| > \Omega$]. The choice of h above is the Nyquist sampling rate, and the Shannon recovery formula

$$x(t) = \sum_{n\in\mathbb{Z}} x(nh) \frac{\sin \pi(t - nh)}{\pi(t - nh)} \tag{2.75}$$

enables us to recover the original analog function $x(t)$. The proof of this theorem is most easily carried out using the Fourier transform and Poisson's sum of Fourier series. We shall defer this proof until Chapter 3.

2.8.2 Linear Shift-Invariant Systems

Let us consider a system characterized by its impulse response $h(n)$. We say that the system is linearly shift invariant if the input $x(n)$ and the output $y(n)$ satisfy the following system relations:

Shift invariance:

$$\begin{aligned} x(n) &\Longrightarrow y(n) \\ x(n-n') &\Longrightarrow y(n-n'). \end{aligned} \tag{2.76}$$

Linearity:

$$\begin{aligned} x_1(n) \Longrightarrow y_1(n) \quad &\text{and} \quad x_2(n) \Longrightarrow y_2(n) \\ x_1(n) + mx_2(n) &\Longrightarrow y_1(n) + my_2(n). \end{aligned} \tag{2.77}$$

In general, a linear shift-invariant system is characterized by

$$x_1(n-n') + mx_2(n-n') \Longrightarrow y_1(n-n') + my_2(n-n'). \tag{2.78}$$

2.8.3 Convolution

Discrete convolution, also known as *moving averages*, defines the input–output relationship of a linear shift-invariant system. The mathematical definition of a linear convolution is

$$\begin{aligned} y(n) &= h(n) * x(n) \\ &= \sum_k h(k-n)x(k) \\ &= \sum_k x(k-n)h(k). \end{aligned} \tag{2.79}$$

We may express the convolution sum in matrix notation as

$$\begin{bmatrix} \cdot \\ y(-1) \\ y(0) \\ y(1) \\ y(2) \\ \cdot \\ \cdot \end{bmatrix} = \begin{bmatrix} \cdot & \cdot & \cdot & \cdot & \cdot & \cdot & \cdot \\ \cdot & \cdot & \cdot & \cdot & \cdot & \cdot & \cdot \\ \cdot & h(1) & h(0) & h(-1) & h(-2) & \cdot & \cdot \\ \cdot & h(2) & h(1) & h(0) & h(-1) & h(-2) & \cdot \\ \cdot & \cdot & h(2) & h(1) & h(0) & h(-1) & h(-2) \\ \cdot & \cdot & \cdot & \cdot & \cdot & \cdot & \cdot \\ \cdot & \cdot & \cdot & \cdot & \cdot & \cdot & \cdot \end{bmatrix} \begin{bmatrix} \cdot \\ x(-1) \\ x(0) \\ x(1) \\ x(2) \\ \cdot \\ \cdot \end{bmatrix}. \tag{2.80}$$

As an example, if $h(n) = \{\frac{1}{4}, \frac{1}{4}, \frac{1}{4}, \frac{1}{4}\}$ and $x(n) = \{1, 0, 1, 0, 1, 0, 1\}$ are causal sequences, the matrix equation for the input–output relations is

$$
\begin{bmatrix} \frac{1}{4} \\ \frac{1}{4} \\ \frac{1}{2} \\ \frac{1}{2} \\ \frac{1}{2} \\ \frac{1}{2} \\ \frac{1}{2} \\ \frac{1}{4} \\ \frac{1}{4} \\ 0 \\ 0 \end{bmatrix} = \begin{bmatrix} \frac{1}{4} & \frac{1}{4} & \frac{1}{4} & \frac{1}{4} & & & & & & & \\ & \frac{1}{4} & \frac{1}{4} & \frac{1}{4} & \frac{1}{4} & & & & & & \\ & & \frac{1}{4} & \frac{1}{4} & \frac{1}{4} & \frac{1}{4} & & & & & \\ & & & \frac{1}{4} & \frac{1}{4} & \frac{1}{4} & \frac{1}{4} & & & & \\ & & & & \frac{1}{4} & \frac{1}{4} & \frac{1}{4} & \frac{1}{4} & & & \\ & & & & & \frac{1}{4} & \frac{1}{4} & \frac{1}{4} & \frac{1}{4} & & \\ & & & & & & \frac{1}{4} & \frac{1}{4} & \frac{1}{4} & \frac{1}{4} & \\ & & & & & & & \frac{1}{4} & \frac{1}{4} & \frac{1}{4} & \frac{1}{4} \\ & & & & & & & & \frac{1}{4} & \frac{1}{4} & \frac{1}{4} \\ & & & & & & & & & \frac{1}{4} & \frac{1}{4} \\ & & & & & & & & & & \frac{1}{4} \end{bmatrix} \begin{bmatrix} x(-2) = 0 \\ x(-1) = 0 \\ x(0) = 1 \\ x(1) = 0 \\ x(2) = 1 \\ x(3) = 0 \\ x(4) = 1 \\ x(5) = 0 \\ x(6) = 1 \\ x(7) = 0 \\ x(8) = 0 \end{bmatrix}. \tag{2.81}
$$

The output signal is seen to be much smoother than the input signal. In fact, the output is very close to the average value of the input. We call this type of filter a *smoothing* or *averaging filter*. In signal processing terms, it is called a *lowpass filter*.

On the other hand, if the impulse response of the filter is $h(n) = \{\frac{1}{4}, -\frac{1}{4}, \frac{1}{4}, -\frac{1}{4}\}$, we have a *differentiating filter* or *highpass filter*. With the input signal $x(n)$ as before, the output signal is

$$
\begin{bmatrix} \frac{1}{4} \\ -\frac{1}{2} \\ \frac{1}{2} \\ -\frac{1}{2} \\ \frac{1}{2} \\ -\frac{1}{2} \\ \frac{1}{2} \\ -\frac{1}{4} \\ \frac{1}{4} \\ 0 \\ 0 \end{bmatrix} = \begin{bmatrix} \frac{1}{4} & -\frac{1}{4} & \frac{1}{4} & -\frac{1}{4} & 0 & 0 & \cdot & \cdot & & & \\ 0 & \frac{1}{4} & -\frac{1}{4} & \frac{1}{4} & -\frac{1}{4} & 0 & 0 & \cdot & \cdot & & \\ 0 & 0 & \frac{1}{4} & -\frac{1}{4} & \frac{1}{4} & -\frac{1}{4} & 0 & 0 & \cdot & \cdot & \\ \cdot & 0 & 0 & \frac{1}{4} & -\frac{1}{4} & \frac{1}{4} & -\frac{1}{4} & 0 & 0 & \cdot & \cdot \\ \cdot & \cdot & 0 & 0 & \frac{1}{4} & -\frac{1}{4} & \frac{1}{4} & -\frac{1}{4} & 0 & 0 & \cdot \\ & \cdot & \cdot & 0 & 0 & \frac{1}{4} & -\frac{1}{4} & \frac{1}{4} & -\frac{1}{4} & 0 & 0 \\ & & \cdot & \cdot & 0 & 0 & \frac{1}{4} & -\frac{1}{4} & \frac{1}{4} & -\frac{1}{4} & 0 \\ & & & \cdot & \cdot & 0 & 0 & \frac{1}{4} & -\frac{1}{4} & \frac{1}{4} & -\frac{1}{4} \\ & & & & \cdot & \cdot & 0 & 0 & \frac{1}{4} & -\frac{1}{4} & \frac{1}{4} \\ & & & & & \cdot & \cdot & 0 & 0 & \frac{1}{4} & -\frac{1}{4} \\ & & & & & & \cdot & \cdot & 0 & 0 & \frac{1}{4} \end{bmatrix} \begin{bmatrix} x(-2) = 0 \\ x(-1) = 0 \\ x(0) = 1 \\ x(1) = 0 \\ x(2) = 1 \\ x(3) = 0 \\ x(4) = 1 \\ x(5) = 0 \\ x(6) = 1 \\ x(7) = 0 \\ x(8) = 0 \end{bmatrix}. \tag{2.82}
$$

The oscillation in the input signal is allowed to pass through the filter, while the average value (dc component) is rejected by this filter. This is evident from the near-zero average of the output, while the average of the input is $\frac{1}{2}$.

2.8.4 z-Transform

The z-transform is a very useful tool for discrete signal analysis. We will use it often in derivations of the wavelet and filter bank algorithms. It is defined by the infinite sum

$$H(z) = \sum_{k \in \mathbb{Z}} h(k) z^{-k}$$
$$= \ldots, h(-1)z^{1} + h(0) + h(1)z^{-1} + h(2)z^{-2} + \cdots \tag{2.83}$$

The variable z^{-1} represents a delay of 1 unit of sampling interval; z^{-M} means a delay of M units. If one replaces z by $e^{j\omega}$, the z-transform becomes the discrete-time Fourier transform, discussed in more detail in Chapter 3:

$$H(z)_{z=e^{j\omega}} = H(e^{j\omega}) = \sum_{k \in \mathbb{Z}} h(k) e^{-jk\omega}. \tag{2.84}$$

We use these notations interchangeably in future discussions. One important property of the z-transform is that the z-transform of a linear discrete convolution becomes a product in the z-transform domain:

$$y(n) = h(n) * x(n) \Longrightarrow Y(z) = H(z)X(z). \tag{2.85}$$

2.8.5 Region of Convergence

The variable z in the z-transform is complex valued. The z-transform, $X(z) = \sum_{n=-\infty}^{\infty} x(n) z^{-n}$, may not converge for some values of z. The *region of convergence* (ROC) of a z-transform indicates the region in the complex plane in which all values of z make the z-transform converge. Two sequences may have the same z-transform but with different regions of convergence.

Example 1: Find the z-transform of $x(n) = a^n \cos(\omega_0 n) u(n)$, where $u(n)$ is the unit step function, defined by

$$u(n) = \begin{cases} 1, & n \geq 0 \\ 0, & \text{otherwise.} \end{cases}$$

SOLUTION: From the definition of z-transform, we have

$$X(z) = \sum_{n=0}^{\infty} a^n \cos(\omega_0 n) z^{-n}$$
$$= \sum_{n=0}^{\infty} a^n \frac{e^{j\omega_0} + e^{-j\omega_0}}{2} z^{-n}$$
$$= \frac{1}{2} \left[\sum_{n=0}^{\infty} \left(a e^{j\omega_0} z^{-1} \right)^n + \sum_{n=0}^{\infty} \left(a e^{-j\omega_0} z^{-1} \right)^n \right]$$
$$= \frac{1}{2} \left(\frac{1}{1 - a e^{j\omega_0} z^{-1}} + \frac{1}{1 - a e^{-j\omega_0} z^{-1}} \right)$$

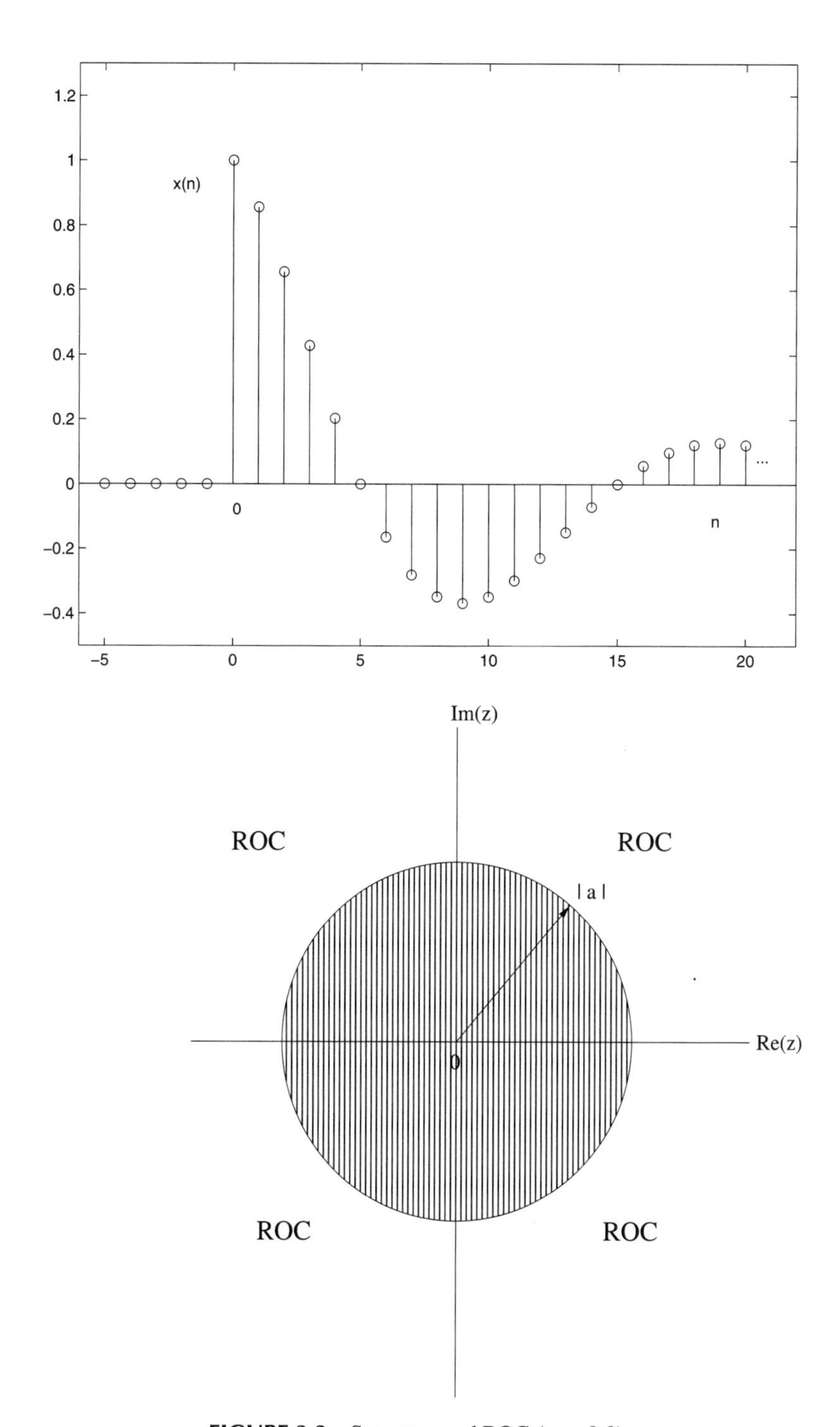

FIGURE 2.3 Sequence and ROC ($a = 0.9$).

$$= \frac{1 - a\cos(\omega_0)z^{-1}}{1 - 2a\cos(\omega_0)z^{-1} + a^2z^{-2}}, \qquad \text{ROC: } |z| > |a|.$$

The case where $a = 0.9$ and $\omega_0 = 10\pi$ is shown in Figure 2.3.

Special cases:

1. If $a = 1$ and $\omega_0 = 0$, we have

$$U(z) = \frac{1 - z^{-1}}{1 - 2z^{-1} + z^{-2}}$$

$$= \frac{1}{1 - z^{-1}}, \qquad \text{ROC: } |z| > 1.$$

2. If $a = 1$, we have

$$X(z) = \frac{1 - \cos(\omega_0)z^{-1}}{1 - 2\cos(\omega_0)z^{-1} + z^{-2}}, \qquad \text{ROC: } |z| > 1.$$

3. If $\omega_0 = 0$, we have

$$X(z) = \frac{1 - az^{-1}}{1 - 2az^{-1} + a^2z^{-2}}$$

$$= \frac{1}{1 - az^{-1}} \qquad \text{ROC: } |z| > |a|.$$

2.8.6 Inverse z-Transform

The formula for recovering the original sequence from its z-transform involves complex integration of the form

$$x(n) = \frac{1}{2\pi j} \oint_c X(z)z^{n-1}\, dz, \tag{2.86}$$

where the contour is taken within the ROC of the transform in the counterclockwise direction. For the purpose of this book, we shall not use (2.86) to recover the sequence. Since the signals and filters that are of interest in this book are rational functions of z, it is more convenient to use partial fractions or long division to recover the sequence.

Example 2: Determine the sequence $x(n)$ corresponding to the following z-transform

$$X(z) = \frac{z - 1}{(z - 0.7)(z - 1.2)} = \frac{z - 1}{z^2 - 1.9z + 0.84}, \qquad \text{ROC: } |z| > 1.2.$$

SOLUTION: Using long division, we have

$$X(z) = z^{-1} + 0.9z^{-2} + 0.87z^{-3} + \frac{0.897z^{-2} - 0.7308z^{-3}}{z^2 - 1.9z + 0.84}.$$

Obviously, $x(0) = 0, x(1) = 1, x(2) = 0.9, x(3) = 0.87, \ldots, x(n)$ is a right-sided infinite sequence since the ROC is outside a circle of radius $r = 1.2$.

If ROC: $|z| < 0.7$, $x(n)$ is a left-sided sequence. We obtain

$$X(z) = \frac{-1 + z}{0.84 - 1.9z + z^2} = -\frac{1}{0.84} + \frac{1}{0.84}\left(1 - \frac{1.9}{0.84}\right)z + \cdots$$

using long division. The sequence $\{x(n)\}$ becomes a sequence of ascending powers of z and is a left-sided sequence where $x(0) = -1/0.84 = -1.19$, $x(-1) = 1/0.84(1 - 1.9/0.84) = -1.5$, $x(-2) = \cdots$.

2.9 EXERCISES

1. Let $u = (-4, -5)$ and $v = (12, 20)$ be two vectors in the two-dimensional space. Find $-5u$, $3u + 2v$, $-v$, and $u + v$. For arbitrary $a, b \in R$, show that $|au| + |bv| \geq |au + bv|$.

2. Expand the function $f(t) = \sin t$ in the polynomial basis set $\{t^n, \} n = 0, 1, 2, \ldots$ Is this an orthogonal set?

3. The following three vectors form a basis set: $e_1 = (1, 2, 1)$; $e_2 = (1, 0, -2)$; $e_3 = (0, 4, 5)$. Is this an orthonormal basis? If not, form an orthonormal basis through a linear combination of e_k, $k = 1, 2, 3$.

4. Let $e_1 = (1, 0)$ and $e_2 = (0, 1)$ be the unit vectors of a two-dimensional Euclidean space. Let $x_1 = (2, 3)$ and $x_2 = (-1, 2)$ be the unit vector of a nonorthogonal basis. If the coordinates of a point w are (3, 1) with respect to the Euclidean space, determine the coordinates of the point with respect to the nonorthogonal coordinate basis.

5. Let $e_1 = (0.5, 0.5)$ and $e_2 = (0, -1)$ be a biorthogonal basis. Determine the dual of this basis.

6. Show that if $\begin{bmatrix} a_{11} & a_{12} \\ a_{21} & a_{22} \end{bmatrix} \begin{bmatrix} b_1 \\ b_2 \end{bmatrix} = \begin{bmatrix} c_{11} & c_{12} \\ c_{21} & c_{22} \end{bmatrix} \begin{bmatrix} b_1 \\ b_2 \end{bmatrix}$ for all $b_1$and b_2 then $\mathbf{A} = \mathbf{C}$.

$$\mathbf{A} = \begin{bmatrix} 1 & 2 \\ 2 & 4 \end{bmatrix}, \mathbf{B} = \begin{bmatrix} -2 & 1 \\ 1 & 3 \end{bmatrix}$$

Form $(\mathbf{AB})^T$and $\mathbf{B}^T\mathbf{A}^T$, and verify that these are the same. Also check if $\mathbf{AB}$ is equal to $\mathbf{BA}$.

7. Find the eigenvalues and the eigenvectors for matrix A.

$$A = \begin{bmatrix} 3 & 1 & 0 \\ 1 & 2 & 2 \\ 0 & 2 & 3 \end{bmatrix}$$

Form the transform matrix P which makes $P^{-1}AP$ a diagonal matrix.

8. Find the convolution of x(n) and $D_2(n)$ where

$$x(n) = (1, 3, 0, 2, 4) \text{ for n} = 0, 1, 2, 3, 4$$

and $D_2(n)$ is given in (2.45). Plot $x(n)$ and $h(n) = x(n) * D_2(n)$ as sequences of n.

9. Find the z-transform of the following sequences and determine the ROC for each of them:

(a) $x(n) = \begin{cases} \cos(n\alpha), & n \geq 0 \\ 0, & n < 0. \end{cases}$

(b) $x(n) = \begin{cases} n, & 1 \leq n \leq m \\ 2m - n, & m + 1 \leq n \leq 2m - 1 \\ 0, & \text{otherwise.} \end{cases}$

10. Find the z-transform of the system function for the following discrete systems:

(a) $y(n) = 3x(n) - 5x(n-1) + x(n-3)$

(b) $y(n) = 4\delta(n) - 11\delta(n-1) + 5\delta(n-4)$, where

$$\delta(n) = \begin{cases} 1, & n = 1 \\ 0, & \text{otherwise.} \end{cases}$$

REFERENCES

1. E. A. Guillemin, *The Mathematics of Circuit Analysis.* New York: John Wiley & Sons, 1949.
2. Ben Noble and James W. Daniel, *Applied Linear Algebra.* Upper Saddle River, N.J.: Prentice Hall, 1988.
3. A. V. Oppenheim and R. W. Schafer, *Discrete-Time Signal Processing.* Upper Saddle River, N.J.: Prentice Hall, 1989.

CHAPTER THREE

Fourier Analysis

Since the days of Joseph Fourier, his analysis has been used in all branches of engineering science and some areas of social science. Simply stated, the Fourier method is the most powerful technique for signal analysis. It transforms the signal from one domain to another domain in which many characteristics of the signal are revealed. One usually refers to this transform domain as the *spectral* or *frequency domain*, while the domain of the original signal is usually the *time* or *spatial domain*. The Fourier analysis includes both the Fourier transform (or Fourier integral) and the Fourier series. The Fourier transform is applicable to functions that are defined on the real line, while the Fourier series is used to analyze functions that are periodic. Since wavelet analysis is similar to Fourier analysis in many aspects, the purpose of this chapter is to provide readers with an overview of Fourier analysis from a signal analysis point of view without going into the mathematical details. Most of the mathematical identities and properties are stated without proof.

3.1 FOURIER SERIES

Fourier series and Fourier transform are often separately treated by mathematicians since they involve two different classes of functions. However, engineers have always been taught that Fourier transform is an extension of Fourier series by allowing the period T of a periodic function to approach infinity. We will follow this route by discussing Fourier series first. The Fourier series representation of a real-valued periodic function $p(t)$ $[p(t) = p(t+T)]$ is given by

$$p(t) = \sum_{k=-\infty}^{\infty} \alpha_k e^{jk\omega_0 t} \tag{3.1}$$

with

$$\alpha_k = \frac{1}{T}\int_{t_0}^{t_0+T} p(t)e^{-jk\omega_0 t}, \tag{3.2}$$

where α_k are the Fourier coefficients and the period $T = 2\pi/\omega_0$ with ω_0 being the *fundamental frequency*. The set of functions $\{e_k\} = \{e^{jk\omega_0 t}\}, k \in \mathbb{Z}$, forms a complete orthogonal basis in $L^2[t_0, t_0 + T]$; that is,

$$\int_{t_0}^{t_0+T} e_k \bar{e}_\ell \, dt = T \delta_{k,\ell}.$$

The coefficient α_k written in the form of an inner product,

$$\alpha_k = \frac{1}{T} \left\langle p(t), e^{jk\omega_0 t} \right\rangle, \tag{3.3}$$

represents the orthogonal component of the function $p(t)$ in $k\omega_0$. Hence the Fourier series is an orthogonal expansion of $p(t)$ with respect to the basis set $\{e_k\}$. The representation in (3.1) is exact. However, if we truncate the series to, say, $\pm N$ terms ($k = -N, \ldots, N$), there will be some error. As described in Section 2.6, being orthogonal projections, the Fourier coefficients minimize the mean square of such error.

A Fourier series may be represented in other forms. Representation using sine and cosine functions is given by

$$p(t) = \frac{a_0}{2} + \sum_{k=1}^{\infty} \left(a_k \cos k\omega_0 t + b_k \sin k\omega_0 t \right), \tag{3.4}$$

in which the a_k and b_k are real quantities. Complex representation using only positive harmonics is written as

$$p(t) = c_0 + \sum_{k=1}^{\infty} c_k \cos \left(k\omega_0 t + \theta_k \right) \tag{3.5}$$

with

$$|c_k| = \sqrt{a_k^2 + b_k^2}, \quad \theta_k = \tan^{-1}\left(-\frac{b_k}{a_k} \right), \tag{3.6}$$

where $c_k = |c_k| \, e^{j\theta_k}$ are complex quantities. Computation formulas for a_k and b_k are

$$a_k = \frac{2}{T} \int_0^T p(t) \cos k\omega_0 t \, dt \tag{3.7}$$

$$b_k = \frac{2}{T} \int_0^T p(t) \sin k\omega_0 t \, dt. \tag{3.8}$$

3.2 EXAMPLES

3.2.1 Rectified Sine Wave

Consider a function $p(t) = |\sin t|$ as shown in Figure 3.1 with the period $T = \pi$ and $\omega_0 = 2\pi/T = 2$ rad/s. Since the function $p(t)$ is an even function with respect to

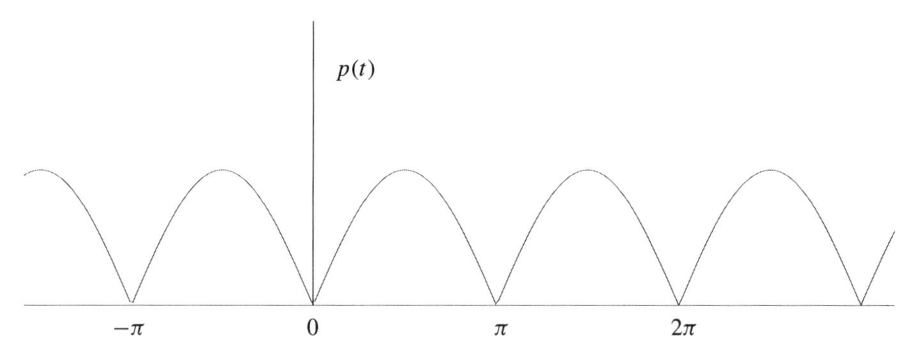

FIGURE 3.1 Rectified sine wave.

$t = 0$ [i.e., $p(-t) = p(t)$], $b_k = 0$ for all k. The coefficients $\{a_k\}$ are computed as

$$\begin{aligned} a_k &= \frac{2}{\pi}\int_0^{\pi} \sin t \cos 2kt \, dt \\ &= \frac{1}{\pi}\int_0^{\pi} [\sin(1-2k)t + \sin(1+2k)t] \, dt \\ &= -\frac{1}{\pi}\left[\frac{\cos(1-2k)t}{1-2k}\bigg|_0^{\pi}\right] - \frac{1}{\pi}\left[\frac{\cos(1+2k)t}{1+2k}\bigg|_0^{\pi}\right] \\ &= -\frac{4}{\pi}\frac{1}{4k^2-1}. \end{aligned} \tag{3.9}$$

Hence the Fourier series of $p(t)$ is given as

$$p(t) = \frac{2}{\pi} - \frac{4}{\pi}\sum_{k=1}^{\infty}\frac{1}{4k^2-1}\cos 2kt.$$

3.2.2 Comb Function and the Fourier Series Kernel $K_N(t)$

In this example we want to find the Fourier series of a periodic impulse train [i.e., a periodic train of the delta function[†] $\delta(t)$]. We write the impulse train with period T as

$$\begin{aligned} I_T(t) &= \sum_{n=-\infty}^{\infty} \delta(t - nT) \\ &= \sum_{k=-\infty}^{\infty} \alpha_k e^{jk\omega_0 t}. \end{aligned} \tag{3.10}$$

[†] This is not a function in the classical sense. It is called a *generalized function* or *distribution*. However, in this book, we refer to this as a *delta function*.

The Fourier coefficients are given by

$$\alpha_k = \frac{1}{T} \int_{-T/2}^{T/2} \sum_{n=-\infty}^{\infty} \delta(t - nT) e^{-jk\omega_0 t}\, dt. \tag{3.11}$$

Since the only n that is within the range of integration is $n = 0$, we find that

$$\alpha_k = \frac{1}{T}, \qquad k \in \mathbb{Z}.$$

Therefore, the Fourier series expansion of an impulse train $I_T(t)$ is written as

$$I_T(t) = \frac{1}{T} \sum_{k=-\infty}^{\infty} e^{jk\omega_0 t}. \tag{3.12}$$

It is instructive to examine the behavior of a truncated version of (3.12). Let $K_N(t)$ be the $(2N+1)$-term finite Fourier sum of $I_T(t)$:

$$K_N(t) = \frac{1}{T} \sum_{k=-N}^{N} e^{jk\omega_0 t}. \tag{3.13}$$

$K_N(t)$ is known as the *Fourier series kernel*. The geometric series sum in (3.13) is carried out to give

$$K_N(t) = \frac{1}{T} \frac{\sin\left(N + \frac{1}{2}\right)\omega_0 t}{\sin(\omega_0 t/2)}.$$

A graph of $K_N(t)$ for $N = 4$ is given in Figure 3.2. We also compute the kernel for $N = 10$ and $N = 15$ but find that the shape of the kernel does not change except that the oscillation frequency is correspondingly increased for higher values of N. The main lobes (main peaks) of the graph become narrower as the value of N increases. The oscillation characteristic of $K_N(t)$ contributes to the Gibbs phenomenon, to be discussed later. These oscillation patterns can be modified by weighting the amplitudes of the coefficients in (3.12). This is a common practice in antenna array design [1].

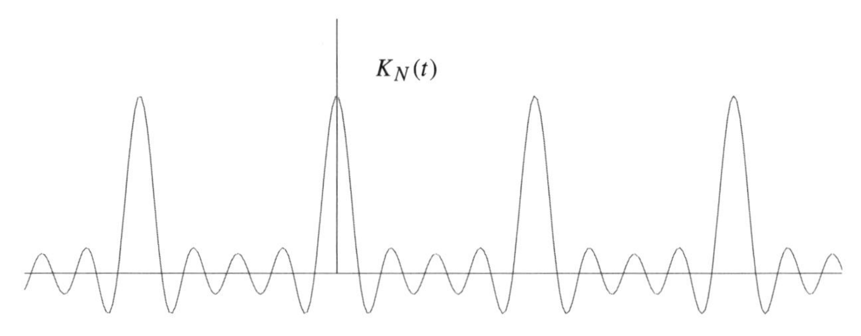

FIGURE 3.2 Fourier series kernel $K_N(t)$.

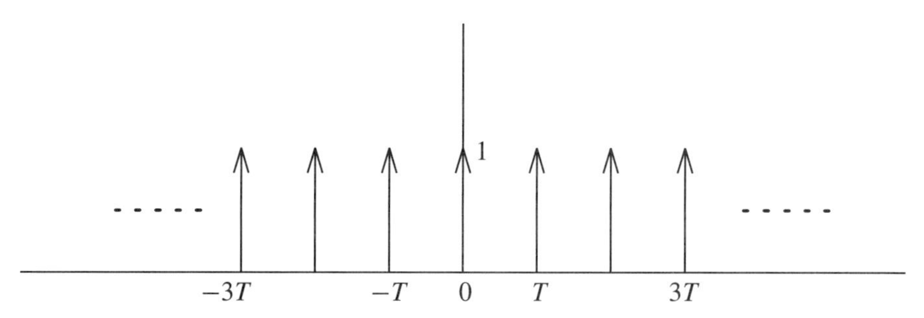

FIGURE 3.3 Comb function.

Since $K_N(t)$ is periodic, we only need to consider the behavior of the kernel within the interval $[-T/2, T/2]$. It is clear that $t/\sin(\omega_0 t/2)$ is bounded in the interval $[-T/2, T/2]$ and $\sin(N + \frac{1}{2})\omega_0 t/t$ approaches $\delta(t)$ as N tends to infinity [2]. Hence

$$\lim_{N\to\infty} K_{N(t)} = \delta(t), \qquad |t| \le \frac{T}{2}.$$

This procedure is applied to all other intervals $[(2k+1)(T/2)]$, $k \in \mathbb{Z}$, and the result is that

$$\lim_{N\to\infty} K_N(t) = I_T(t) = \sum_{k\in\mathbb{Z}} \delta(t - kT). \tag{3.14}$$

The impulse train of (3.12) is called the *comb function* in the engineering literature [3] (see Figure 3.3).

Several important properties of Fourier series, such as Poisson's sum formula, partial sum, and sampling theorem, require the use of Fourier transform for efficient derivation. We consider these topics later in this chapter.

3.3 FOURIER TRANSFORM

To extend the Fourier series to Fourier transform, let us consider (3.1) and (3.2).

The time function $p(t)$ in (3.1) can be expressed using (3.2) as

$$\begin{aligned} p(t) &= \sum_{k=-\infty}^{\infty} \left[\frac{1}{T} \int_{-T/2}^{T/2} p(t') e^{-jk\omega_0 t'}\, dt' \right] e^{jk\omega_0 t} \\ &= \frac{1}{2\pi} \sum_{k=-\infty}^{\infty} \omega_0 \left[\int_{-T/2}^{T/2} p(t') e^{-jk\omega_0 t'}\, dt' \right] e^{jk\omega_0 t}. \end{aligned} \tag{3.15}$$

We extend the period T to infinity so that ω_0 approaches $d\omega$ and $k\omega_0$ approaches ω. The summation in (3.15) becomes an integral,

$$p(t) = \frac{1}{2\pi} \int_{-\infty}^{\infty} \left[\int_{-\infty}^{\infty} p(t') e^{-j\omega t'} \, dt' \right] e^{j\omega t} \, d\omega. \tag{3.16}$$

The integral inside the brackets is represented by a function $\widehat{p}(\omega)$:

$$\widehat{p}(\omega) = \int_{-\infty}^{\infty} p(t') e^{-j\omega t'} \, dt', \tag{3.17}$$

and (3.16) becomes

$$p(t) = \frac{1}{2\pi} \int_{-\infty}^{\infty} \widehat{p}(\omega) e^{j\omega t} \, d\omega. \tag{3.18}$$

Equations (3.17) and (3.18) are known as the *Fourier transform pair*.

From here on we use $f(t)$ to represent a time-domain function, while $p(t)$ is restricted to representing periodic time functions. Let's rewrite (3.17) in new notation.

The *Fourier transform* of a finite-energy function $f(t) \in L^2(\mathbb{R})$ of a real variable t is defined by the integral

$$\widehat{f}(\omega) = \int_{-\infty}^{\infty} f(t) e^{-j\omega t} \, dt. \tag{3.19}$$

In inner product notation, described in Chapter 2, the Fourier transform can also be expressed as

$$\widehat{f}(\omega) = \left\langle f(t), \, e^{j\omega t} \right\rangle. \tag{3.20}$$

We should emphasize the fact that $\widehat{f}(\omega)$ is a complex-valued function that can be expressed in terms of amplitude and phase by

$$\widehat{f}(\omega) = \left| \widehat{f}(\omega) \right| \, e^{j\phi(\omega)}. \tag{3.21}$$

However, the mapping from the domain of $f(t)$ to that of $\widehat{f}(\omega)$ is from $\mathbb{R}$ to $\mathbb{R}$ (i.e., from the t-axis to the ω-axis), even though the real-valued function $f(t)$ is mapped to a complex-valued function $\widehat{f}(\omega)$.

The interpretation of (3.20) is very important. This equation states that for an ω_1, $\widehat{f}(\omega_1)$ represents the component of $f(t)$ at ω_1. If we can determine all the components of $f(t)$ on the ω-axis, a superposition of these components should give back (reconstruct) the original function $f(t)$:

$$f(t) = \frac{1}{2\pi} \int_{-\infty}^{\infty} \widehat{f}(\omega) \, e^{j\omega t} \, d\omega. \tag{3.22}$$

Hence (3.22) can be viewed as a superposition integral that produces $f(t)$ from its components. The integral is referred to as the *inverse Fourier transform* of $\widehat{f}(\omega)$. If the variable t represents *time*, $\widehat{f}(\omega)$ is called the *spectrum* of $f(t)$. If t represents space, $\widehat{f}(\omega)$ is called the *spatial spectrum*.

The Fourier transform is very important in the development of wavelet analysis and will be used often in subsequent chapters. We will use it as an example to present some of the properties of the δ-function.

Let us recall that

$$\int_{-\infty}^{\infty} f(t)\,\delta(t-y)\,dt = f(y). \tag{3.23}$$

Consequently, the Fourier transform of $\delta(t)$,

$$\widehat{\delta}(\omega) = \int_{-\infty}^{\infty} \delta(t)\,e^{-j\omega t}\,dt = e^{-j\omega 0} = 1. \tag{3.24}$$

From the inverse transform of $\widehat{\delta}(\omega)$, the identity

$$\delta(t) = \frac{1}{2\pi}\int_{-\infty}^{\infty} e^{j\omega t}\,d\omega \tag{3.25}$$

is established. The inverse transform in (3.22) can now be shown to be

$$\begin{aligned}\frac{1}{2\pi}\int_{-\infty}^{\infty} \widehat{f}(\omega)\,e^{j\omega t}\,d\omega &= \frac{1}{2\pi}\int_{-\infty}^{\infty} e^{j\omega t}\,d\omega \int_{-\infty}^{\infty} f(t')\,e^{-j\omega t'}\,dt' \\ &= \frac{1}{2\pi}\int_{-\infty}^{\infty} f(t') \int_{-\infty}^{\infty} e^{j\omega(t-t')}\,d\omega\,dt' \\ &= \int_{-\infty}^{\infty} f(t')\,\delta(t-t')\,dt' = f(t).\end{aligned}$$

Since the Fourier transform is unique, we may write

$$f(t) \Leftrightarrow \hat{f}(\omega),$$

meaning that for each function $f(t)$ there is a unique Fourier transform corresponding to that function, and vice versa.

3.4 PROPERTIES OF THE FOURIER TRANSFORM

Since the focus of this chapter is not a detailed exposition of the Fourier analysis, only the properties that are relevant to wavelet analysis will be discussed.

3.4.1 Linearity

If $f(t) = \alpha f_1(t) + \beta f_2(t)$ for some constants α and β, the Fourier transform

$$\begin{aligned}\widehat{f}(\omega) &= \int_{-\infty}^{\infty} f(t)\,e^{-j\omega t}\,dt = \alpha\int_{-\infty}^{\infty} f_1(t)\,e^{-j\omega t}\,dt + \beta\int_{-\infty}^{\infty} f_2(t)\,e^{-j\omega t}\,dt \\ &= \alpha\,\widehat{f_1}(\omega) + \beta\,\widehat{f_2}(\omega).\end{aligned} \tag{3.26}$$

The extension of (3.26) to the finite sum of functions is trivial.

3.4.2 Time Shifting and Time Scaling

Let the function $f(t)$ be shifted by an amount t_0. The spectrum is changed by a phase shift. Indeed, the spectrum of the shifted function $f_0(t) := f(t - t_0)$ is expressed by

$$\begin{aligned}\widehat{f_0}(\omega) &= \int_{-\infty}^{\infty} f(t - t_0)\, e^{-j\omega t}\, dt = \int_{-\infty}^{\infty} f(u)\, e^{-j\omega(u+t_0)}\, du \\ &= e^{-j\omega t_0}\, \widehat{f}(\omega) = \left|\widehat{f}(\omega)\right| e^{j\phi(\omega) - j\omega t_0}, \end{aligned} \tag{3.27}$$

where $\phi(\omega)$ is the phase of the original function $f(t)$. The magnitude of the spectrum remains unchanged for a shifted signal. The shifting is incorporated into the phase term of the spectrum.

Let a be a nonzero constant; the spectrum of $f_a(t) := f(at)$ is given by

$$\widehat{f_a}(\omega) = \int_{-\infty}^{\infty} f(at)\, e^{-j\omega t}\, dt \tag{3.28}$$

$$= \int_{-\infty}^{\infty} f(u)\, e^{-j\omega(u/a)}\, d\left(\frac{u}{a}\right) \tag{3.29}$$

$$= \frac{1}{|a|}\, \widehat{f}\left(\frac{\omega}{a}\right). \tag{3.30}$$

Depending on whether a is greater or smaller than 1, the spectrum is expanded or contracted, respectively. We shall see this important property occur frequently later in the development of wavelet analysis.

3.4.3 Frequency Shifting and Frequency Scaling

The results for frequency shifting and scaling follow in a similar way. If $\widehat{f_0}(\omega) := \widehat{f}(\omega - \omega_0)$, then

$$f_0(t) = f(t)e^{j\omega_0 t}, \tag{3.31}$$

and if $\widehat{f_a}(\omega) := f(a\omega)$ for a nonzero value of a, then

$$f_a(t) = \frac{1}{|a|} f\left(\frac{t}{a}\right). \tag{3.32}$$

3.4.4 Moments

The nth-order moment of a function is defined as

$$M_n := \int_{-\infty}^{\infty} t^n f(t)\, dt. \tag{3.33}$$

The first-order moment,

$$M_1 = \int_{-\infty}^{\infty} t\, f(t)\, dt = (-j)^{-1} \frac{d}{d\omega} \int_{-\infty}^{\infty} f(t)\, e^{-j\omega t}\, dt \bigg|_{\omega=0}$$

$$= (-j)^{-1} \frac{d\widehat{f}(\omega)}{d\omega}\bigg|_{\omega=0}. \tag{3.34}$$

The extension of this formula to the nth-order moment results in

$$M_n = (-j)^{-n} \frac{d^n \widehat{f}(\omega)}{d\omega^n}\bigg|_{\omega=0}. \tag{3.35}$$

3.4.5 Convolution

The convolution of two functions $f_1(t)$ and $f_2(t)$ is defined by

$$f(t) = \int_{-\infty}^{\infty} f_1(y)\, f_2(t-y)\, dy. \tag{3.36}$$

We write (3.36) symbolically by

$$f(t) = f_1(t) * f_2(t). \tag{3.37}$$

Notice that if $f_2(t)$ is $\delta(t)$, the convolution integral recovers the function $f_1(t)$. It is well known that a linear system represented symbolically by the block diagram in Figure 3.4 has the input–output relation given by

$$O(t) = h(t) * i(t), \tag{3.38}$$

where $h(t)$ is the system response function. Hence if $i(t)$ is a delta function, the output function $O(t)$ is the same as $h(t)$. For an arbitrary input function $f(t)$ the convolution integral

$$O(t) = \int_{-\infty}^{\infty} h(\tau)\, i(t-\tau)\, d\tau \tag{3.39}$$

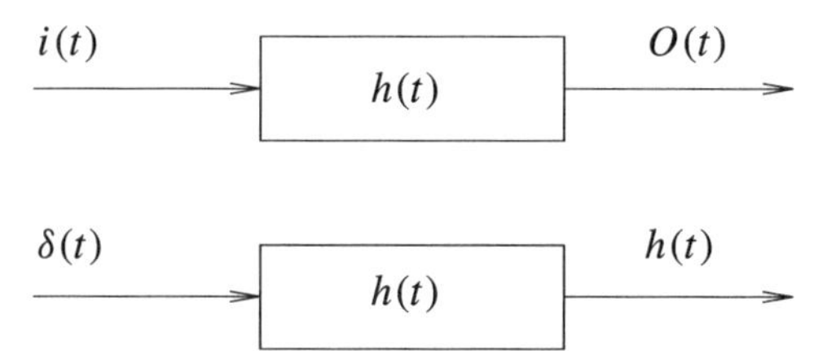

FIGURE 3.4 Linear system.

represents a superposition of the output due to a series of input delta functions whose amplitudes are modulated by the input signal. It is easy to show that the spectral domain representation of the convolution integral of (3.36) is given by

$$\widehat{f}(\omega) = \widehat{f}_1(\omega)\widehat{f}_2(\omega).$$

3.4.6 Parseval's Theorem

Parseval's theorem states that

$$\int_{-\infty}^{\infty} |f(t)|^2 \, dt = \frac{1}{2\pi} \int_{-\infty}^{\infty} \left|\widehat{f}(\omega)\right|^2 d\omega. \tag{3.40}$$

Two functions, $f(t)$ and $g(t)$, are related to their Fourier transform $\widehat{f}(\omega)$ and $\widehat{g}(\omega)$ via the Parseval's identity for Fourier transform, given as

$$\langle f(t), g(t)\rangle = \frac{1}{2\pi} \left\langle \widehat{f}(\omega), \widehat{g}(\omega)\right\rangle. \tag{3.41}$$

This can be shown from

$$\begin{aligned}
\langle f(t), g(t)\rangle &= \int_{-\infty}^{\infty} f(t)\overline{g(t)}\, dt \\
&= \frac{1}{2\pi} \int_{-\infty}^{\infty} \left(\int_{-\infty}^{\infty} \widehat{f}(\omega) e^{j\omega t}\, d\omega \right) \overline{g(t)}\, dt \\
&= \frac{1}{2\pi} \int_{-\infty}^{\infty} \widehat{f}(\omega) \left(\overline{\int_{-\infty}^{\infty} g(t) e^{-j\omega t}\, dt} \right) d\omega \\
&= \frac{1}{2\pi} \int_{-\infty}^{\infty} \widehat{f}(\omega)\overline{\widehat{g}(\omega)}\, d\omega \\
&= \frac{1}{2\pi} \left\langle \widehat{f}(\omega), \widehat{g}(\omega)\right\rangle.
\end{aligned} \tag{3.42}$$

In particular, with $g(t) = f(t)$, we get Parseval's theorem, given in (3.40). Equation (3.40) is a statement about the energy content in the signal. It states that the total energy computed in the time-domain $\left[\int_{-\infty}^{\infty} |f(t)|^2 \, dt\right]$ is equal to the total energy computed in the spectral domain $[(1/2\pi) \int_{-\infty}^{\infty} \left|\widehat{f}(\omega)\right|^2 d\omega]$. The Parseval theorem allows the energy of the signal to be considered in either the spectral domain or the time domain and can be interchanged between domains for convenience of computation.

3.5 EXAMPLES OF THE FOURIER TRANSFORM

We evaluate the Fourier transforms of several functions that will occur frequently in various applications. For this purpose we may use the definition given in Section 3.4 directly or use the properties of Fourier transform.

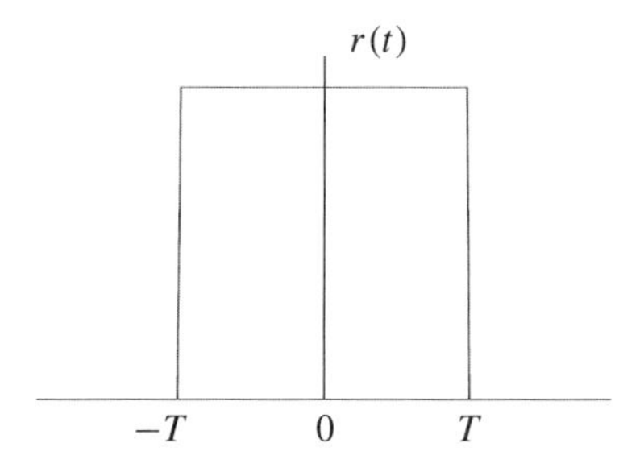

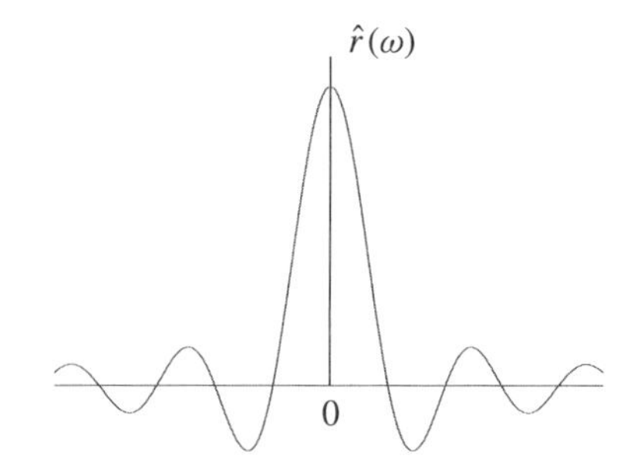

FIGURE 3.5 Rectangular pulse and its Fourier transform.

3.5.1 Rectangular Pulse

The rectangular pulse (Figure 3.5), $r(t)$, is defined by

$$r(t) = u(t+T) - u(t-T) \tag{3.43}$$

$$= \begin{cases} 1, & |t| < T \\ 0, & \text{otherwise.} \end{cases} \tag{3.44}$$

We obtain

$$\widehat{r}(\omega) = \int_{-\infty}^{\infty} r(t)\, e^{-j\omega t}\, dt = \int_{-T}^{T} e^{-j\omega t}\, dt = \frac{2T \sin \omega T}{\omega T}. \tag{3.45}$$

The function $(\sin \omega T)/\omega T$, called the *sinc function*, is the Fourier transform of a rectangular pulse. We remark here that for $T = \frac{1}{2}$, the function $r\big(t - \frac{1}{2}\big) = \chi_{[0,1)}(t)$ is called the *characteristic function* or the *first-order B-spline*. This is an important function to remember and will be recalled later in the development of wavelet theory.

3.5.2 Triangular Pulse

By convolving two rectangular pulses, we obtain a triangular pulse (Figure 3.6), which is expressed by

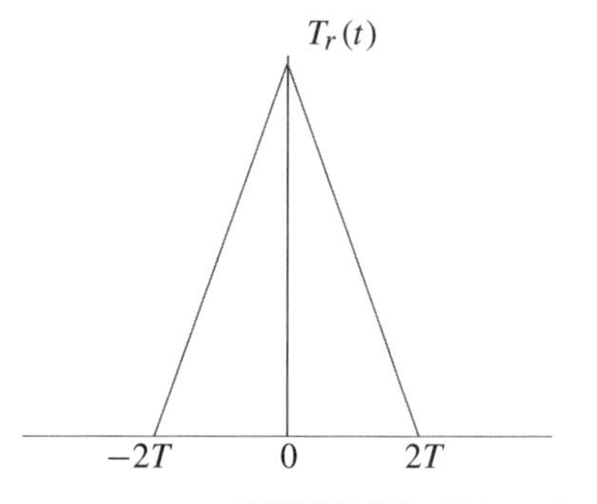

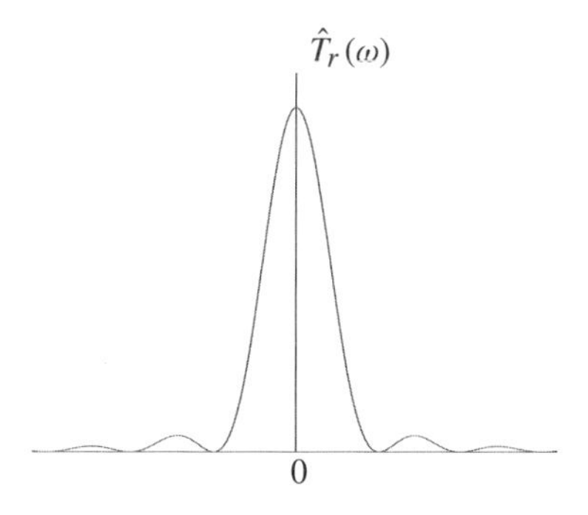

FIGURE 3.6 Triangular function and its Fourier transform.

$$T_r(t) = r(t) * r(t) \tag{3.46}$$

$$= \begin{cases} 2T\left(1+\dfrac{t}{2T}\right), & -2T \le t \le 0 \\ 2T\left(1-\dfrac{t}{2T}\right), & 0 \le t \le 2T \\ 0, & \text{otherwise.} \end{cases} \tag{3.47}$$

By the convolution theorem we have

$$\widehat{T}_r(\omega) = \left(\frac{2T\ \sin\omega T}{\omega T}\right)^2 \tag{3.48}$$

$$= 4T^2\,\frac{\sin^2(\omega T)}{(\omega T)^2}. \tag{3.49}$$

If $T = \frac{1}{2}$,

$$T_r(t) = \begin{cases} 1+t, & -1 \le t \le 0 \\ 1-t, & 0 \le t \le 1 \\ 0, & \text{otherwise,} \end{cases} \tag{3.50}$$

and $\widehat{T}_r(\omega) = [\sin(\omega/2)/(\omega/2)]^2$. The triangular function with $T = 1/2$, called the *second-order B-spline*, plays an important role as a scaling function of the spline wavelet.

3.5.3 Gaussian Function

The Gaussian function is one of the most important functions in probability theory and the analysis of random signals. It plays a central role in the Gabor transform, to be developed later. The Gaussian function with unit amplitude is expressed as

$$g(t) = e^{-\alpha t^2}. \tag{3.51}$$

$\widehat{g}(\omega)$ can be computed easily as

$$\begin{aligned} \widehat{g}(\omega) &= \int_{-\infty}^{\infty} e^{-\alpha t^2} e^{-j\omega t}\,dt \\ &= \int_{-\infty}^{\infty} e^{-\alpha\left[t^2 + j(\omega/\alpha)t - \omega^2/4\alpha\right] - \omega^2/4\alpha}\,dt \\ &= e^{-\omega^2/4\alpha} \int_{-\infty}^{\infty} e^{-\alpha[t + j(\omega/2\alpha)]^2}\,dt \\ &= \sqrt{\frac{\pi}{\alpha}}\, e^{-\omega^2/4\alpha}. \end{aligned} \tag{3.52}$$

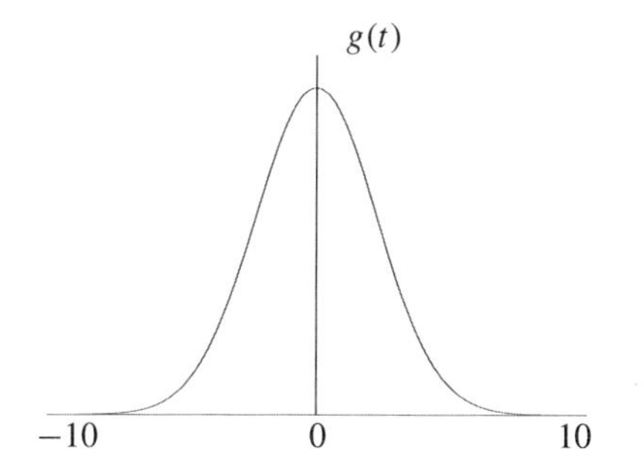

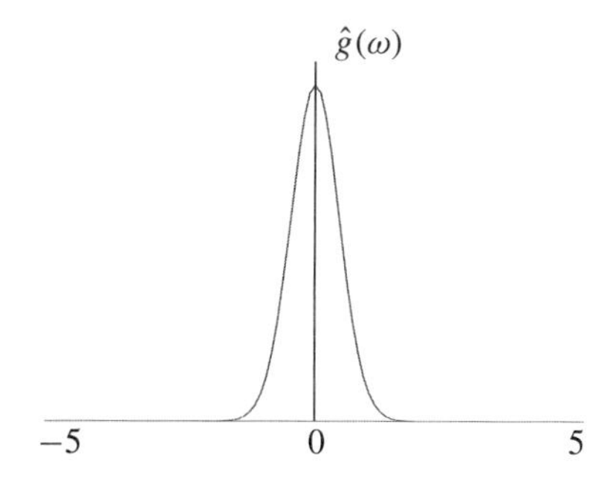

FIGURE 3.7 Gaussian function and its Fourier transform.

It is interesting to note that the Fourier transform of a Gaussian function is also a Gaussian function. The waveform and its transform are shown in Figure 3.7.

The parameter α can be used to control the width of the Gaussian pulse. It is evident from (3.51) and (3.52) that a large value of α produces a narrow pulse, but its spectrum spreads wider on the ω-axis.

3.6 POISSON'S SUM

We now return to the Fourier series and discuss Poisson's sum, whose derivation is made much simpler by using some properties of the Fourier transform. In many applications it is necessary to form a periodic function from a nonperiodic function with finite energy for the purpose of analysis.

Poisson's sum formula is useful in relating the time-domain information of such a function with its spectrum. Let $f(t) \in L^2(\mathbb{R})$. The periodic version of $f(t)$, called $f_p(t)$, is obtained by

$$f_p(t) := \sum_{n=-\infty}^{\infty} f(t + 2\pi n), \tag{3.53}$$

where we have assumed $T = 2\pi$ to be the period of $f_p(t)$. Consequently, $\omega_0 = 2\pi/T = 1$, and the Fourier series representation of $f_p(t)$ is

$$f_p(t) = \sum_{k=-\infty}^{\infty} c_k e^{jkt} \tag{3.54}$$

with the coefficient c_k given by

$$\begin{aligned} c_k &= \frac{1}{2\pi} \int_0^{2\pi} f_p(t) e^{-jkt}\, dt \\ &= \frac{1}{2\pi} \int_0^{2\pi} \sum_{n \in \mathbb{Z}} f(t + 2\pi n) e^{-jkt}\, dt \end{aligned}$$

$$= \frac{1}{2\pi} \sum_{n \in \mathbb{Z}} \int_0^{2\pi} f(t + 2\pi n) e^{-jkt}\, dt$$

$$= \frac{1}{2\pi} \sum_{n \in \mathbb{Z}} \int_{2\pi n}^{2\pi(n+1)} f(\xi) e^{-jk(\xi - 2\pi n)} d\xi, \tag{3.55}$$

where a change of variable $\xi = t + 2\pi n$ has been used. Since the summation and the integration limits effectively extend the integration over the entire real line $\mathbb{R}$, we may write

$$c_k = \frac{1}{2\pi} \int_{-\infty}^{\infty} f(\xi) e^{-jk\xi} d\xi$$

$$= \frac{1}{2\pi} \widehat{f}(k), \tag{3.56}$$

where the definition of the Fourier transform has been used. Combining (3.53), (3.54), and (3.56), we have *Poisson's sum formula,*

$$\sum_{n=-\infty}^{\infty} f(t + 2\pi n) = \frac{1}{2\pi} \sum_{k=-\infty}^{\infty} \widehat{f}(k) e^{jkt}. \tag{3.57}$$

For an arbitrary period T, the formula is generalized to

$$\sum_{n=-\infty}^{\infty} f(t + nT) = \frac{1}{T} \sum_{k=-\infty}^{\infty} \widehat{f}(k\omega_0) e^{jk\omega_0 t}. \tag{3.58}$$

If $g(t)$ is a scaled version of $f(t)$, that is,

$$g(t) = f(at), \qquad a > 0, \tag{3.59}$$

we have

$$\widehat{g}(\omega) = \frac{1}{a} \widehat{f}\left(\frac{\omega}{a}\right). \tag{3.60}$$

Poisson's sum formula for $f(at)$ is

$$\sum_{n=-\infty}^{\infty} f(at + 2\pi an) = \frac{1}{2\pi a} \sum_{k=-\infty}^{\infty} \widehat{f}\left(\frac{k}{a}\right) e^{jkt}. \tag{3.61}$$

If at is renamed as t, we have

$$\sum_{n=-\infty}^{\infty} f(t + 2\pi an) = \frac{1}{2\pi a} \sum_{k=-\infty}^{\infty} \widehat{f}\left(\frac{k}{a}\right) e^{jkt/a}. \tag{3.62}$$

Two other forms of Poisson's sum will be needed for derivations in subsequent sections. They are stated here without proof. The proofs are left as exercises.

$$\sum_{k\in\mathbb{Z}} \widehat{f}(\omega + 2\pi k) = \sum_{k\in\mathbb{Z}} f(k)e^{-jk\omega} \tag{3.63}$$

$$\frac{1}{a}\sum_{k\in\mathbb{Z}} \widehat{f}\left(\frac{\omega + 2\pi k}{a}\right) = \sum_{k\in\mathbb{Z}} f(ak)e^{-jk\omega} \tag{3.64}$$

3.6.1 Partition of Unity

A direct consequence of Poisson's sum is that a basis may be found so that unity is expressed as a linear sum of the basis. We call this property the *partition of unity*. Let a be $1/2\pi$ in (3.62). Poisson's sum formula becomes

$$\sum_{n=-\infty}^{\infty} f(t+n) = \sum_{k=-\infty}^{\infty} \widehat{f}(2\pi k)e^{j2\pi kt}. \tag{3.65}$$

If the spectrum of a function $f(t) \in L^2(\mathbb{R})$ is such that

$$\widehat{f}(2\pi k) = \delta_{0,k} \qquad \text{for } k \in \mathbb{Z}, \tag{3.66}$$

that is,

$$\widehat{f}(0) = 1$$

and

$$\widehat{f}(2\pi k) = 0 \qquad k \in \mathbb{Z} \setminus \{0\},$$

it follows from (3.65) that

$$\sum_{n\in\mathbb{Z}} f(t+n) \equiv 1. \tag{3.67}$$

The first and second orders of B-splines are good examples of functions satisfying this property.

First-Order B-Spline

$$\begin{aligned}
N_1(t) &:= \chi_{[0,1)}(t) \\
\widehat{N}_1(\omega) &= \int_0^1 e^{-j\omega t}\,dt = \frac{1-e^{-j\omega}}{j\omega} \\
\widehat{N}_1(0) &= \lim_{\omega\to 0} \frac{1-e^{-j\omega}}{j\omega} = 1 \\
\widehat{N}_1(2\pi k) &= 0, \qquad k \in \mathbb{Z} \setminus \{0\}.
\end{aligned}$$

Hence

$$\sum_{k\in\mathbb{Z}} N_1(t+k) \equiv 1. \tag{3.68}$$

Second-Order B-Spline

$$N_2(t) = N_1(t) * N_1(t) \tag{3.69}$$

$$= \begin{cases} t, & t \in [0,1) \\ 2-t, & t \in [1,2) \\ 0, & \text{otherwise.} \end{cases} \tag{3.70}$$

From the convolution property, we have

$$\widehat{N}_2(\omega) = \left[\widehat{N}_1(\omega)\right]^2 \tag{3.71}$$

$$\widehat{N}_2(\omega) = \left(\frac{1-e^{-j\omega}}{j\omega}\right)^2. \tag{3.72}$$

Again, we find here that

$$\widehat{N}_2(0) = 1 \tag{3.73}$$

$$\widehat{N}_2(2\pi k) = 0, \qquad k \in \mathbb{Z}\setminus\{0\}. \tag{3.74}$$

Consequently, $N_2(t)$ also satisfies the conditions for partition of unity. In fact, from the recursive relation of the B-spline,

$$N_m(t) = N_{m-1}(t) * N_1(t) \tag{3.75}$$

$$= \int_0^1 N_{m-1}(t-\tau)\,d\tau, \tag{3.76}$$

we have $\widehat{N}_m(\omega) = [(1-e^{-j\omega})/j\omega]^m$, which satisfies the requirement for partition of unity. Hence B-splines of arbitrary orders all have that property. Graphic illustrations for the partition of unity are shown in Figure 3.8.

3.7 SAMPLING THEOREM

The sampling theorem is fundamentally important to digital signal analysis. It states that if a signal $f(t)$ is bandlimited with bandwidth 2Ω, the signal $f(t)$ can be reconstructed exactly from its sampled values at equidistant grid points. The distance between adjacent sample points, called the sampling period h, should not exceed π/Ω. The function $f(t)$ is recovered by using the formula

$$f(t) = \sum_{k\in\mathbb{Z}} f(kh)\frac{\sin[\Omega(t-kh)]}{\pi(t-kh)}, \qquad k \in \mathbb{Z}. \tag{3.77}$$

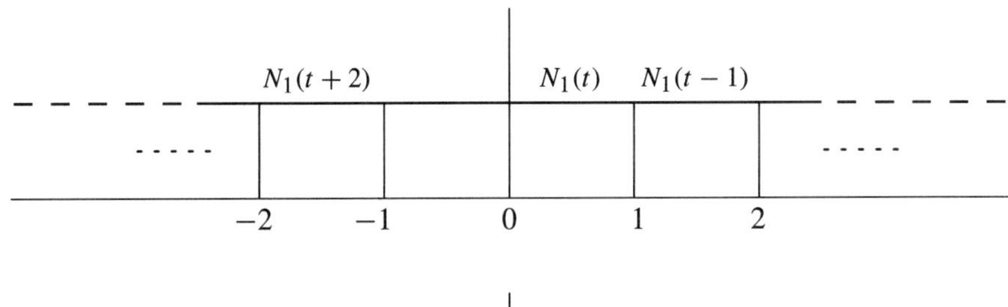

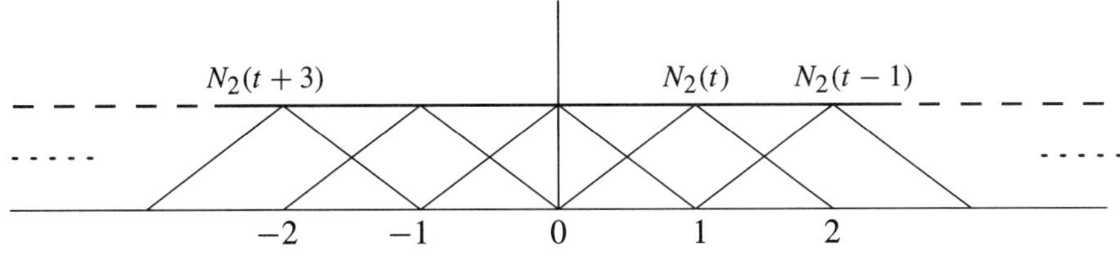

FIGURE 3.8 Partition of unity.

If $h = \pi/\Omega$, the sampling frequency $f_s = 1/h = \Omega/\pi$ is called the *Nyquist rate*. Theoretically, $f(t)$ can always be reconstructed perfectly from samples if $h < \pi/\Omega$. In practice, however, we cannot recover $f(t)$ without error due to the infinite nature of the sinc function.

Let $\widehat{f}(\omega)$ be the Fourier transform of $f(t)$:

$$\widehat{f}(\omega) = \int_{-\infty}^{\infty} f(t)e^{-j\omega t}\, dt.$$

The integral can be approximated using Simpson's rule as

$$\widehat{f}(\omega) \cong \widehat{F}(\omega) = h \sum_{k\in\mathbb{Z}} f(kh)e^{-j\omega kh}. \tag{3.78}$$

Using Poisson's sum formula in (3.64), we can rewrite $\widehat{F}(\omega)$ as

$$\begin{aligned}\widehat{F}(\omega) &= h \sum_{k\in\mathbb{Z}} f(kh)e^{-jk\omega h} \\ &= \sum_{k\in\mathbb{Z}} \widehat{f}\left(\frac{\omega h + 2\pi k}{h}\right) \\ &= \widehat{f}(\omega) + \sum_{k\in\mathbb{Z}\setminus\{0\}} \widehat{f}\left(\omega + \frac{2\pi k}{h}\right).\end{aligned} \tag{3.79}$$

Hence $\widehat{F}(\omega)$ contains $\widehat{f}(\omega)$ plus infinitely many copies of $\widehat{f}(\omega)$ shifted along the ω-axis. In order for $\widehat{f}(\omega)$ to be disjointed with its copies, the amount of shift, $2\pi/h$,

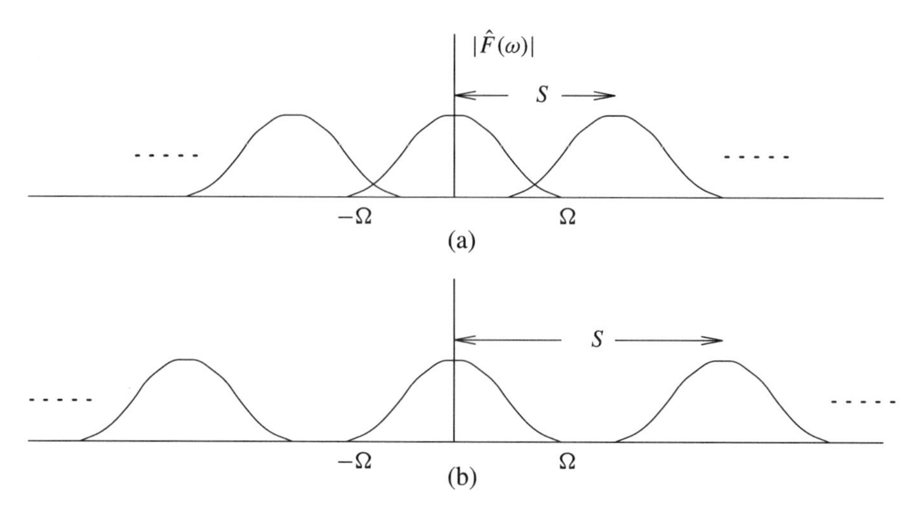

FIGURE 3.9 (*a*) Undersampling ($S = 2\pi/h < 2\Omega$); (*b*) oversampling ($S = 2\pi/h > 2\Omega$).

must be at least 2Ω (see Figure 3.9); that is,

$$\frac{2\pi}{h} \geq 2\Omega, \qquad h \leq \frac{\pi}{\Omega}. \tag{3.80}$$

To recover the original function, we introduce a spectral window

$$\widehat{W}(\omega) = \begin{cases} 1, & |\omega| \leq \Omega \\ 0, & \text{otherwise} \end{cases} \tag{3.81}$$

and recover $\widehat{f}(\omega)$ by

$$\widehat{f}(\omega) = \widehat{F}(\omega)\widehat{W}(\omega). \tag{3.82}$$

From the convolution theorem we obtain $f(t)$:

$$f(t) = F(t) * W(t). \tag{3.83}$$

Since $W(t) = \sin \Omega t / \pi t$ is well known, we compute $F(t)$ from the inverse Fourier transform

$$\begin{aligned} F(t) &= \frac{h}{2\pi} \int_{-\infty}^{\infty} \sum_{k\in\mathbb{Z}} f(kh) e^{-jkh\omega} e^{j\omega t}\, d\omega \\ &= \frac{h}{2\pi} \sum_{k\in\mathbb{Z}} f(kh) \int_{-\infty}^{\infty} e^{j\omega(t-kh)}\, d\omega \\ &= h \sum_{k\in\mathbb{Z}} f(kh)\delta(t - kh), \end{aligned} \tag{3.84}$$

where we have used (3.25). The function $f(t)$ is recovered by using the convolution formula

$$\begin{aligned} f(t) &= h \sum_{k \in \mathbb{Z}} f(kh) \int_{-\infty}^{\infty} \delta(\tau - kh) W(t - \tau)\, d\tau \\ &= h \sum_{k \in \mathbb{Z}} f(kh) W(t - kh) \\ &= h \sum_{k \in \mathbb{Z}} f(kh) \frac{\sin[\Omega(t - kh)]}{\pi(t - kh)} \\ &= \sum_{k \in \mathbb{Z}} f(kh) \frac{\sin(\Omega t - k\pi)}{\Omega t - k\pi}, \end{aligned} \tag{3.85}$$

where we have used $\Omega h = \pi$. We remark here that (3.85) represents an interpolation formula. Since $\sin[\Omega(t - kh)]/[\Omega(t - kh)]$ is unity at $t = kh$ and zero at all other sampling points, the function value at kh is not influenced by other sampled values:

$$f(kh) = \sum_{k \in \mathbb{Z}} f(kh) \frac{\sin(0)}{0} = f(kh). \tag{3.86}$$

Hence the function $f(t)$ is reconstructed through interpolation of its sampled values with the sinc function as the interpolation kernel.

3.8 PARTIAL SUM AND THE GIBBS PHENOMENON

The partial sum of a Fourier series is a least-square-error approximation to the original periodic function. Let $p_M(t)$ be the $(2M + 1)$-term partial sum of the Fourier series of a periodic function $p(t)$ with period T:

$$p_M(t) = \sum_{k=-M}^{M} \alpha_k e^{jk\omega_0 t}, \tag{3.87}$$

with the Fourier coefficients given by

$$\alpha_k = \frac{1}{T} \int_{-T/2}^{T/2} p(t) e^{-jk\omega_0 t}\, dt. \tag{3.88}$$

Putting α_k back into (3.87), we have the partial sum

$$p_M(t) = \sum_{k=-M}^{M} \frac{1}{T} \int_{-T/2}^{T/2} p(\tau) e^{-jk\omega_0 \tau} e^{jk\omega_0 t}\, d\tau. \tag{3.89}$$

On interchanging the order of summation and integration, we obtain

$$p_M(t) = \frac{1}{T}\int_{-T/2}^{T/2} p(\tau) \sum_{k=-M}^{M} e^{jk\omega_0(t-\tau)}\, d\tau$$

$$= \frac{1}{T}\int_{-T/2}^{T/2} p(\tau) \frac{\sin\left(M+\frac{1}{2}\right)(t-\tau)\omega_0}{\sin\frac{1}{2}(t-\tau)\omega_0}\, d\tau, \tag{3.90}$$

which is the convolution between the orginial periodic function with the Fourier series kernel discussed in Section 3.2.2. We can easily see that the oscillatory characteristic of K_N is carried into the partial sum. If $p(t)$ is a rectangular pulse train or a periodic function with jump discontinuities, the partial Fourier series will exhibit oscillation around the discontinuities. This is known as the Gibbs phenomenon. The percentage of overshoot remains constant regardless of the number of terms taken for the approximation. As $M \to \infty$, the sum converges to the midpoint at the discontinuity [4].

3.9 FOURIER ANALYSIS OF DISCRETE-TIME SIGNALS

Since computation of Fourier series coefficients and Fourier transform requires integration, the function must be describable analytically by elementary functions such as sine and cosine functions, exponential functions, and terms from a power series. In general, most signals we encounter in real life are not representable by elementary functions. We must use numerical algorithms to compute the spectrum. If the signals are sampled signals, the discrete Fourier series and discrete-time Fourier transform are computable directly. They produce an approximate spectrum of the original analog signal.

3.9.1 Discrete Fourier Basis and Discrete Fourier Series

For a given periodic sequence with periodicity N, we have

$$f_p(n+mN) = f_p(n), \qquad m \in \mathbb{Z}. \tag{3.91}$$

The Fourier basis for this periodic sequence has only N basis functions, namely,

$$\mathbf{e}_k(n) = e^{(j2\pi/N)kn}, \qquad k = 0, 1, \ldots, N-1. \tag{3.92}$$

We can easily show the periodicity of the basis set

$$\begin{aligned} \mathbf{e}_{k+N} &= e^{(j2\pi/N)(k+N)n} \\ &= e_k \cdot e^{j2\pi n} \\ &= e_k \end{aligned} \tag{3.93}$$

since $e^{j2\pi n} = 1$ for integer n. Therefore, the expansion of $f_p(n)$ is in the form

$$f_p(n) = \sum_{k=0}^{N-1} \alpha_k \mathbf{e}_k(n) \tag{3.94}$$

$$= \sum_{k=0}^{N-1} \alpha_k e^{(j2\pi/N)\,kn}, \tag{3.95}$$

and then we can compute the coefficients by

$$\alpha_k = \left\langle f_p(n),\ e^{(j2\pi k/N)\,n} \right\rangle$$

$$= \frac{1}{N} \sum_{k=0}^{N-1} f_p(n)\, e^{(-j2\pi k/N)\,n}. \tag{3.96}$$

Equations (3.94) and (3.96) form a transform pair for discrete periodic sequences and their discrete spectra. It is quite easy to see from (3.96) that the Fourier coefficients $\{\alpha_k\}$ are also periodic with N.

$$\alpha_k = \alpha_{k+mN}, \qquad m \in \mathbb{Z}.$$

Example 1: Find the Fourier series coefficients for the sequence

$$f(n) = \cos(\sqrt{5}\pi n).$$

SOLUTION: The given sequence is not a periodic sequence since we cannot find an integer N such that $f(n+N) = f(n)$. Consequently, $f(n)$ does not have a discrete Fourier series representation.

Example 2: Find the Fourier series representation of

(a) $f(n) = \cos(n\pi/5)$, and
(b) $f(n) = \{1, 1, 0, 0\}$.

SOLUTION:

(a) Instead of computing the coefficients directly using (3.96), we may represent the cosine function in its exponential form,

$$f(n) = \frac{1}{2}\left(e^{j(2\pi/10)n} + e^{-j(2\pi/10)n}\right). \tag{3.97}$$

The periodicity of this sequence is seen as $N = 10$. Since (3.97) is already in the form of an exponential series as in (3.95), we conclude that

$$\alpha_k = \begin{cases} \frac{1}{2}, & k = 1 \\ \frac{1}{2}, & k = 9 \\ 0, & \text{otherwise.} \end{cases} \tag{3.98}$$

(b) We compute the Fourier coefficients using (3.96) to obtain

$$\alpha_k = \frac{1}{4}\left(1 + e^{-j(2\pi/4)k}\right), \qquad k = 0, 1, 2, 3.$$

We have

$$\alpha_k = \begin{cases} \frac{1}{2}, & k = 0 \\ \frac{1}{4}(1 - j), & k = 1 \\ 0, & k = 2 \\ \frac{1}{4}(1 + j), & k = 3. \end{cases} \tag{3.99}$$

The sequence and its magnitude spectrum are shown in Figure 3.10.

3.9.2 Discrete-Time Fourier Transform

If a discrete signal is aperiodic, we may consider it to be a periodic signal with period $N = \infty$. In this case we extend the discrete Fourier series analysis to the discrete-time Fourier transform (DTFT), similar to the extension in the analog domain. In DTFT, the time variable (n) is discretized while the frequency variable (ω) is continuous since

$$\Delta\omega = \lim_{N\to\infty} \frac{2\pi}{N} \to \omega.$$

The DTFT pair is given explicitly by

$$\widehat{f}(\omega) = \sum_{n=-\infty}^{\infty} f(n)e^{-jn\omega} \tag{3.100}$$

$$f(n) = \frac{1}{2\pi}\int_{-\pi}^{\pi} \widehat{f}(\omega)e^{jn\omega}\, d\omega. \tag{3.101}$$

Example 3: Determine the spectrum of the exponential sequence

$$f(n) = a^n, \qquad \forall\, n \in \mathbb{Z}^+ := \{0, 1, \ldots\}, \qquad |a| < 1.$$

SOLUTION: Using (3.100) yields

$$\widehat{f}(\omega) = \sum_{n=0}^{\infty} a^n e^{-jn\omega}$$

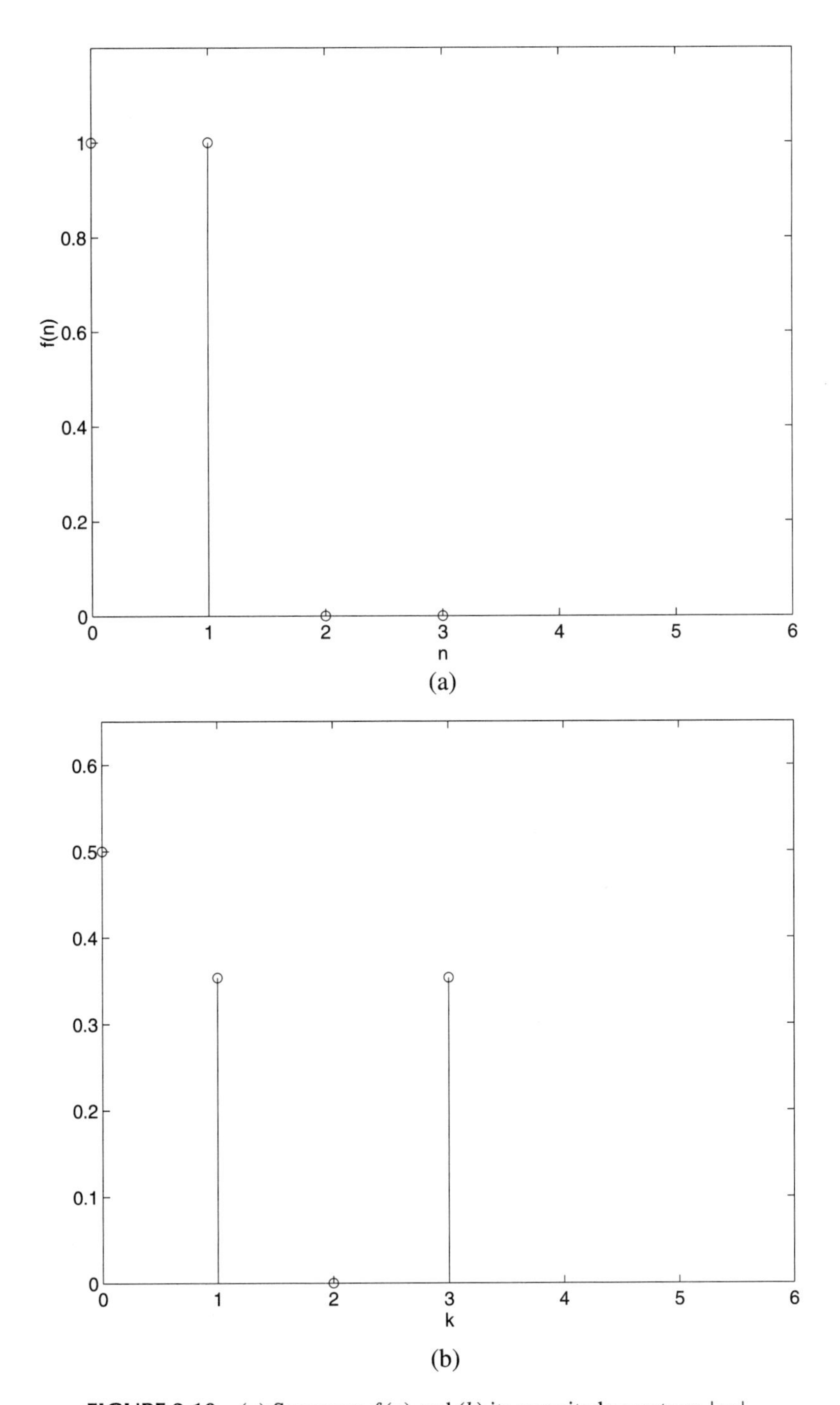

FIGURE 3.10 (*a*) Sequency $f(n)$ and (*b*) its magnitude spectrum $|\alpha_k|$.

$$= \sum_{n=0}^{\infty} (ae^{-j\omega})^n$$

$$= \frac{1}{1-(ae^{-j\omega})}. \tag{3.102}$$

We have pointed out in Chapter 2 that the DTFT can be obtained from the z-transform by replacing the variable z with $e^{j\omega}$. For this example the z-transform is

$$F(z) = \sum_{n=0}^{\infty} a^n z^{-n}$$

$$= \sum_{n=0}^{\infty} (az^{-1})^n$$

$$= \frac{1}{1-(az^{-1})}.$$

Therefore, replacing the variable z with $e^{j\omega}$ yields

$$F(z)|_{z=e^{j\omega}} = \frac{1}{1-(ae^{-j\omega})} = \widehat{f}(\omega).$$

The z-transform $F(z)$ and the DTFT $\left[\widehat{f}(\omega) = F(z)|_{z=e^{j\omega}}\right]$ will be used interchangeably in future derivations and discussions of wavelet construction.

3.10 DISCRETE FOURIER TRANSFORM

The integral in the inverse DTFT discussed in Section 3.9 must be evaluated to recover the original discrete-time signal. Instead of evaluating the integral, we can obtain a good approximation by a discretization on the frequency (ω) axis.

Since the function $f(t)$ is bandlimited (if it is not, we make it so by passing it through a lowpass filter with sufficiently large width), we need to discretize the interval $[-\Omega, \Omega]$ only, namely

$$\omega_n = \frac{2\pi n}{Nh}, \qquad n = -\frac{N}{2}, \ldots, \frac{N}{2}. \tag{3.103}$$

The integral in (3.19) can now be approximated as a series sum, namely

$$\widehat{f}(\omega_n) \approx h \sum_{k=0}^{N-1} f(kh) e^{-j\omega_n kh} = h\widehat{f}(n), \tag{3.104}$$

where

$$\widehat{f}(n) = \sum_{k=0}^{N-1} f(kh) e^{-j(2\pi kn/N)}. \tag{3.105}$$

We can easily verify that evaluation of the discrete Fourier transform using (3.105) is an $O(N^2)$ process. We can compute the discrete Fourier transform (DFT) with an $O(N \log_2 N)$ operation with the well-known algorithm of the *fast Fourier transform* (FFT). One of the commonly used FFT algorithms is that of Danielson and Lanczos, according to which, assuming N to be such that it is continuously divisible by 2, a DFT of data length N can be written as a sum of two discrete Fourier transforms, each of length $N/2$. This process can be used recursively until we arrive at the DFT of only two data points. This is known as the radix-2 FFT algorithm. Without getting into many details of the algorithm, which the interested reader can obtain from many excellent books available on these topics, we simply mention here that by appropriately arranging the data of length N, where N is an integer power of 2 (known as decimation-in-time and decimation-in-frequency arrangements), we can compute the discrete Fourier transform in an $O(N \log_2 N)$ operation. If N is not an integer power of 2, we can always make it so by padding the data sequence with zeros.

3.11 EXERCISES

1. Verify that the order of taking the complex conjugate and the Fourier transform of a function $f \in L^2(-\infty, \infty)$ can be reversed as follows:

$$\widehat{\overline{F}}(\eta) = \overline{\widehat{F}}(-\eta)$$

for any $\eta \in R$.

2. Check that the condition

$$\frac{d^j}{d\omega^j} \widehat{\psi}(\omega) \mid_{\omega=0} \mid = 0$$

is equivalent to the moment condition

$$\int_{-\infty}^{\infty} t^j \psi(t)\, dt = 0$$

for any positive integer number j.

3. Show that the Dirichlet kernel

$$D_n(u) = \frac{1}{\pi}\left(\frac{1}{2} + \sum_{k=1}^{n} \cos ku\right) = \frac{\sin(n+\frac{1}{2})u}{2\pi \sin(u/2)}.$$

Plot the kernel for $n = 6$.

4. Find the Fourier series of $f(t) = e^{jxt}$, $-\pi \le t < \pi$, and $T = 2\pi$.

5. Determine the energy-normalized constant A of the Gaussian function $g_\alpha(t) = Ae^{-\alpha t^2}$ and derive the expression of the Fourier transform.

6. Extend Poisson's sum formula to arbitrary period T.

7. Derive the following Poisson's sum formulas in the spectral domain:

$$\sum_k \widehat{f}(\omega + 2\pi k) = \sum_k f(k)e^{-jk\omega}$$

$$\frac{1}{a}\sum_k \widehat{f}\left(\frac{\omega + 2\pi k}{a}\right) = \sum_k f(ak)e^{-jk\omega}.$$

REFERENCES

1. R. S. Elliott, *The Theory of Antenna Arrays.* Vol. 2 of *Microwave Scanning Antennas*, R. C. Hansen (Ed.). San Diego, Calif.: Academic Press, 1966.
2. A. Papoulis, *The Fourier Integral and Its Applications.* New York: McGraw-Hill Book Company, 1962.
3. K. Iizuka, *Engineering Optics*, 2nd ed. New York: Springer-Verlag, 1986.
4. W. W. Harman and D. W. Lytle, *Electrical and Mechanical Networks, An Introduction to Their Analysis.* New York: McGraw-Hill Book Company, 1962.

CHAPTER FOUR

Time–Frequency Analysis

We summarized the Fourier analysis rather briefly in Chapter 3 to refresh the memory of the reader and to point out a few important concepts in the analysis that will be useful when we discuss time–frequency analysis. We observe from the definition of the Fourier transform (3.19) that the integration cannot be carried out until the entire waveform in the whole of the real line $(-\infty, \infty)$ is known. This is because the functions $e^{j\omega t}$ or $\cos \omega t$ and $\sin \omega t$ are *global functions*. By this we mean that a small perturbation of the function at any point along the t-axis influences every point on the ω-axis, and vice versa. If we imagine the signal $f(t)$ as the modulating function for $e^{j\omega t}$, a perturbation at any point on the t-axis will propagate through the entire ω-axis. Another observation we make on the Fourier transform is that the integral can be evaluated at only one frequency at a time. This is quite inconvenient from a signal processing point of view. Although there are fast algorithms to compute the transform digitally, it cannot be carried out in real time. All necessary data must be stored in the memory before the discrete or fast Fourier transform can be computed.

Although unquestionably the most versatile method, Fourier analysis becomes inadequate when one is interested in the local frequency contents of a signal. In other words, the Fourier spectrum does not provide any time-domain information about the signal. To demonstrate this point, let us examine the function shown in Figure 4.1a, which represents a truncated sinusoid of frequency 4 Hz in the time domain with perturbations near $t = 0.7$ and $t = 1.3$ s. We have seen in Chapter 3 that a sinusoid in the time domain will appear as a delta function in the frequency domain, and vice versa. Observe that the frequency spread near 4 Hz in Figure 4.1b is due to truncation of the sinusoid. We conclude from the Fourier spectrum shown in Figure 4.1b that the sharp pulse near 4 Hz comes *primarily* from the sinusoid of 4 Hz, and the small ripples that appear throughout the frequency axis are due *primarily* to some delta functions (sharp changes) in the time domain. However, we are unable to point out the locations of these delta functions in the time axis by observing the spectrum of Figure 4.1b. This can be explained simply by the Fourier representation of a delta function (3.25). The delta function requires an infinite number of sinusoidal functions that combine constructively at $t = 0$ while interfering with one another de-

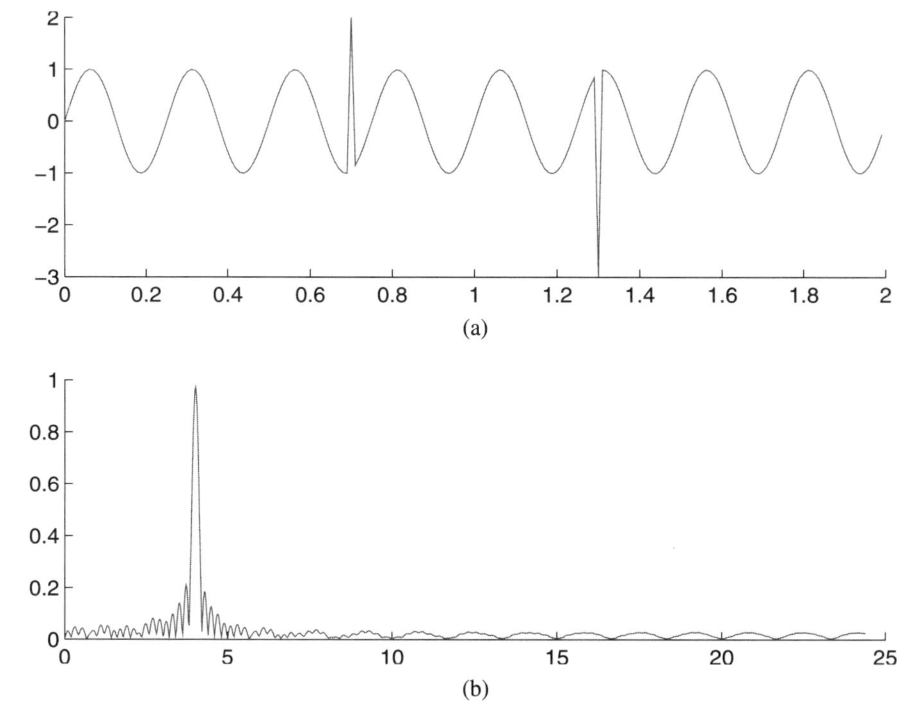

FIGURE 4.1 (*a*) Sinusoid signal with perturbation at $t = 0.7$ and $t = 1.3$, and (*b*) its magnitude spectrum.

structively to produce zero at all points $t \neq 0$. This shows the extreme cumbersomeness and ineffectiveness of using global functions $e^{j\omega t}$ to represent local functions. To correct this deficiency, a local analysis is needed to combine both the time- and frequency-domain analyses to achieve *time–frequency analysis,* by means of which we can extract the local frequency contents of a signal. This is very important, since in practice we may be interested in some particular portion of the spectrum only and, therefore, we may like to know which portion of the time-domain signal is primarily responsible for a given characteristic in the spectrum.

Common sense dictates that to know the local frequency contents of a signal, we should first "remove" the portion desired from the signal given and then take the Fourier transform of the part removed. Such a time–frequency analysis method is referred to as *short-time Fourier transform* (STFT). Before we discuss STFT, let us discuss the notion of *window function,* by means of which the desired portion of a given signal can be removed.

4.1 WINDOW FUNCTION

A desired portion of a signal can be removed from the main signal by multiplying the original signal by another function, which is zero outside the interval desired.

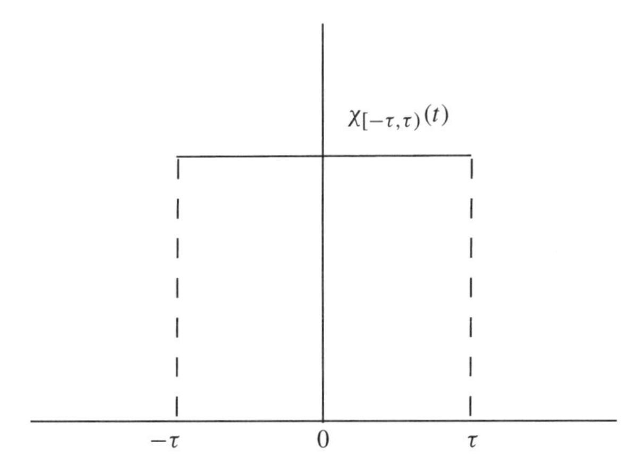

FIGURE 4.2 Characteristic function.

Let $\phi(t) \in L^2(\mathbb{R})$ be a real-valued window function. Then the product $f(t)\phi(t-b) =: f_b(t)$ will contain the information of $f(t)$ near $t = b$. In particular, if $\phi(t) = \chi_{[-\tau,\tau)}(t)$, as shown in Figure 4.2, then

$$f_b(t) = \begin{cases} f(t), & t \in [b-\tau, b+\tau) \\ 0, & \text{otherwise.} \end{cases} \tag{4.1}$$

By changing the parameter b we can slide the window function along the time axis to analyze the local behavior of the function $f(t)$ in different intervals.

The two most important parameters for a window function are its center and width; the latter is usually twice the radius. It is clear that the center and the standard width of the window function in Figure 4.2 are 0 and 2τ, respectively. For a general window function $\phi(t)$, we define its center t^* as

$$t^* := \frac{1}{\|\phi\|^2} \int_{-\infty}^{\infty} t\, |\phi(t)|^2 \, dt \tag{4.2}$$

and the root-mean-square (RMS) radius Δ_ϕ as

$$\Delta_\phi := \frac{1}{\|\phi\|} \left[\int_{-\infty}^{\infty} (t-t^*)^2 \, |\phi(t)|^2 \, dt \right]^{1/2}. \tag{4.3}$$

For the particular window of Figure 4.2, it is easy to verify that $t^* = 0$ and $\Delta_\phi = \tau/\sqrt{3}$. Therefore, the RMS width is smaller than the standard width by $1/\sqrt{3}$.

The function $\phi(t)$ described above with finite Δ_ϕ is called a *time window*. Similarly, we can have a frequency window $\widehat{\phi}(\omega)$ with center ω^* and the RMS radius $\Delta_{\widehat{\phi}}$ defined analogous to (4.2) and (4.3) as

$$\omega^* := \frac{1}{\|\widehat{\phi}\|^2} \int_{-\infty}^{\infty} \omega \left|\widehat{\phi}(\omega)\right|^2 d\omega, \tag{4.4}$$

$$\Delta_{\widehat{\phi}} := \frac{1}{\|\widehat{\phi}\|}\left[\int_{-\infty}^{\infty}(\omega-\omega^*)^2\left|\widehat{\phi}(\omega)\right|^2 d\omega\right]^{1/2}. \tag{4.5}$$

As wc know, theoretically a function cannot be limited in time and frequency simultaneously. However, we can have $\phi(t)$ such that both Δ_ϕ and $\Delta_{\widehat{\phi}}$ are both finite; in such a case the function $\phi(t)$ is called a *time–frequency window*. It is easy to verify that for the window of Figure 4.2, $\omega^* = 0$ and $\Delta_\phi = \infty$. This window is the best (ideal) time window but the worst (unacceptable) frequency window.

A figure of merit for the time–frequency window is its time–frequency-width product $\Delta_\phi \Delta_{\widehat{\phi}}$, which is bounded from below by the *uncertainty principle* and is given by

$$\Delta_\phi \Delta_{\widehat{\phi}} \geq \frac{1}{2}, \tag{4.6}$$

where the equality holds only when ϕ is of the Gaussian type (see Section 3.5.3).

4.2 SHORT-TIME FOURIER TRANSFORM

In the beginning of this chapter we indicated that we could obtain the approximate frequency contents of a signal $f(t)$ in the neighborhood of some desired location in time, say $t = b$, by first windowing the function using an appropriate window function $\phi(t)$ to produce the windowed function $f_b(t) = f(t)\phi(t-b)$ and then taking the Fourier transform of $f_b(t)$. This is the short-time Fourier transform (STFT). Formally, we can define the STFT of a function $f(t)$ with respect to the window function $\phi(t)$ evaluated at the location (b, ξ) in the time–frequency plane as

$$G_\phi f(b,\xi) := \int_{-\infty}^{\infty} f(t)\overline{\phi_{b,\xi}(t)}\,dt \tag{4.7}$$

where

$$\phi_{b,\xi}(t) := \phi(t-b)e^{j\xi t}. \tag{4.8}$$

The window function $\phi(t)$ in (4.7) is allowed to be complex and satisfies the condition

$$\widehat{\phi}(0) = \int_{-\infty}^{\infty}\phi(t)dt \neq 0.$$

In other words, $\widehat{\phi}(\omega)$ behaves as a lowpass filter. That is, the spectrum is nonzero at $\omega = 0$. Because of the windowing nature of the STFT, this transform is also referred to as the *windowed Fourier transform* or *running-window Fourier transform*.

Unlike the case of Fourier transform, in which the function $f(t)$ must be known for the entire time axis before its spectral component at any single frequency can be computed, STFT needs to know $f(t)$ only in the interval in which $\phi(t-b)$ is nonzero. In other words, $G_\phi f(b,\xi)$ gives the approximate spectrum of f near $t=b$.

If the window function $\phi(t-b)$ in (4.7) is considered as the modulating function of the sinusoid $e^{-j\xi t}$, the STFT can be written as

$$G_\phi f(b,\xi) = \left\langle f(t), \phi(t-b)e^{j\xi t} \right\rangle. \tag{4.9}$$

The function $\phi_{b,\xi}(t) = \phi(t-b)e^{j\xi t}$ behaves like a *packet of waves,* where the sinusoidal wave oscillates inside the envelope function $\phi(t)$. In addition, (4.8) indicates that each of these packets of waves behaves like a basis function, so that the STFT may be interpreted as the components of the function $f(t)$ with respect to this basis in the time–frequency plane.

4.2.1 Inversion Formula

One can recover the time function $f_b(t)$ by taking the inverse Fourier transform of $(G_\phi f)(b,\xi)$:

$$f_b(t) = \phi(t-b)f(t) = \frac{1}{2\pi}\int_{-\infty}^{\infty} G_\phi f(b,\xi)e^{j\xi t}\,d\xi. \tag{4.10}$$

The original $f(t)$ is obtained by multiplying (4.10) with $\overline{\phi(t-b)}$ and integrating with respect to b. The final recovery formula is

$$f(t) = \frac{1}{2\pi\,\|\phi(t)\|}\int_{-\infty}^{\infty} d\xi\, e^{j\xi t}\int_{-\infty}^{\infty} G_\phi f(b,\xi)\overline{\phi(t-b)}\,db. \tag{4.11}$$

One may observe a similar symmetric property between (4.7) and (4.11), and that of the Fourier transforms in (3.19) and (3.22).

4.2.2 Gabor Transform

The Gabor transform was developed by D. Gabor [1], who used the Gaussian function

$$g_\alpha(t) = \frac{1}{2\pi\alpha}e^{-t^2/4\alpha}, \qquad \alpha > 0 \tag{4.12}$$

as the window function. The Fourier transform of (4.12) is

$$\widehat{g}_\alpha(\omega) = e^{-\alpha\omega^2}, \qquad \alpha > 0. \tag{4.13}$$

The window property of $g_\alpha(t)$ can be computed using the formulas in Section 4.1 to give $t^* = \omega^* = 0$, $\Delta_{g_\alpha} = \sqrt{\alpha}$, and $\Delta\widehat{g}_\alpha = 1/2\sqrt{\alpha}$. Observe that $\Delta g_\alpha\,\Delta\widehat{g}_\alpha = 0.5$ attains the lower bound of the *uncertainty principle.*

4.2.3 Time–Frequency Window

Let us consider the window function $\phi(t)$ in (4.7). If t^* is the center and Δ_ϕ the radius of the window function, then (4.7) gives the information of the function $f(t)$ in the time window:

$$\left[t^* + b - \Delta_\phi, t^* + b + \Delta_\phi\right]. \tag{4.14}$$

To derive the corresponding window in the frequency domain, apply Parseval's identity (3.41) to (4.7). We have

$$G_\phi f(b,\xi) = \int_{-\infty}^{\infty} f(t)\overline{\phi(t-b)}e^{-j\xi t}\,dt \tag{4.15}$$

$$= \frac{1}{2\pi}e^{-j\xi b}\int_{-\infty}^{\infty} \widehat{f}(\omega)\overline{\widehat{\phi}(\omega-\xi)e^{jb\omega}}\,d\omega$$

$$= e^{-j\xi b}\left[\widehat{f}(\omega)\overline{\widehat{\phi}(\omega-\xi)}\right]^{\vee}(b), \tag{4.16}$$

where the symbol "$\vee$" represents the inverse Fourier transform. Observe that (4.15) has a form similar to (4.7). If ω^* is the center and $\Delta\widehat{\phi}$ is the radius of the window function $\widehat{\phi}(\omega)$, then (4.15) gives us information about the function $\widehat{f}(\omega)$ in the interval

$$\left[\omega^* + \xi - \Delta\widehat{\phi}, \omega^* + \xi + \Delta\widehat{\phi}\right]. \tag{4.17}$$

Because of the similarity of representations in (4.7) and (4.15), the STFT gives information about the function $f(t)$ in the time–frequency window:

$$\left[t^* + b - \Delta\phi, t^* + b + \Delta\phi\right] \times \left[\omega^* + \xi - \Delta\widehat{\phi}, \omega^* + \xi + \Delta\widehat{\phi}\right]. \tag{4.18}$$

Figure 4.3 represents graphically the notion of the time–frequency window given by (4.17). Here we have assumed that $t^* = \omega^* = 0$.

4.2.4 Properties of STFT

Linearity Let $f(t) = \alpha f_1(t) + \beta f_2(t)$ be a linear combination of two functions $f_1(t)$ and $f_2(t)$ with weights α and β independent of t. Then the STFT of $f(t)$,

$$G_\phi f(b,\xi) = \alpha(G_\phi f_1)(b,\xi) + \beta(G_\phi f_2)(b,\xi), \tag{4.19}$$

is the linear sum of the STFT of the individual function. Hence STFT is a linear transformation.

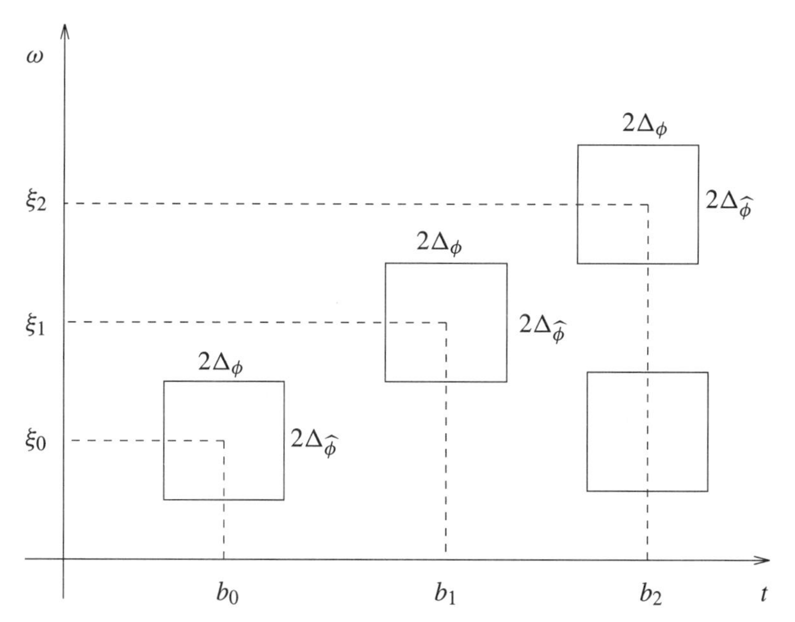

FIGURE 4.3 Time–frequency window for short-time Fourier transform ($t^* = \omega^* = 0$).

Time Shift Letting $f_0(t) = f(t - t_0)$, then

$$\begin{aligned} G_\phi f_0(b, \xi) &= \int_{-\infty}^{\infty} f(t - t_0)\phi(t - b)e^{-j\xi t}\, dt \\ &= \int_{-\infty}^{\infty} f(t)\phi(t - (b - t_0))e^{-j\xi t}e^{-j\xi t_0}\, dt \\ &= e^{-j\xi t_0}(G_\phi f)(b - t_0, \xi). \end{aligned} \tag{4.20}$$

Equation (4.20) simply means that if the original function $f(t)$ is shifted by an amount t_0 in the time axis, the location of STFT in the time–frequency domain will shift by the same amount in time while the frequency location will remain unchanged. Apart from the change in position, there is also a change in the phase of the STFT which is directly proportional to the time shift.

Frequency Shift Letting $f(t)$ be the modulation function of a carrier signal $e^{j\omega_0 t}$ such that

$$f_0(t) = f(t)e^{j\omega_0 t}; \tag{4.21}$$

then the STFT of $f_0(t)$ is given by

$$G_\phi f_0(b,\xi) = \int_{-\infty}^{\infty} f(t)e^{j\omega_0 t}\phi(t-b)e^{-j\xi t}\,dt$$
$$= G_\phi f(b, \xi - \omega_0). \tag{4.22}$$

Equation (4.22) implies that both the magnitude and phase of the STFT of $f_0(t)$ remain the same as those of $f(t)$, except that the new location in the t–ω domain is shifted along the frequency axis by the carrier frequency ω_0.

4.3 DISCRETE SHORT-TIME FOURIER TRANSFORM

Similar to the discussion of Section 3.10, we can efficiently evaluate the integral of (4.7) as a series sum by appropriately sampling the function $f(t)$ and the window function $\phi(t)$. In its discrete form, the short-time Fourier transform can be represented as

$$G_\phi f(b_n, \xi_n) \approx h \sum_{k=0}^{N-1} f(t_k)\phi(t_k - b_n)e^{-j\xi_n t_k}, \tag{4.23}$$

where

$$t_k = b_k = kh, \qquad k = 0, \ldots, N-1 \tag{4.24}$$

and

$$\xi_n = \frac{2\pi n}{Nh}, \qquad n = -\frac{N}{2}, \ldots, \frac{N}{2}. \tag{4.25}$$

In particular, when $h = 1$, we have

$$G_\phi f(n, \xi_n) \approx \sum_{k=0}^{N-1} f(k)\phi(k-n)e^{-j(2\pi kn/N)}. \tag{4.26}$$

4.3.1 Examples of STFT

We use an example similar to the one used in [2] to show the computation of the STFT and the effect of the window width with respect to resolution. The signal

$$f(t) = \sin 2\pi\nu_1 t + \sin 2\pi\nu_2 t + K\left[\delta(t-t_1) + \delta(t-t_2)\right] \tag{4.27}$$

consists of two sinusoids at frequencies of $\nu_1 = 500$ Hz and $\nu_2 = 1000$ Hz and two delta functions occurring at $t_1 = 192$ and $t_2 = 196$ ms. We arbitrarily choose $K = 3$. We apply a rectangular window to the function and compute the STFT for four different window sizes. The signal and the window function are both sampled at 8 kHz. The window size varies from 16 to 2 ms and the corresponding number

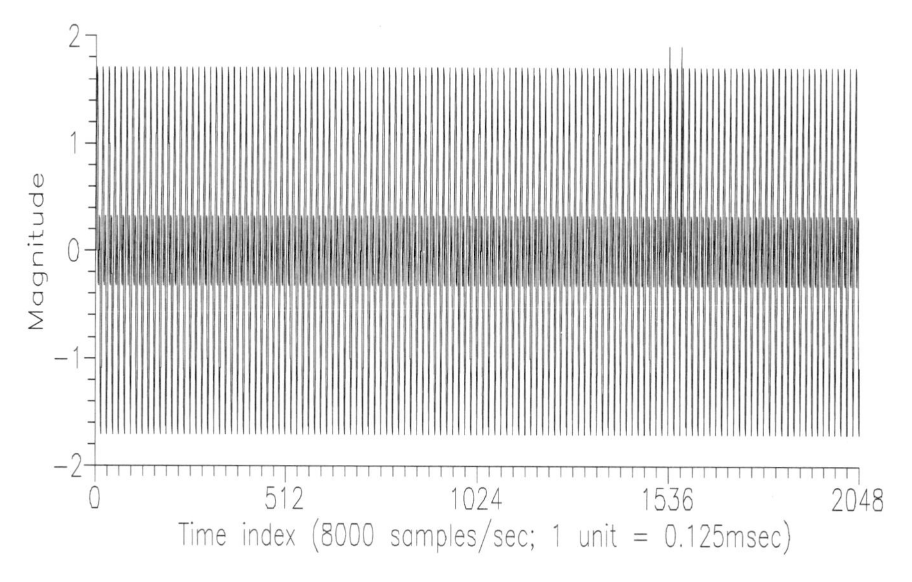

FIGURE 4.4 Signal for which the STFT is shown in Figure 4.5.

of samples in the windows are 128, 64, 32, and 16, respectively. Since the delta functions are separated by 32 samples, window sizes equal to or greater than 32 samples are not narrow enough to resolve the delta functions.

To compute the STFT, we apply the FFT algorithm to the product of the signal and the window function. We compute a 128-point FFT each time the window is moved to the right by one sample. Figure 4.4 shows the function $f(t)$, and the results of these STFTs are given in Figure 4.5.

Initially, when the time window is wide, the delta functions are not resolvable at all. However, the two frequencies are well distinguished by the high resolution of the window in the spectral domain. As the window size gets smaller, we begin to see the two delta functions while the frequency resolution progressively worsens. At a window size of 16 samples we can distinguish the delta functions quite easily, but the two frequencies cannot be resolved accurately. To resolve events in the frequency and time axes, we must compute the STFT every time we change the window size. Computation load is a serious issue in using STFT for signal processing.

4.4 DISCRETE GABOR REPRESENTATION

Formally writing the Gabor transform given in Section 4.2.2, we obtain

$$\begin{aligned} G_{g_a} f(b, \xi) &:= \int_{-\infty}^{\infty} f(t) \overline{g_a(t-b) e^{-j\xi t}}\, dt \\ &= \frac{1}{2\alpha\pi} \int_{-\infty}^{\infty} f(t) e^{-(t-b)^2/4\alpha} e^{-j\xi t}\, dt \end{aligned} \tag{4.28}$$

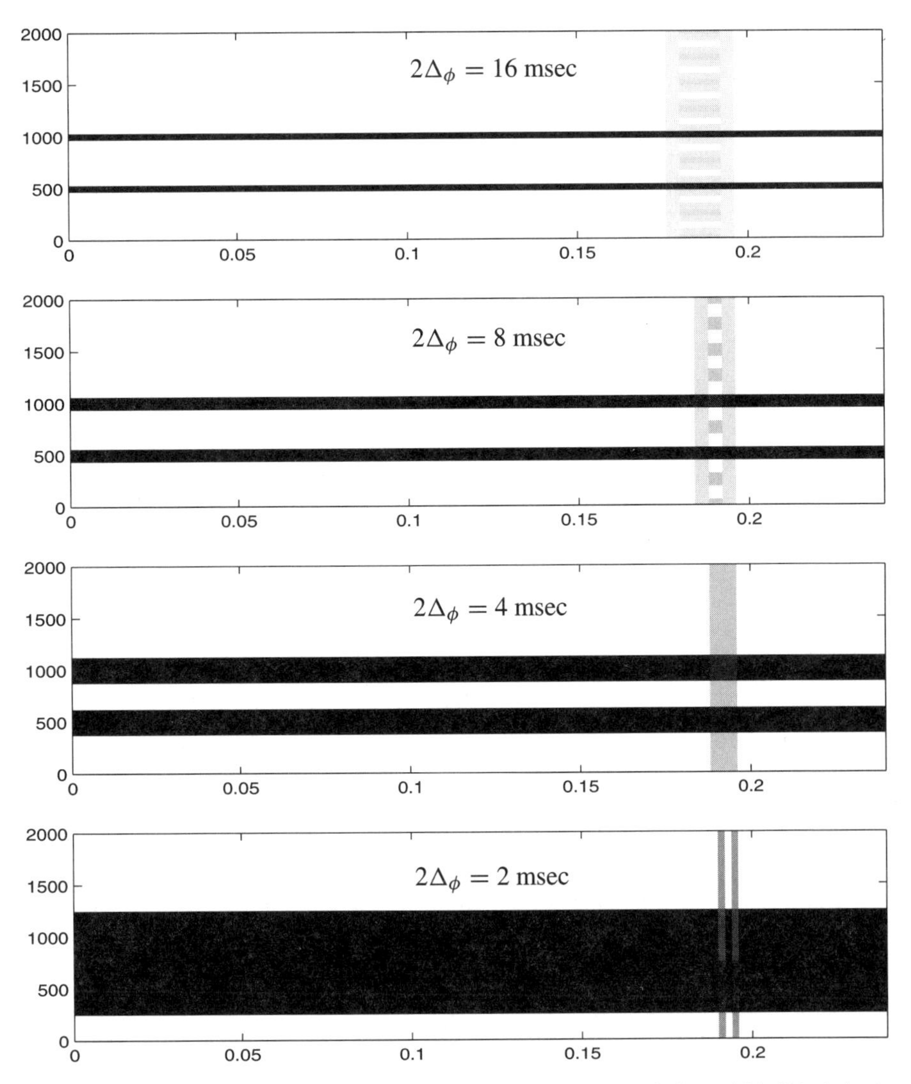

FIGURE 4.5 STFT of signal shown in Figure 4.4 with different window width ($2\Delta_\phi$); horizontal axis is time (second) and the vertical axis is frequency (Hz).

for $-\infty \leq b, \xi \leq \infty$. The Gabor transform is dense over the t–f plane. Computation load for the Gabor transform in (4.28) is quite heavy. We may, instead of (4.28), compute the discretized version of (4.28). That is, we compute (4.28) only at a set of points on the t–f plane:

$$\begin{aligned} G_{g_a} f(b_n, \xi_k) &= \int_{-\infty}^{\infty} f(t)\overline{g_a(t - b_n)e^{-j\xi_k t}}\, dt \\ &= \left\langle f(t), g_a(t - b_n)e^{j\xi_k t} \right\rangle \\ &= \left\langle f(t), \phi_{n,k}(t) \right\rangle . \end{aligned} \tag{4.29}$$

The last expression of (4.29) is the inner product of the function with the function $\phi_{n,k}(t) = g_a(t - b_n)e^{j\xi_k t}$. The function $f(t)$ may be recovered under the restricted condition [3]

$$f(t) = \sum_n \sum_k G_{g_a} f(b_n, \xi_k) g_a(t - b_n) e^{j\xi_k t}. \tag{4.30}$$

Equation (4.30) is known as the *Gabor expansion,* in which $(G_{g_a} f)(b_n, \xi_k)$ play the role of coefficients in the recovery formula

$$f(t) = \sum_n \sum_k G_{g_a} f(b_n, \xi_k) \phi_{n,k}(t). \tag{4.31}$$

The function $\phi_{n,k}(t)$ is a Gaussian-modulated sinusoid. Spread of the function is controlled by α, while the oscillation frequency is controlled by ξ_k. These "bullets" of the t–f plane form the basis of the Gabor expansion. Since the Gaussian function has the minimum size of a time–frequency window, it has the highest concentration of energy in the t–f plane. The Gabor basis $\phi_{n,k}(t)$ appears to be a useful basis for signal representation. However, it lacks basic properties such as orthogonality, completeness, and independence needed to achieve simple representations and efficient computation.

4.5 CONTINUOUS WAVELET TRANSFORM

The STFT discussed in Section 4.4 provides one of many ways to generate a time–frequency analysis of signals. Another linear transform that provides such analyses is the integral (or continuous) wavelet transform. The terms *continuous wavelet transform* (CWT) and *integral wavelet transform* (IWT) will be used interchangeably throughout this book. Fixed time–frequency resolution of the STFT poses a serious constraint in many applications. In addition, developments on the discrete wavelet transform (DWT) and the wavelet series (WS) make the wavelet approach more suitable than the STFT for signal and image processing. To clarify our points, let us observe that the radii Δ_ϕ and $\Delta_{\widehat{\phi}}$ of the window function for STFT do not depend upon location in the t–ω plane. For instance, if we choose $\phi(t) = g_\alpha(t)$ as in the Ga-

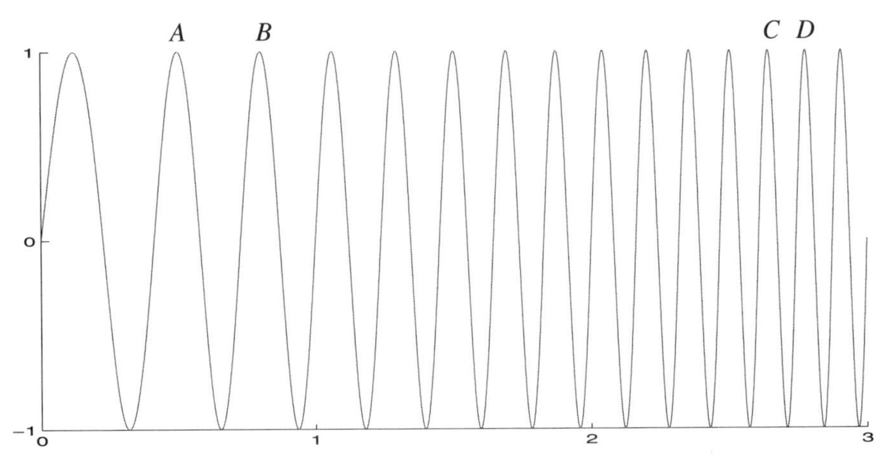

FIGURE 4.6 Chirp signal with frequency changing linearly with time.

bor transform (see Section 4.2.2), once α is fixed, so are Δg_α and $\Delta \widehat{g}_\alpha$, regardless of the window location in the t–ω plane. A typical STFT time–frequency window has been shown in Figure 4.3. Once the window function is chosen, the time–frequency resolution is fixed throughout the processing. To understand the implications of such a fixed resolution, let us consider the chirp signal shown in Figure 4.6, in which the frequency of the signal increases with time.

If we choose the parameters of the window function $\phi(t)$ [α in the case of $g_\alpha(t)$] such that $\Delta\phi$ is approximately equal to AB, the STFT as computed using (4.7) will be able to resolve the low-frequency portion of the signal better, while there will be poor resolution of the high-frequency portion. On the other hand, if Δ_ϕ is approximately equal to CD, the low frequency will not be resolved properly. Observe that if Δ_ϕ is very small, $\Delta_{\widehat{\phi}}$ will be proportionally large, and hence the low-frequency part will be blurred.

Our objective is to devise a method that can give good time–frequency resolution at an arbitrary location in the t–ω plane. In other words, we must have a window function whose radius increases in time (reduces in frequency) while resolving the low-frequency contents, and decreases in time (increases in frequency) while resolving the high-frequency contents of a signal. This objective leads us to the development of wavelet functions $\psi(t)$.

The integral wavelet transform of a function $f(t) \in L^2$ with respect to some analyzing wavelet ψ is defined as

$$W_\psi f(b,a) := \int_{-\infty}^{\infty} f(t)\overline{\psi_{b,a}(t)}\,dt, \tag{4.32}$$

where

$$\psi_{b,a}(t) = \frac{1}{\sqrt{a}}\psi\frac{t-b}{a}, \qquad a > 0. \tag{4.33}$$

The parameters b and a are called *translation* and *dilation parameters*, respectively. The normalization factor $a^{-1/2}$ is included so that $\|\psi_{b,a}\| = \|\psi\|$.

For ψ to be a window function and to recover $f(t)$ from its IWT, $\psi(t)$ must satisfy the following condition:

$$\widehat{\psi}(0) = \int_{-\infty}^{\infty} \psi(t) dt = 0. \tag{4.34}$$

In addition to satisfying (4.34), a wavelet is constucted so that it has a higher order of *vanishing moments*. A wavelet is said to have vanishing moments of order m if

$$\int_{-\infty}^{\infty} t^p \psi(t)\, dt = 0, \qquad p = 0, \ldots, m-1. \tag{4.35}$$

Strictly speaking, integral wavelet transform provides time-scale analysis and not time–frequency analysis. However, by proper scale-to-frequency transformation (discussed later), one can get an analysis that is very close to time–frequency analysis. Observe that in (4.33), by reducing a, the support of $\psi_{b,a}$ is reduced in time and hence covers a larger frequency range, and vice versa. Therefore, $1/a$ is a measure of frequency. The parameter b, on the other hand, indicates the location of the wavelet window along the time axis. Thus, by changing (b, a), $W_\psi f$ can be computed on the entire time–frequency plane. Furthermore, because of the condition (4.34), we conclude that all wavelets must oscillate, giving them the nature of small waves and hence the name *wavelets*. Recall that such an oscillation is not required for the window function in STFT. Compared with the definition of STFT in (4.7), the wavelet $\psi_{b,a}(t)$ takes the place of $\phi_{b,\xi}$. Hence a wavelet also behaves like a window function. The behavior and measures of wavelet windows are discussed in more detail in Section 4.5.2.

4.5.1 Inverse Wavelet Transform

Since the purpose of the inverse transform is to reconstruct the original function from its integral wavelet transform, it involves a two-dimensional integration over the scale parameter a and the translation parameter b. The expression for the inverse wavelet transform is

$$f(t) = \frac{1}{C_\psi} \int_{-\infty}^{\infty} db \int_{-\infty}^{\infty} \frac{1}{a^2} \left[W_\psi f(b,a) \right] \psi_{b,a}(t)\, da, \tag{4.36}$$

where C_ψ is a constant that depends on the choice of wavelet and is given by

$$C_\psi = \int_{-\infty}^{\infty} \frac{|\widehat{\psi}(\omega)|^2}{|\omega|}\, d\omega < \infty. \tag{4.37}$$

The condition (4.37), known as the *admissibility condition*, restricts the class of functions that can be wavelets. In particular, it implies that all wavelets must have

$\widehat{\psi}(0) = 0$ [see (4.34)] in order to make the left-hand side of (4.37) a finite number. For a proof of (4.36), readers may refer to [2, Chap. 2].

Equation (4.36) is essentially a superposition integral. Integration with respect to a sums all the contributions of the wavelet components at location b, while the integral with respect to b includes all locations along the b-axis. Since computation of the inverse wavelet transform is quite cumbersome and the inverse wavelet transform is used only for synthesizing the original signal, it is not used as frequently as the integral wavelet transform for the analysis of signals. In subsequent sections where the discrete wavelet transform (DWT) is introduced, the inverse of the DWT is very useful in data communication and signal processing.

4.5.2 Time–Frequency Window

The definitions of the frequency-domain center and radius discussed in Section 4.1 do not apply to wavelet windows because unlike the window of STFT in which $\widehat{\phi}(0) = 1$, here the wavelet window $\widehat{\psi}(0) = 0$. In other words, $\widehat{\psi}(\omega)$ exhibits band-pass filter characteristics. Consequently, we have two centers and two radii for $\widehat{\psi}(\omega)$. We are interested only in the positive frequencies. Let us, therefore, define the center ω_+^* and the radius $\Delta_{\widehat{\psi}}^+$ on the positive-frequency axis as

$$\omega_+^* := \frac{\int_0^\infty \omega \left|\widehat{\psi}(\omega)\right|^2 d\omega}{\int_0^\infty \left|\widehat{\psi}(\omega)\right|^2 d\omega} \tag{4.38}$$

$$\Delta_{\widehat{\psi}}^+ := \left[\frac{\int_0^\infty (\omega - \omega_+^*)^2 \left|\widehat{\psi}(\omega)\right|^2 d\omega}{\int_0^\infty \left|\widehat{\psi}(\omega)\right|^2 d\omega}\right]^{1/2}. \tag{4.39}$$

The definitions for t^* and Δ_ψ remain the same as those in Section 4.1, with $\phi(t)$ replaced by $\psi(t)$. For wavelets the uncertainty principle gives

$$\Delta_\psi \Delta_{\widehat{\psi}}^+ > \frac{1}{2}. \tag{4.40}$$

If t^* is the center and Δ_ψ is the radius of $\psi(t)$, then $W_\psi f(b, a)$ contains the information of $f(t)$ in the time window

$$\left[at^* + b - a\Delta_\psi, at^* + b + a\Delta_\psi\right]. \tag{4.41}$$

Let us apply Parseval's identity to (4.32) to get an idea of the frequency window:

$$W_\psi f(b, a) = \frac{1}{\sqrt{a}} \int_{-\infty}^{\infty} f(t) \overline{\psi\left(\frac{t-b}{a}\right)} \, dt \tag{4.42}$$

$$= \frac{\sqrt{a}}{2\pi} \int_{-\infty}^{\infty} \widehat{f}(\omega) \overline{\widehat{\psi}(a\omega)} e^{jb\omega} \, d\omega. \tag{4.43}$$

From (4.43) it is clear that the frequency window is

$$\left[\frac{1}{a}(\omega_+^* - \Delta_{\widehat{\psi}}^{+}), \frac{1}{a}(\omega_+^* + \Delta_{\widehat{\psi}}^{+})\right]. \tag{4.44}$$

The time–frequency window product $= 2a\Delta_\psi \times \frac{2}{a}\Delta_{\widehat{\psi}}^{+} = 4\Delta_\psi\Delta_{\widehat{\psi}}^{+} =$ constant.

Figure 4.7 graphically represents the notion of the time–frequency window for the wavelet transform. Comparing Figure 4.7 with the corresponding Figure 4.3 for STFT, we observe the flexible nature of the window in the wavelet transform. For higher frequency $(1/a_2)$, the time window is small, whereas for lower frequency $(1/a_0)$, the time window is large. For a fixed frequency level, $(1/a_0)$, for example, both the time and frequency windows are fixed. Recall that in STFT the time–frequency window is fixed regardless of the frequency level.

Example: We perform a continuous wavelet transform on the same function used for computing the STFT. We choose the complex Morlet wavelet, given by

$$\psi(t) = e^{-t^2/2}e^{-j5.336t} \tag{4.45}$$

to compute the CWT

$$W_f(b,a) = \frac{1}{\sqrt{a}}\int_{-\infty}^{\infty} f(t)\overline{\psi}\,\frac{t-b}{a}\,dt.$$

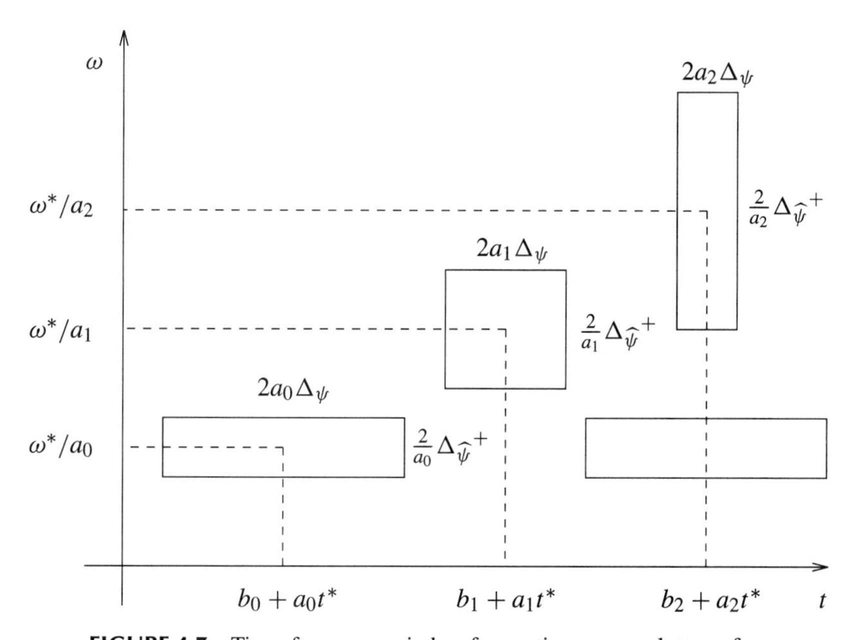

FIGURE 4.7 Time–frequency window for continuous wavelet transform.

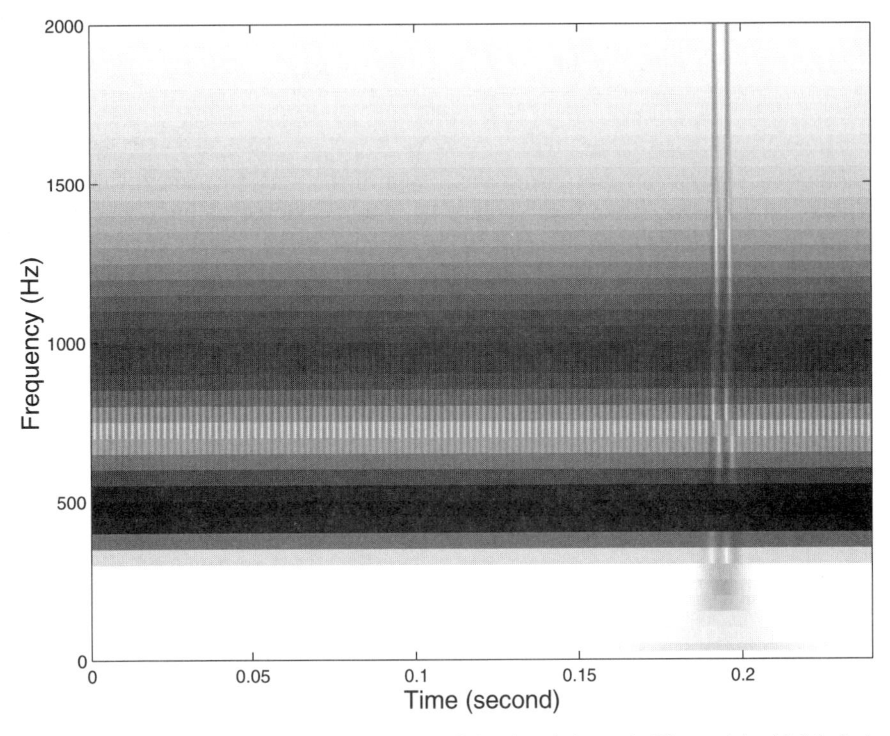

FIGURE 4.8 Continuous wavelet transform of the signal shown in Figure 4.4 with Morlet's wavelet.

The results are shown in Figure 4.8. The figure indicates good resolution of the events in both the time and frequency axes. If we choose an appropriate range for a, the transform needs to be computed only once to capture most, if not all, of the events occurring in the time and frequency domains.

4.6 DISCRETE WAVELET TRANSFORM

Similar to the discrete Fourier transform and discrete short-time Fourier transform, we have the discrete wavelet transform (DWT). However, unlike the discretized time and frequency axes shown earlier in Fourier analysis, here we take the discrete values of the scale parameter a and the translation parameter b in a different way. The interest here is to introduce the DWT and show the relationship between DWT and IWT. A detailed discussion of the DWT is presented in Chapter 7. Here we just mention that we will take a to be of the form 2^{-s} and b to be of the form $k2^{-s}$, where $k, s \in \mathbb{Z}$. With these values of a and b, the integral of (4.32) becomes

$$W_\psi f(k2^{-s}, 2^{-s}) = 2^{s/2} \int_{-\infty}^{\infty} f(t)\psi(2^s t - k)dt. \tag{4.46}$$

Let us now discretize the function $f(t)$. For simplicity, assume the sampling rate to be 1. In that case, the integral of (4.46) can be written as

$$W_\psi f(k2^{-s}, 2^{-s}) \approx 2^{s/2} \sum_n f(n)\psi(2^s n - k). \tag{4.47}$$

To compute the wavelet transform of a function at some point in the time-scale plane, we do not need to know the function values for the entire time axis. All we need is the function at those values of time at which the wavelet is nonzero. Consequently, evaluation of the wavelet transform can be done almost in real time. We discuss algorithms to compute the wavelet transform in later chapters.

One of the important observations about (4.47) is its time-variant nature. The DWT of a function shifted in time is quite different from the DWT of the original function. To explain it further, let

$$f_m(t) = f(t - t_m). \tag{4.48}$$

This gives

$$\begin{aligned} W_\psi f_m(k2^{-s}, 2^{-s}) &= 2^{s/2} \int_{-\infty}^{\infty} f_m(t)\psi(2^s t - k)\, dt \\ &\approx 2^{s/2} \sum_n f(n - m)\psi(2^s n - k) \\ &= 2^{s/2} \sum_n f(n)\psi\left[2^s n - (k - m2^s)\right] \\ &\approx W_\psi f\left[(k - m2^s)2^{-s}, 2^{-s}\right]. \end{aligned} \tag{4.49}$$

Therefore, we see that for DWT, a shift in time of a function manifests itself in a rather complicated way. Recall that a shift in time of a function appears as a shift in time location by an exact amount in the case of STFT, with an additional phase shift. Also in the Fourier transform, the shift appears only as a phase change in the frequency domain.

4.7 WAVELET SERIES

Analogous to the Fourier series, we have the wavelet series. Recall that the Fourier series exists for periodic functions only. Here, for any function $f(t) \in L^2$, we have its wavelet series representation, given as

$$f(t) = \sum_s \sum_k w_{k,s}\psi_{k,s}(t), \tag{4.50}$$

where

$$\psi_{k,s}(t) = 2^{s/2}\psi(2^s t - k). \tag{4.51}$$

The double summation in (4.50) is due to the fact that wavelets have two parameters: the translation and scale parameters. For a periodic function $p(t)$, its Fourier series is given by

$$p(t) = \sum_k c_k e^{jkt}. \tag{4.52}$$

Since $\{e^{jkt} : k \in \mathbb{Z}\}$ is an orthogonal basis of $L^2(0, 2\pi)$, we can obtain c_k as

$$c_k = \frac{1}{2\pi} \left\langle p(t), e^{jkt} \right\rangle. \tag{4.53}$$

On a similar line, if $\{\psi_{k,s}(t) : k, s \in \mathbb{Z}\}$ forms an orthonormal basis of $L^2(R)$, we can get

$$w_{k,s} = \left\langle f(t), \psi_{k,s}(t) \right\rangle nn \tag{4.54}$$

$$= 2^{s/2} W_\psi f \left(\frac{k}{2^s}, \frac{1}{2^s} \right). \tag{4.55}$$

Therefore, the coefficients $\{w_{k,s}\}$ in the wavelet series expansion of a function are nothing but the integral wavelet transform of the function evaluated at certain dyadic points $(k/2^s, 1/2^s)$. No such relationship exists between Fourier series and Fourier transform, which are applicable to different classes of functions; Fourier series applies to functions that are square integrable in $[0, 2\pi]$, whereas Fourier transform is for functions that are in $L^2(\mathbb{R})$. Both wavelet series and wavelet transform, on the other hand, are applicable to functions in $L^2(\mathbb{R})$.

If $\left\{\psi_{k,s}(t)\right\}$ is not an orthonormal basis, we can obtain $w_{k,s}$ using the dual wavelet $\left\{\widetilde{\psi}_{k,s}(t)\right\}$ as $w_{k,s} = \left\langle f(t), \widetilde{\psi}_{k,s}(t) \right\rangle$. The concept of dual wavelets will appear in subsequent chapters.

4.8 INTERPRETATIONS OF THE TIME–FREQUENCY PLOT

Let us briefly discuss the significance of a surface over the time–frequency plane. Usually, the height of a point on the surface represents the magnitude of the STFT or the IWT. Suppose the given function is such that its frequency does not change with time; then we should expect a horizontal line parallel to the time axis in the time–frequency plot, corresponding to the frequency of the function. However, because of the finite support of the window function and the truncation of the sinusoid, instead of getting a line we will see a (widened line) band near the frequency. To understand it more clearly, let us consider a truncated sinusoid of frequency ω_0. We assume, for the purpose of explaining the time–frequency plot here, that even though the sinusoid is truncated, its Fourier transform is represented as $\widehat{\delta}(\omega - \omega_0)$.

By replacing $\widehat{f}(\omega) = \widehat{\delta}(\omega - \omega_0)$ in (4.16) and (4.43), respectively, we obtain

$$\left| G_\phi f(b, \xi) \right| = \frac{1}{2\pi} \left| \widehat{\phi}(\omega_0 - \xi) \right| \tag{4.56}$$

$$\left|W_\psi f(b,a)\right| = \frac{\sqrt{a}}{2\pi}\left|\widehat{\psi}(a\omega_0)\right|. \tag{4.57}$$

It is clear from (4.56) and (4.57) that $\left|G_\phi f(b,\xi)\right|$ and $\left|W_\psi f(b,\xi)\right|$ do not depend upon b. On the frequency axis, since $\left|\widehat{\phi}(0)\right| = 1$, and assuming that $\left|\widehat{\phi}(\omega)\right| \le 1,\ \omega \ne 0$, we will get the maximum magnitude of STFT at $\xi = \omega_0$. Then there will be a band around $\xi = \omega_0$, the width of which will depend on $\Delta_{\widehat{\phi}}$, the radius of $\widehat{\phi}(\omega)$.

Interpretation of (4.57) is a little complicated since, unlike STFT, wavelet transform does not give a time–frequency plot directly. Let us consider a point ω' on the frequency axis such that

$$\left|\widehat{\psi}(\omega')\right| = \max\left\{\left|\widehat{\psi}(\omega)\right|;\quad \omega \in (0,\infty)\right\}. \tag{4.58}$$

For all practical purposes, we may take $\omega' = \omega_+^*$.

Now if we consider a variable $\xi = \omega_+^*/a$ and rewrite (4.57) in terms of the new variable ξ, we have

$$\left|W_\psi f(b, \frac{\omega_+^*}{\xi})\right| = \frac{1}{2\pi}\sqrt{\frac{\omega_+^*}{\xi}}\left|\widehat{\psi}\left(\frac{\omega_+^*}{\xi}\omega_0\right)\right|. \tag{4.59}$$

Therefore, the maximum value of the wavelet transform (4.57) will occur at $\xi = \omega_0$ with a band around $\xi = \omega_0$, depending on the radius $\Delta_{\widehat{\psi}}^{+}$ of the wavelet $\widehat{\psi}(\omega)$.

As our next example, let $f(t) = \delta(t - t_0)$. Since this function has all the frequency components, we should expect a vertical line in the time–frequency plane at $t = t_0$. Substituting $f(t) = \delta(t - t_0)$ in (4.7) and (4.32), we obtain

$$\left|G_\phi f(b,\xi)\right| = \left|\phi(t_0 - b)\right| \tag{4.60}$$

$$\left|W_\psi f(b,a)\right| = \left|\frac{1}{\sqrt{a}}\right|\left|\psi\frac{t_0 - b}{a}\right|. \tag{4.61}$$

Explanation of the STFT is straightforward. As expected, it does not depend on ξ. We get a vertical line parallel to the frequency axis near $b = t_0$ with the time spread determined by Δ_ϕ. For wavelet transform we observe that it depends on the scale parameter a. Rewriting (4.61) in terms of the new variable ξ, we have

$$\left|W_\psi f\left(b, \frac{\omega_+^*}{\xi}\right)\right| = \sqrt{\frac{\xi}{\omega_+^*}}\left|\psi\left(\frac{\xi}{\omega_+^*}(t_0 - b)\right)\right|. \tag{4.62}$$

Although all the frequency contents of the delta function in time are indicated by (4.62), it is clear that as we reduce ξ, the time spread increases. Furthermore, the location of the maximum will depend on the shape of $\psi(t)$. Readers are referred to [4] for more information on the interpretation of time–frequency plots.

4.9 WIGNER–VILLE DISTRIBUTION

We have considered in previous sections linear time–frequency representations of a signal. Thc STFT and CWT are linear transforms because they satisfy the linear superposition theorem. That is,

$$T[\alpha_1 f_1 + \alpha_2 f_2] = \alpha_1 T[f_1] + \alpha_2 T[f_2], \tag{4.63}$$

where T may represent either the STFT or the CWT, and $f_1(\tau)$ and $f_2(\tau)$ are two different signals in the same class with coefficients α_1 and α_2. These transforms are important because they provide an interpretation to the *local spectrum* of a signal at the vicinity of time t. In addition, easy implementation and high computation efficiency of their algorithms add to their advantages. On the other hand, these linear transforms do not provide instantaneous energy information of the signal at a specific instant of time. Intuitively, we want to consider a transform of the type

$$\int_{-\infty}^{\infty} |f(\tau - t)|^2 e^{-j\omega\tau}\, d\tau = \int_{-\infty}^{\infty} f(\tau - t)\overline{f(\tau - t)} e^{-j\omega\tau}\, d\tau.$$

Since it is not easy to determine the energy of a signal at a given time, it is more meaningful to consider the energy within an interval $(t-\tau/2, t+\tau/2)$ that is centered around the time location t. For this purpose, the Wigner–Ville distribution (WVD) is defined by

$$\mathbf{W}_f(t, \omega) = \frac{1}{2\pi} \int_{-\infty}^{\infty} f\left(t + \frac{\tau}{2}\right) \overline{f\left(t - \frac{\tau}{2}\right)} e^{-j\omega\tau}\, d\tau. \tag{4.64}$$

The constant $1/2\pi$ is a normalization factor for simplicity of computation. We should note that the linearity property no longer holds for the equation above. The Wigner–Ville distribution is a nonlinear (or bilinear) time–frequency transform because the signal enters the integral more than once. One may also observe that the Wigner–Ville distribution at a given time t looks symmetrically to the left and right sides of the signal at a distance $\tau/2$. Computation of $\mathbf{W}_f(t, \omega)$ requires signal information at $t \pm \tau/2$ and cannot be carried out in real time.

Example 1: Let us consider a chirp signal that is modulated by a Gaussian envelope:

$$f(t) = \left(\frac{a}{\pi}\right)^{1/4} \exp\left(\frac{-at^2}{2} + j\frac{bt^2}{2} + j\omega_0 t\right), \tag{4.65}$$

where $\exp(-at^2/2)$ is the Gaussian term, $\exp(-jbt^2/2)$ is the chirp signal, and $e^{j\omega_0 t}$ is a frequency-shifting term. The Wigner–Ville distribution from (4.64) yields

$$\mathbf{W}_f(t, \omega) = \frac{1}{2\pi}\left(\frac{a}{\pi}\right)^{1/2} \int_{-\infty}^{\infty} \exp\left[\frac{-a(t+\tau/2)^2}{2} + j\frac{b(t+\tau/2)^2}{2} + j\omega_0\left(t + \frac{\tau}{2}\right)\right]$$

$$\times \exp\left[\frac{-a(t-\tau/2)^2}{2} - j\frac{-b(t-\tau/2)^2}{2} - j\omega_0\left(t-\frac{\tau}{2}\right) - j\omega\tau\right] d\tau$$

$$= \frac{1}{2\pi}\left(\frac{a}{\pi}\right)^{1/2} e^{-at^2} \int_{-\infty}^{\infty} \exp\left(\frac{-a\tau^2}{4} + jbt\tau + j\omega_0\tau - j\omega\tau\right) d\tau. \tag{4.66}$$

Using the Fourier transform of a Gaussian function as given in Chapter 3, the WVD of a Gaussian sinusoid-modulated chirp is

$$\mathbf{W}_f(t,\omega) = \frac{1}{\pi}\exp\left[-\frac{at^2}{2} - \frac{(\omega-\omega_0-bt)^2}{a}\right]. \tag{4.67}$$

The function and its WVD are shown in Figure 4.9.

Example 2: A sinusoidal modulated chirp signal is given by

$$f(t) = \exp\left(j\frac{bt^2}{2} + j\omega_0 t\right). \tag{4.68}$$

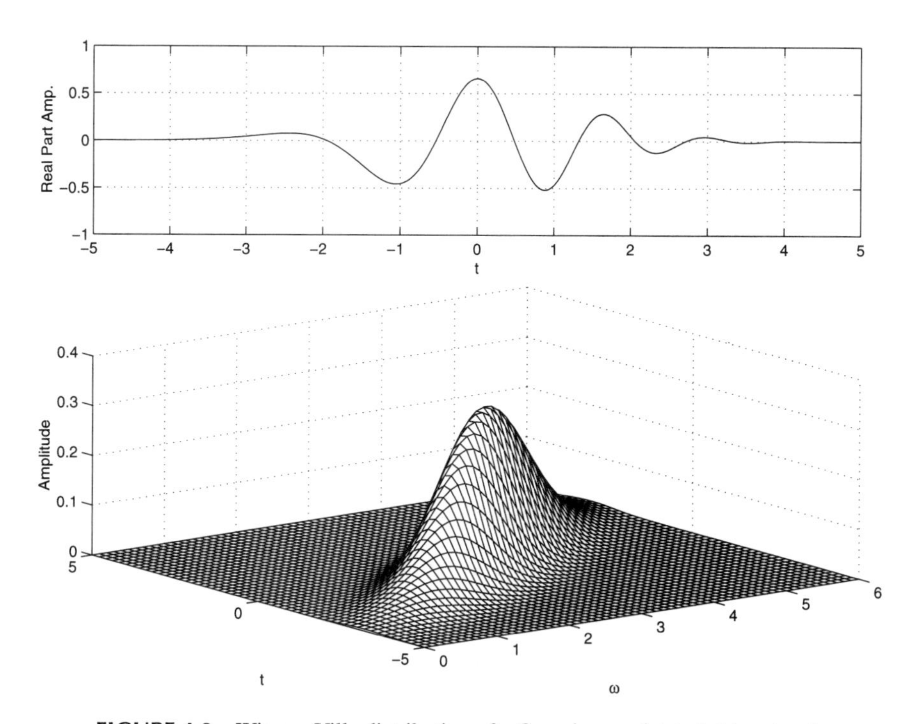

FIGURE 4.9 Wigner–Ville distribution of a Gaussian-modulated chirp signal.

We compute the WVD straightforwardly to obtain

$$\begin{aligned}
\mathbf{W}_f(t,\omega) &= \frac{1}{2\pi}\int_{-\infty}^{\infty} \exp\left[j\frac{b(t+\tau/2)^2}{2} + j\omega_0\left(t+\frac{\tau}{2}\right)\right] \\
&\quad \times \exp\left[-j\frac{b(t-\tau/2)^2}{2} - j\omega_0\left(t-\frac{\tau}{2}\right) - j\omega\tau\right] d\tau \\
&= \frac{1}{2\pi}\int_{-\infty}^{\infty} \exp(jbt\tau + j\omega_0\tau - j\omega\tau)d\tau \\
&= \delta(\omega - \omega_0 - bt).
\end{aligned} \tag{4.69}$$

Example 3: We compute the WVD of a pure sinusoidal signal $e^{j\omega_0 t}$ by setting the chirp parameter b to zero. Therefore,

$$e^{j\omega_0 t} \Longleftrightarrow \delta(\omega - \omega_0). \tag{4.70}$$

The WVDs of (4.67) and (4.69) on the time–frequency plane are a straight line with a slope b and a straight line parallel to the time axis, respectively. They are given in Figures 4.10 and 4.12. Figure 4.11 shows the WVD of a Gaussian-modulated sinusoidal function.

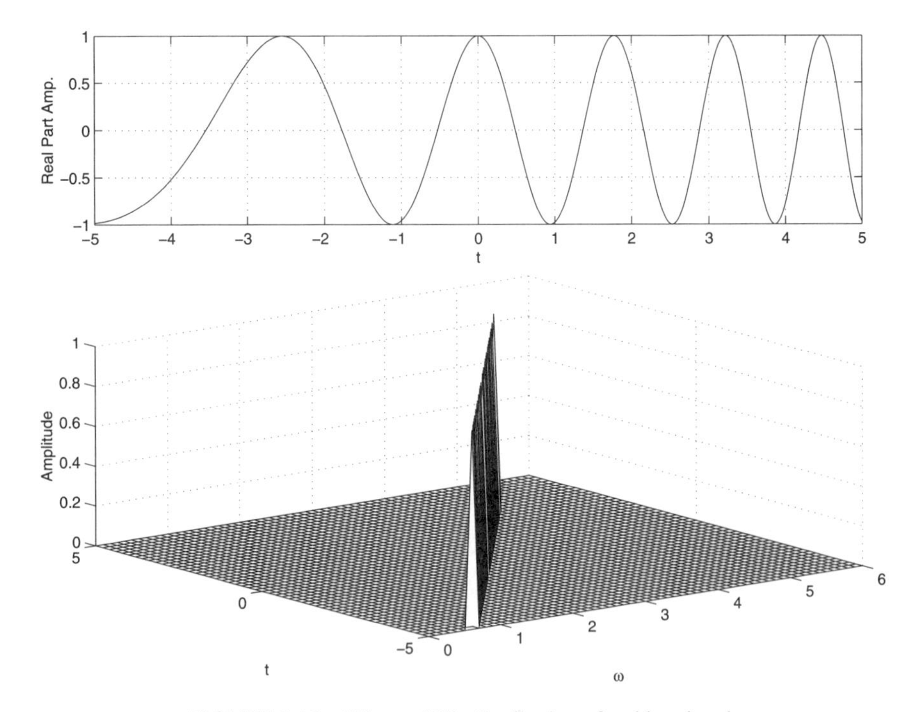

FIGURE 4.10 Wigner–Ville distribution of a chirp signal.

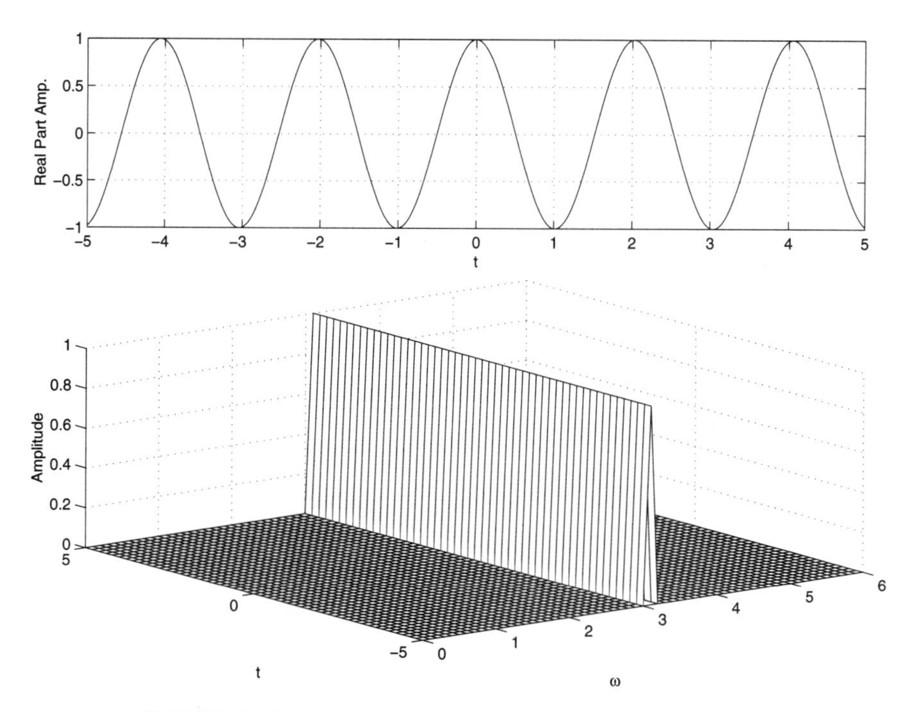

FIGURE 4.11 Wigner–Ville distribution of a sinusoidal function.

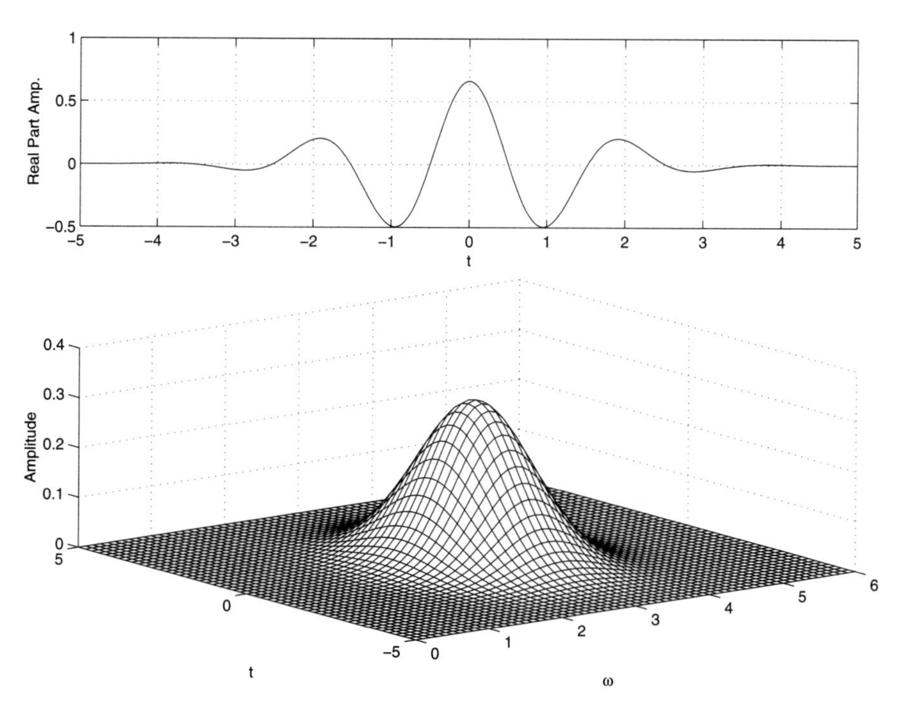

FIGURE 4.12 Wigner–Ville distribution of a Gaussian-modulated sinusoid.

4.10 PROPERTIES OF THE WIGNER–VILLE DISTRIBUTION

There are several general properties of WVD that are important for signal representation in signal processing. Some of them are discussed in this section. It has been shown [5] that the Wigner–Ville distribution has the highest concentration of signal energy in the time–frequency plane. Any other distribution that has higher energy concentration than WVD will be in violation of the uncertainty principle. Furthermore, it cannot satisfy the *marginal properties* discussed in this section.

4.10.1 A Real Quantity

The Wigner–Ville distribution is always real, regardless of whether the signal is real or complex. This can be seen by considering the complex conjugate of the Wigner–Ville distribution:

$$\begin{aligned}\overline{\mathbf{W}_s(t,\omega)} &= \frac{1}{2\pi}\int_{-\infty}^{\infty} s\left(t-\frac{\tau}{2}\right)\overline{s\left(t+\frac{\tau}{2}\right)}e^{j\omega\tau}\,d\tau \\ &= \frac{1}{2\pi}\int_{-\infty}^{\infty} s\left(t+\frac{\tau}{2}\right)\overline{s\left(t-\frac{\tau}{2}\right)}e^{-j\omega\tau}\,d\tau \\ &= \mathbf{W}_s(t,\omega). \end{aligned} \tag{4.71}$$

Wigner–Ville distribution is always real but not always positive. Figure 4.13 shows the WVD of a function that becomes negative near the center. Consequently, WVD may not be used as a measure of energy density or probability density.

4.10.2 Marginal Properties

Of particular concern to signal processing is the energy conservation. This is expressed by the marginal properties of the distribution:

$$\int_{-\infty}^{\infty} \mathbf{W}_f(t,\omega)\,d\omega = |f(t)|^2 \tag{4.72}$$

$$\int_{-\infty}^{\infty} \mathbf{W}_f(t,\omega)\,dt = \left|\widehat{f}(\omega)\right|^2. \tag{4.73}$$

Marginal (density) expresses the energy density in terms of one of the two variables alone. If we wish to find the energy density in terms of t, we simply integrate (sum up) the distribution with respect to ω, and vice versa. The total energy of the signal can be computed by a two-dimensional integration of the Wigner–Ville distribution over the entire time–frequency plane.

$$E = \int_{-\infty}^{\infty} |f(t)|^2\,dt = \int_{-\infty}^{\infty} \left|\widehat{f}(\omega)\right|^2 d\omega = \int_{-\infty}^{\infty}\int_{-\infty}^{\infty} \mathbf{W}_f(t,\omega)\,d\omega\,dt = 1$$

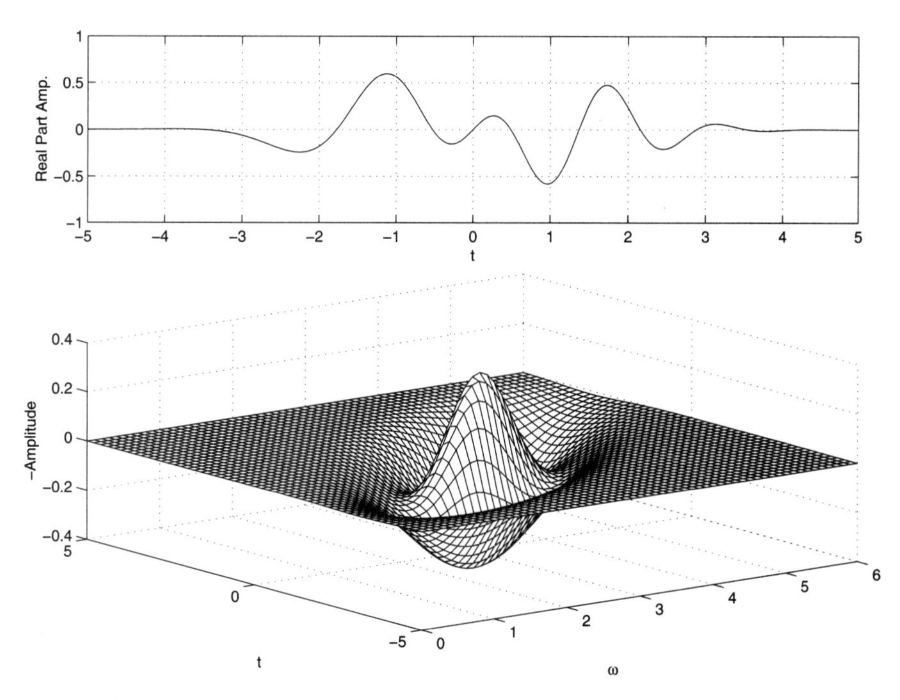

FIGURE 4.13 Plot indicating that Wigner–Ville distribution may be negative.

4.10.3 Correlation Function

We can compute the correlation functions in the time or frequency domains easily from the marginals. That is,

$$\gamma_t(t') = \int_{-\infty}^{\infty} f(\tau)\overline{f(\tau + t')}\, d\tau = \mathbf{W}_f(t', 0) \tag{4.74}$$

$$\gamma_\omega(\omega') = \int_{-\infty}^{\infty} f(\omega)\overline{f(\omega + \omega')}\, d\omega = \mathbf{W}_f(0, \omega'). \tag{4.75}$$

4.11 QUADRATIC SUPERPOSITION PRINCIPLE

We recall that WVD is a nonlinear distribution where the linear superposition principle does not apply. For instance, let a multicomponent signal be

$$f(t) = \sum_{k=1}^{m} f_k(t). \tag{4.76}$$

The Wigner–Ville distribution of this signal is

$$\mathbf{W}_f(t,\omega) = \sum_{k=1}^{m} \mathbf{W}_{f_k}(t,\omega) + \sum_{k=1}^{m} \sum_{\ell=1,\ell\neq k}^{m} \mathbf{W}_{f_k,f_\ell}(t,\omega), \tag{4.77}$$

where $\mathbf{W}_{f_k}(t,\omega)$ is called the auto-term of the WVD, while $\mathbf{W}_{f_k,\ell}(t,\omega)$ is a cross-term defined by

$$\mathbf{W}_{f_k,\ell}(t,\omega) = \frac{1}{2\pi}\int_{-\infty}^{\infty} f_k\left(t+\frac{\tau}{2}\right)\overline{f_\ell\left(t-\frac{\tau}{2}\right)}e^{-j\omega\tau}\,d\tau. \tag{4.78}$$

These cross-terms of the WVD are also called interference terms, which represent the cross coupling of energy between two components of a multicomponent signal. These interference terms are undesirable in most signal processing applications, and much research effort has been devoted to reducing the contribution of these terms. We must remember that these cross-terms [6, 7] are necessary for perfect reconstruction of the signal. In signal detection and identification applications, we are interested in discovering only those signal components that have significant energy. The cross-terms are rendered unimportant since reconstruction of the signal is not necessary.

In radar signal processing and radar imaging, the signals to be processed have a time-varying spectrum like that of a linear chirp or quadratic chirp. Using either the STFT or WT to represent a chirp signal loses resolution in the time–frequency plane. However, the WVDs of these signals produce a well-defined concentration of energy in the time–frequency plane, as shown in Figure 4.10. For multicomponent signals, the energy concentration of the WVD will be far apart in the time–frequency plane if the bandwidths of the components are not overlapped too much (see Figure 4.14).

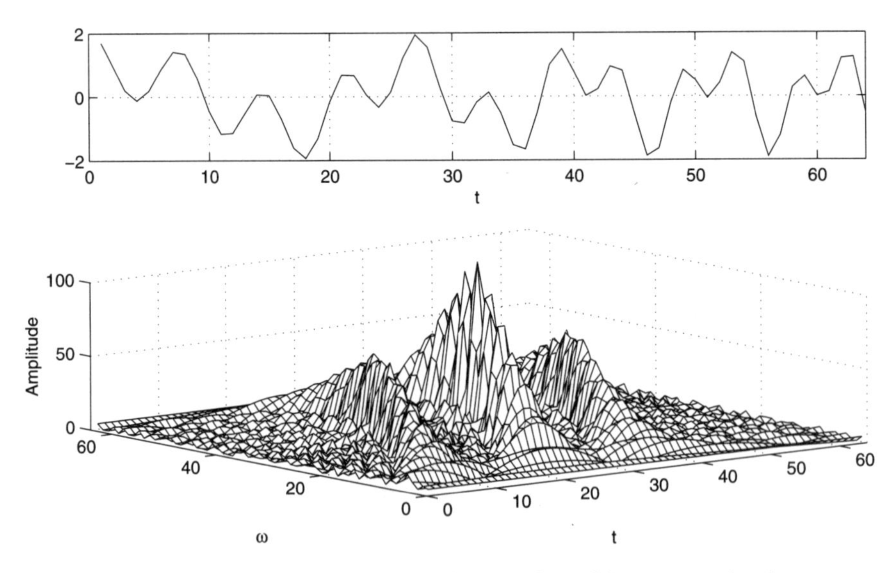

FIGURE 4.14 Wigner–Ville distribution of a multicomponent signal.

However, if this is not the case, certain cross-interference reduction techniques must be applied, and that leads to the reduction of resolution.

4.12 AMBIGUITY FUNCTION

The ambiguity function (AF) is the characteristic function of the Wigner–Ville distribution, defined mathematically as

$$\mathbf{A}_f(u, v) = \int_{-\infty}^{\infty}\int_{-\infty}^{\infty} e^{jut+jv\omega}\mathbf{W}_f(t, \omega)\, dt\, d\omega. \tag{4.79}$$

While the Wigner–Ville distribution is a time–frequency function that measures the energy density of the signal on the time–frequency plane, the ambiguity function is a distribution that measures the energy distribution over a frequency-shift (Doppler) and time-delay plane. This is a very important function in radar signal processing, particularly in the area of waveform design. We shall see some applications of this function toward the end of this book.

Apart from a complex constant, we may express the AF in terms of the signal as

$$\mathbf{A}_f(\tau, \zeta) = K\int_{-\infty}^{\infty} f\left(t - \frac{\tau}{2}\right)\overline{f\left(t + \frac{\tau}{2}\right)}e^{-j\zeta t}\, dt, \tag{4.80}$$

where K is a complex constant. The proof of this relationship can be found in [8]. For further information on the ambiguity function, readers are referred to [9]. Fig-

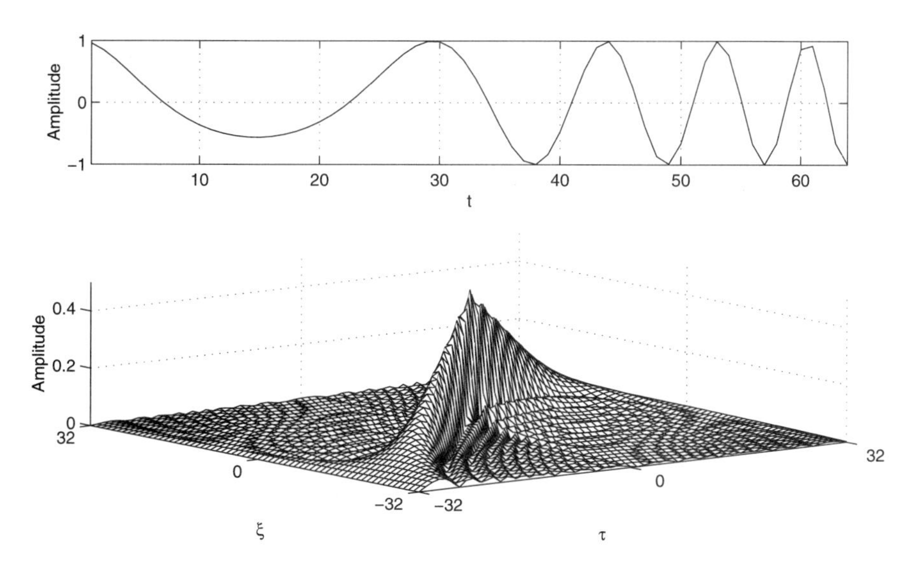

FIGURE 4.15 Ambiguity function of a chirp signal.

ure 4.15 shows the AF for a chirp signal. This time–frequency representation will be revisited in Chapter 9, where a combination of *wavelet packets* and the Wigner–Ville distribution is applied to radar signal detection.

4.13 EXERCISES

1. Verify that for any function $\psi \in L^2(-\infty, \infty)$, the normalized function given by $\psi_{k,s}(t) = 2^{s/2}\psi(2^s t - k)$ for $k, s \in \mathbb{Z}$, $t \in \mathbb{R}$, has the same L^2 norm as ψ :

$$\int_{-\infty}^{\infty} |\psi(t)|^2 \, dt = \int_{-\infty}^{\infty} \left|\psi_{k,s}(t)\right|^2 dt, \qquad k, s \in \mathbb{Z}.$$

2. Consider the window function $g_a(t) = e^{-at^2}, a > 0$. Compute the window widths in the time and frequency domains and verify the uncertainty principle.

3. The *hat function* N_2 is defined by

$$N_2(t) = \begin{cases} t, & \text{for } 0 \le t < 1 \\ 2 - t, & \text{for } 1 \le t < 2 \\ 0, & \text{otherwise.} \end{cases}$$

   Compute the time–frequency window for $N_2(t)$.

4. Show that $\|f(t)\|^2 = \frac{1}{2\pi \|\phi\|^2} \int\int \left|G_\phi f(b, \xi)\right|^2 dbd\xi$.

5. Given that $f(t) = \sin \pi t^2$, and using the raised cosine as the window function

$$\phi(t) = \begin{cases} 1 + \cos(\pi t), & |t| \le 1 \\ 0, & \text{otherwise,} \end{cases}$$

   plot the window-shifted time functions $f_3(t) = \overline{\phi(t - 3)f(t)}$ and $f_7(t)$ and their spectra.
   Consider the time–frequency atoms or the kernel

$$\begin{cases} \operatorname{Re}\left[\phi(t-4)e^{j4\pi t} + \phi(t-6)e^{j8\pi t}\right] \\ \operatorname{Re}\left[\phi(t-4)e^{j4\pi t} + \phi(t-6)e^{j6\pi t}\right]. \end{cases}$$

   Plot the spectral energy density of the two time–frequency atoms. Comment on the time–frequency resolution of the two atoms.

6. In the CWT, show that the normalization constant $1/\sqrt{a}$ is needed to give $\|\psi(t)\| = \left\|\psi_{b,a}(t)\right\|$.

7. Show that the energy conservation principle in the CWT implies that

$$\int_{-\infty}^{\infty} f(t)\overline{g(t)} \, dt = \frac{1}{C_\psi} \int_{-\infty}^{\infty} \int_{-\infty}^{\infty} W_\psi f(b, a)\overline{W_\psi g(b, a)} \, db \frac{da}{a^2}.$$

8. Show that the frequency window width of a wavelet ψ is $[(1/a)(\omega_+^* - \Delta\widehat{\psi}), (1/a)(\omega_+^* + \Delta\widehat{\psi})]$.

9. Identify the reason for dividing the frequency axis by 2 in the program wvd.m.

4.14 COMPUTER PROGRAMS

4.14.1 Short-Time Fourier Transform

```
%
% PROGRAM stft.m
%
% Short-time Fourier Transform using  Rectangular window [0,1]
% generates Figure 4.5
%

% Signal

v1 = 500; % frequency
v2 = 1000;
r = 8000;  %sampling rate
t1 = 0.192; % location of the delta function
t2 = 0.196;

k = 1:2048;
t = (k-1)/r;
f = sin(2*pi*v1*t) + sin(2*pi*v2*t);

k = t1 * r;
f(k) = f(k) + 3;
k = t2 * r;
f(k) = f(k) + 3;

plot(t,f)
axis([0 0.24 -2 2])

figure(2)

% STFT computation

N = 16 % rectangular window width
bot = 0.1;
hi = 0.175;

for kk = 1:4
  Nb2 = N / 2;
  for b = 1:2048-N+1
    fb = f(b:b+N-1);
```

```
    fftfb = abs(fft(fb));
    STFT(b,:) = fftfb(1:Nb2);
  end

% Plot

  NColor = 256;
  colormap(gray(NColor));
  STFT_min = min(min(STFT));
  STFT_max = max(max(STFT));
  STFT = (STFT - STFT_max) * NColor / (STFT_min - STFT_max);
  time=(0:2048-N)/r;
  freq = (0:Nb2-1) * r / N;

  axes('position',[0.1 bot 0.8 hi])
  image(time,freq,STFT')
  axis([0 0.24 0 2000])
  YTickmark = [0 500 1000 1500 2000];
  set(gca,'YDir','normal','ytick',YTickmark)

  hold on;
  N = N * 2
  bot = bot + 0.225;
  clear STFT; clear time; clear freq;
end

set(gcf,'paperposition',[0.5 0.5 7.5 10])
```

4.14.2 Wigner–Ville Distribution

```
%
% PROGRAM wvd.m
%
% Computes Wigner-Ville Distribution
%

% Signal

r = 4000; % sampling rate
t = (0:255) / r;
omega1 = 2.0 * pi * 500.0;
f = exp(i*omega1 * t) ;

% WVD Computation

N=length(f);

if (mod(N,2) == 1) ;
f = [f 0];
```

```
N = N + 1;
end

N2m1 = 2 * N - 1;
Nb2 = N / 2;

for m = 1:N
  s = zeros(1,N2m1);
  s(N-(m-1):N2m1-(m-1)) = f;
  s = conj(fliplr(s)).*s;
  s = s(Nb2:N2m1-Nb2);
  shat = abs(fft(s));
%
% Normalize with the number of overlapping terms
%
  if m <= Nb2
    shat = shat / (2 * m - 1);
  else
    shat = shat / (2 * N - 2 * m + 1);
  end

wvd(m,:)=shat(1:Nb2);
end

% Plot

time = (0:N-1) / r;
freq = (0:Nb2-1) * r / N / 2;

NColor = 256;
colormap(gray(NColor));
wvd_min = min(min(wvd));
wvd_max = max(max(wvd));
wvd = (wvd - wvd_max) * NColor / (wvd_min - wvd_max);

image(time,freq,wvd');

% Because of the finite support of the signal, there will
% end effects

xlabel('Time (seconds)');
ylabel('Frequency (Hz)');
set(gca,'YDir','normal')
```

REFERENCES

1. D. Gabor, "Theory of communication," *J. IEE (London)*, **93**, pp. 429–457, 1946.
2. I. Daubechies, *Ten Lectures on Wavelets,* CBMS-NSF Ser. Appl. Math. # 61. Philadelphia: SIAM, 1992.
3. J. B. Allan and L. R. Rabiner, "A unified approach to STFT analysis and synthesis," *Proc. of IEEE*, **65**, pp. 1558–1564, November 1977.
4. A. Grossmann, R. Kronland-Martinet, and J. Morlet, "Reading and understanding continuous wavelet transform," in *Wavelets, Time–Frequency Methods and Phase Space*, J. M. Combes, A. Grossmann, and Ph. Tchamitchian (Eds). Berlin: Springer-Verlag, 1989, pp. 2–20.
5. T. A. C. M. Claasen and W. F. G. Mecklenbrauker, "The Wigner–Ville distribution: a tool for time–frequency signal analysis: I. Continuous time signals," *Philips J. Res.*, **35**, pp. 217–250, 1980.
6. A. Moghaddar and E. K. Walton, "Time–frequency distribution analysis of scattering from waveguide cavities," *IEEE Trans. Antennas Propag.*, **41**, pp. 677–680, May 1993.
7. L. Cohen, "Time–frequency distributions: A review," *Proc. IEEE*, **77**, pp. 941–981, 1989.
8. Boualem Boshash, "Time–frequency signal analysis," in *Advances in Spectral Estimation and Array Processing*, Vol. 1, S. Haykin, (Ed.). Upper Saddle River, N.J.: Prentice Hall, 1991, Chap. 9.
9. C. E. Cook and M. Bernfeld, *Radar Signals*. San Diego, Calif.: Academic Press, 1967.

CHAPTER FIVE

Multiresolution Analysis

Multiresolution analysis (MRA) forms the most important building block for the construction of scaling functions and wavelets (Chapter 6) and the development of algorithms (Chapters 7 and 8). As the name suggests, in multiresolution analysis, a function is viewed at various levels of approximations or resolutions. The idea was developed by Meyer [1] and Mallat [2,3]. By applying the MRA we can divide a complicated function into several simpler ones and study them separately. To understand the notion of MRA, let us consider a situation where a function consists of slowly varying and rapidly varying segments, as illustrated in Figure 5.1. If we want to represent this function at a single level of approximation, we have to discretize it using step size (h), determined by the rapidly varying segment. This leads to a large number of data points. By representing the function using several discretization steps (resolutions) we can significantly reduce the number of data points required for accurate representation. The coarsest approximation of the function together with the details at every level completely represent the original function. Observe that with every level (scale), the step size is doubled. This corresponds to octave-level representation, familiar in audio signal processing. In addition to this specific example, there are many situations in signal processing as well as in computational electromagnetics, where multiresolution analysis can be very useful.

In this chapter we begin with an understanding of the requirements of MRA. Two-scale relations and decomposition relations are explained. Cardinal B-splines, discussed in Section 5.5, generate an MRA and form the basis of most of the wavelets discussed in this book and elsewhere. Finally, in Section 5.6 we discuss how to map a given function into an appropriate subspace before starting an MRA.

5.1 MULTIRESOLUTION SPACES

Let us go back to Figure 5.1. Every time we go down one level by doubling the step size, we remove certain portions of the function, shown on the right-hand-side plots. Then there are the "leftover" parts which are further decomposed. In Figure 5.1

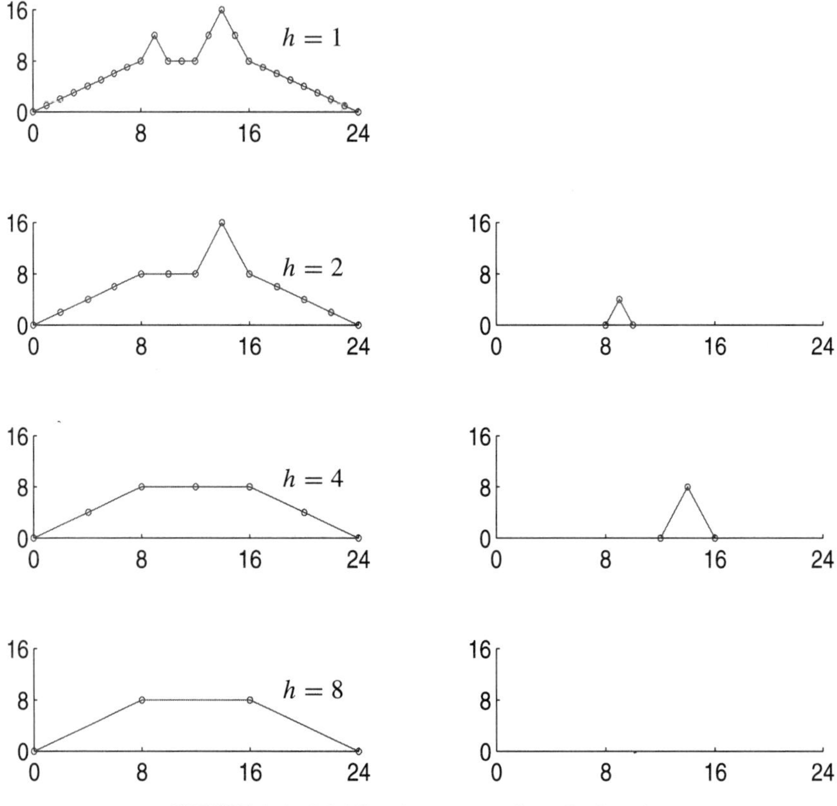

FIGURE 5.1 Multilevel representation of a function.

we assign all the functions on the left-hand side to $\mathbf{A}_s$ and those on the right-hand side to $\mathbf{W}_s$, where s represents individual scales. Let $\mathbf{A}_s$ be generated by the bases $\{\phi_{k,s} : 2^{s/2}\phi(2^s t - k); k \in \mathbb{Z}\}$ and $\mathbf{W}_s$ by $\{\psi_{k,s} : 2^{s/2}\psi(2^s t - k); k \in \mathbb{Z}\}$. In other words, any function $x_s(t)$ and $y_s(t)$ can be represented as the linear combinations of $\phi_{k,s}(t)$ and $\psi_{k,s}(t)$, respectively.

Observe that the functions $x_{s-1}(t) \in \mathbf{A}_{s-1}$ and $y_{s-1}(t) \in \mathbf{W}_{s-1}(t)$ are both derived from $x_s \in \mathbf{A}_s$. Therefore, we should expect that the bases $\phi_{k,s-1}$ of $\mathbf{A}_{s-1}$ and $\psi_{k,s-1}$ of $\mathbf{W}_{s-1}$ should somehow be related to the bases $\phi_{k,s}$ of $\mathbf{A}_s$. Such a relationship will help in devising an algorithm to obtain the functions x_{s-1} and y_{s-1} from x_s more efficiently.

To achieve a multiresolution analysis of a function as shown in Figure 5.1, we must have a finite-energy function $\phi(t) \in L^2(\mathbb{R})$, called a *scaling function*, that generates a nested sequence $\{\mathbf{A}_j\}$, namely

$$\{0\} \leftarrow \cdots \subset \mathbf{A}_{-1} \subset \mathbf{A}_0 \subset \mathbf{A}_1 \subset \cdots \rightarrow L^2,$$

and satisfies a dilation (refinement) equation

$$\phi(t) = \sum_k g_0[k]\, \phi(at - k)$$

for some $a > 0$ and coefficients $\{g_0[k]\} \in \ell^2$. We will consider $a = 2$, which corresponds to octave scales. Observe that the function $\phi(t)$ is represented as a superposition of a scaled and translated version of itself—hence the name *scaling function*. More precisely, $\mathbf{A}_0$ is generated by $\{\phi(\cdot - k) : k \in \mathbb{Z}\}$ and, in general, $\mathbf{A}_s$, by $\{\phi_{k,s} : k, s \in \mathbb{Z}\}$. Consequently, we have the following two obvious results:

$$x(t) \in \mathbf{A}_s \Leftrightarrow x(2t) \in \mathbf{A}_{s+1} \tag{5.1}$$

$$x(t) \in \mathbf{A}_s \Leftrightarrow x(t + 2^{-s}) \in \mathbf{A}_s. \tag{5.2}$$

There are many functions that generate nested sequences of subspaces. But the properties (5.1) and (5.2), and the dilation equation are unique to MRA.

For each s, since $\mathbf{A}_s$ is a proper subspace of $\mathbf{A}_{s+1}$, there is some space left in $\mathbf{A}_s$, called $\mathbf{W}_s$, which when combined with $\mathbf{A}_s$ gives us $\mathbf{A}_{s+1}$. This space $\{\mathbf{W}_s\}$ is called the *wavelet subspace* and is complementary to $\mathbf{A}_s$ in $\mathbf{A}_{s+1}$, meaning that

$$\mathbf{A}_s \cap \mathbf{W}_s = \{0\}, \qquad s \in \mathbb{Z} \tag{5.3}$$

$$\mathbf{A}_s \oplus \mathbf{W}_s = \mathbf{A}_{s+1}. \tag{5.4}$$

With the condition (5.3), the summation in (5.4) is referred to as a *direct sum* and the decomposition in (5.4) as a *direct-sum decomposition.*

Subspaces $\{\mathbf{W}_s\}$ are generated by $\psi(t) \in L^2$, called the *wavelet*, in the same way as $\{\mathbf{A}_s\}$ is generated by $\phi(t)$. In other words, any $x_s(t) \in \mathbf{A}_s$ can be written as

$$x_s(t) = \sum_k a_{k,s}\, \phi(2^s t - k), \tag{5.5}$$

and any function $y_s(t) \in \mathbf{W}_s$ can be written as

$$y_s(t) = \sum_k w_{k,s}\, \psi(2^s t - k) \tag{5.6}$$

for some coefficients $\{a_{k,s}\}_{k\in\mathbb{Z}}$, $\{w_{k,s}\}_{k\in\mathbb{Z}} \in \ell^2$.

Since

$$\begin{aligned} \mathbf{A}_{s+1} &= \mathbf{W}_s \oplus \mathbf{A}_s \\ &= \mathbf{W}_s \oplus \mathbf{W}_{s-1} \oplus \mathbf{A}_{s-1} \\ &= \mathbf{W}_s \oplus \mathbf{W}_{s-1} \oplus \mathbf{W}_{s-2} \oplus \cdots, \end{aligned} \tag{5.7}$$

we have

$$\mathbf{A}_s = \bigoplus_{\ell=-\infty}^{s-1} \mathbf{W}_\ell.$$

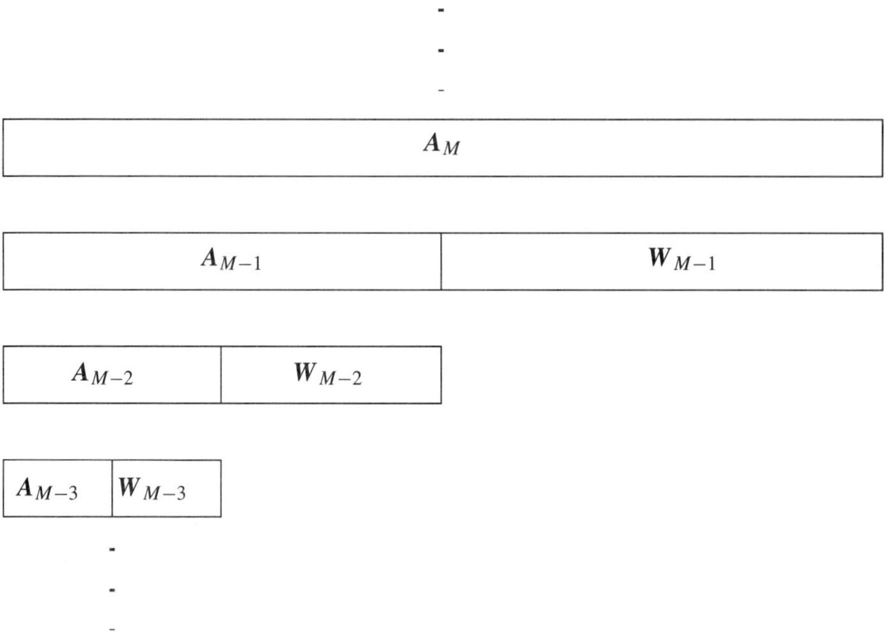

A_M

A_{M-1} W_{M-1}

A_{M-2} W_{M-2}

A_{M-3} W_{M-3}

FIGURE 5.2 Splitting of MRA subspaces.

Observe that the $\{\mathbf{A}_s\}$ are nested while the $\{\mathbf{W}_s\}$ are mutually orthogonal. Consequently, we have

$$\begin{cases} \mathbf{A}_\ell \cap \mathbf{A}_m = \mathbf{A}_\ell, & m > \ell \\ \mathbf{W}_\ell \cap \mathbf{W}_m = \{0\}, & \ell \neq m \\ \mathbf{A}_\ell \cap \mathbf{W}_m = \{0\}, & \ell \leq m. \end{cases}$$

A schematic representation of the hierarchical nature of $\mathbf{A}_s$ and $\mathbf{W}_s$ is shown in Figure 5.2.

5.2 ORTHOGONAL, BIORTHOGONAL, AND SEMIORTHOGONAL DECOMPOSITION

In Section 5.1, the only requirement we had for the wavelet subspace $\mathbf{W}_s$ was that it be complementary to $\mathbf{A}_s$ in $\mathbf{A}_{s+1}$. In addition to this, if we also require that $\mathbf{W}_s \perp \mathbf{A}_s$, such a decomposition is called an *orthogonal decomposition*. Let us explain the orthogonality of $\mathbf{A}_s$ and $\mathbf{W}_s$ a little further. For simplicity, let $s = 0$. For this case, $\{\phi(t-k) : k \in \mathbb{Z}\}$ spans $\mathbf{A}_0$; similarly, $\{\psi(t-k) : k \in \mathbb{Z}\}$ spans $\mathbf{W}_0$. Then $\mathbf{A}_0 \perp \mathbf{W}_0$ implies that

$$\int_{-\infty}^{\infty} \phi(t)\psi(t-\ell)\,dt = 0 \qquad \text{for all } \ell \in \mathbb{Z}. \tag{5.8}$$

In general, $\{\phi(t-k) : k \in \mathbb{Z}\}$ and $\{\psi(t-k) : k \in \mathbb{Z}\}$ need not be orthogonal in themselves; that is,

$$\int_{-\infty}^{\infty} \phi(t)\phi(t-\ell)\,dt \neq 0 \tag{5.9}$$

$$\int_{-\infty}^{\infty} \psi(t)\psi(t-\ell)\,dt \neq 0. \tag{5.10}$$

Let us relax the condition that $\mathbf{A}_s$ and $\mathbf{W}_s$ be orthogonal to each other and assume that the wavelet $\psi_{k,s} \in \mathbf{W}_s$ has a dual, $\widetilde{\psi}_{k,s} \in \widetilde{\mathbf{W}}_s$. *Duality* implies that the biorthogonality condition is satisfied, namely

$$\left\langle \psi_{j,k}, \widetilde{\psi}_{\ell,m} \right\rangle = \delta_{j,\ell} \cdot \delta_{k,m}, \qquad j,k,\ell,m \in \mathbb{Z}. \tag{5.11}$$

Although we do not require that $\mathbf{W}_s \perp \mathbf{A}_s$, we do need $\widetilde{\mathbf{W}}_s \perp \mathbf{A}_s$, the importance of which will become clear later. Similar to the dual wavelet $\widetilde{\psi}_{k,s}$, we also consider a dual scaling function $\widetilde{\phi}_{k,s}$ that generates another MRA $\{\widetilde{\mathbf{A}}_s\}$ of L^2. In other words, $\phi_{k,s}$ and $\psi_{k,s}$ are associated with the MRA $\{\mathbf{A}_s\}$, and $\widetilde{\phi}_{k,s}$ and $\widetilde{\psi}_{k,s}$ are associated with the MRA $\{\widetilde{\mathbf{A}}_s\}$.

Let us summarize our results so far before proceeding to explain their importance.

MRA $\{\mathbf{A}_s\}$:

$$\begin{aligned}
\mathbf{A}_{s+1} &= \mathbf{A}_s + \mathbf{W}_s, && m > \ell\\
\mathbf{A}_\ell \cap \mathbf{A}_m &= \mathbf{A}_\ell, && m > \ell\\
\mathbf{W}_\ell \cap \mathbf{W}_m &= \{0\}, && \ell \neq m\\
\mathbf{A}_\ell \cap \mathbf{W}_m &= \{0\}, && \ell \leq m.
\end{aligned}$$

MRA $\{\widetilde{\mathbf{A}}_s\}$:

$$\begin{aligned}
\widetilde{\mathbf{A}}_{s+1} &= \widetilde{\mathbf{A}}_s + \widetilde{\mathbf{W}}_s, && m > \ell\\
\widetilde{\mathbf{A}}_\ell \cap \widetilde{\mathbf{A}}_m &= \widetilde{\mathbf{A}}_\ell, && m > \ell\\
\widetilde{\mathbf{W}}_\ell \cap \widetilde{\mathbf{W}}_m &= \{0\}, && \ell \neq m\\
\widetilde{\mathbf{A}}_\ell \cap \widetilde{\mathbf{W}}_m &= \{0\}, && \ell \leq m.
\end{aligned}$$

$$\begin{aligned}
\mathbf{W}_s \perp \widetilde{\mathbf{A}}_s &\Rightarrow \widetilde{\mathbf{A}}_\ell \cap \mathbf{W}_m = \{0\} && \text{for } \ell \leq m\\
\widetilde{\mathbf{W}}_s \perp \mathbf{A}_s &\Rightarrow \mathbf{A}_\ell \cap \widetilde{\mathbf{W}}_m = \{0\} && \text{for } \ell \leq m.
\end{aligned}$$

The decomposition process discussed so far is called *biorthogonal decomposition*. To understand its importance, let us briefly point out the procedure of decomposing a function into scales, as shown in Figure 5.1. The details are left for Chapter 7.

Given a function $x(t) \in L^2$, the decomposition into various scales begins by mapping the function into a sufficiently high resolution subspace $\mathbf{A}_M$, that is,

$$L^2 \ni x(t) \longmapsto x_M = \sum_k a_{k,M}\phi(2^M t - k) \in \mathbf{A}_M. \tag{5.12}$$

Now since

$$\begin{aligned}\mathbf{A}_M &= \mathbf{W}_{M-1} + \mathbf{A}_{M-1} \\ &= \mathbf{W}_{M-1} + \mathbf{W}_{M-2} + \mathbf{A}_{M-2} \\ &= \sum_{n=1}^{N} \mathbf{W}_{M-n} + \mathbf{A}_{M-N},\end{aligned} \tag{5.13}$$

we can write

$$x_M(t) = \sum_{n=1}^{N} y_{M-n} + x_{M-N}, \tag{5.14}$$

where $x_{M-N}(t)$ is the coarsest approximation of $x_M(t)$ and

$$x_s(t) = \sum_k a_{k,s}\phi(2^s t - k) \in \mathbf{A}_s \tag{5.15}$$

$$y_s(t) = \sum_k w_{k,s}\psi(2^s t - k) \in \mathbf{W}_s. \tag{5.16}$$

Now the importance of dual wavelets becomes clear. By using the biorthogonality condition (5.11), we can obtain the coefficients $\{w_{k,s}\}$ as

$$w_{k,s} = 2^s \int_{-\infty}^{\infty} y_s(t)\widetilde{\psi}(2^s t - k)\, dt. \tag{5.17}$$

Recall that $\widetilde{\psi}(2^s t - k) \in \widetilde{\mathbf{W}}_s$ and $\mathbf{A}_\ell \perp \widetilde{\mathbf{W}}_s$ for $\ell \leq s$. Therefore, by taking the inner product of (5.14) with $\widetilde{\psi}_{k,s}(t)$ and by using the condition (5.11), we have

$$\begin{aligned}w_{k,s} &= 2^s \int_{-\infty}^{\infty} x_M(t)\widetilde{\psi}(2^s t - k)\, dt \\ &= 2^{s/2}\, W_{\widetilde{\psi}} x_M\left(\frac{k}{2^s}, \frac{1}{2^s}\right).\end{aligned} \tag{5.18}$$

The dual wavelet $\widetilde{\psi}$ can be used to analyze a function x_M by computing its integral wavelet transform at a desired time-scale location, while ψ can be used to obtain its function representation at any scale. Therefore, we call $\widetilde{\psi}$ an *analyzing wavelet*, while ψ is called a *synthesis wavelet*.

Of course, if we have orthogonal decomposition $\mathbf{A}_s \perp \mathbf{W}_s$ with orthonormal bases $\{\phi, \psi\}$, the analyzing and synthesis wavelets are the same. Observe that when we say orthonormal wavelets, this implies that the wavelets are orthonormal with respect to scale as well as with respect to translation in a given scale. But orthonormal scaling function implies that the scaling functions are orthonormal only with respect to translation in a given scale, not with respect to the scale because of the nested nature of the MRA.

A question that arises is: Why do we need biorthogonal wavelets? One of the attractive features in delegating the responsibilities of analysis and synthesis to two different functions in the biorthogonal case as opposed to a single function in the orthonormal case is that in the former, we can have compactly supported symmetric analyzing and synthesis wavelets and scaling functions, something that a continuous orthonormal basis cannot achieve. Furthermore, orthonormal scaling functions and wavelets have poor time-scale localization.

In some applications discussed in later chapters, we need to interchange the roles of the analysis and synthesis pairs, $\{\phi, \psi\}$ and $\{\widetilde{\phi}, \widetilde{\psi}\}$, respectively. In biorthogonal decomposition, we cannot do so easily since ϕ and $\widetilde{\phi}$ generate two different MRAs, $\mathbf{A}$ and $\widetilde{\mathbf{A}}$, respectively. For such an interchange, we need to map the given function $x \longmapsto \widetilde{x}_M \in \widetilde{\mathbf{A}}_M$, and then we can use ψ as analyzing and $\widetilde{\psi}$ as synthesizing wavelets.

In addition to biorthogonal and orthonormal decomposition, there is another class of decomposition, called *semiorthogonal decomposition*, for which $\mathbf{A}_s \perp \mathbf{W}_s$. Since in this system, the scaling function and wavelets are nonorthogonal, we still need their duals, $\widetilde{\phi}$ and $\widetilde{\psi}$. However, unlike the biorthogonal case, there is no dual space. That is, $\phi, \widetilde{\phi} \in \mathbf{A}_s$ and $\psi, \widetilde{\psi} \in \mathbf{W}_s$, for some appropriate scale s. In this system it is very easy to interchange the roles of ϕ, ψ with those of $\widetilde{\phi}, \widetilde{\psi}$.

For semiorthogonal scaling functions and wavelets, we have

$$\left\langle \phi(t-k), \widetilde{\phi}(t-\ell) \right\rangle = \delta_{k,\ell}, \qquad k, \ell \in \mathbb{Z} \tag{5.19}$$

and

$$\left\langle \psi_{j,k}, \widetilde{\psi}_{\ell,m} \right\rangle = 0 \qquad \text{for } j \neq \ell \quad \text{and} \quad j, k, \ell, m \in \mathbb{Z}. \tag{5.20}$$

The wavelets $\{\phi, \psi\}$ are related to $\{\widetilde{\phi}, \widetilde{\psi}\}$ as

$$\hat{\widetilde{\phi}}(\omega) = \frac{\hat{\phi}(\omega)}{E_\phi(e^{j\omega})} \tag{5.21}$$

and

$$\hat{\widetilde{\psi}}(\omega) = \frac{\hat{\psi}(\omega)}{E_\psi(e^{j\omega})} \tag{5.22}$$

with

$$E_x(e^{j\omega}) := \sum_{k=-\infty}^{\infty} |\hat{x}(\omega + 2\pi k)|^2 = \sum_{k=-\infty}^{\infty} A_x(k) e^{jk\omega}, \tag{5.23}$$

where $A_x(t)$ is the autocorrelation function of $x(t)$. For a proof of (5.23), see Section 7.6.1. Observe that the relation above is slightly different from the orthonormalization relation (2.35) in that here we do not have a square root in the denominator. In Chapter 6 we discuss the construction of all the scaling functions and wavelets that we have discussed.

5.3 TWO-SCALE RELATIONS

Two-scale relations relate the scaling function and the wavelets at a given scale with the scaling function at the next-higher scale. Since

$$\phi(t) \in \mathbf{A}_0 \subset \mathbf{A}_1 \tag{5.24}$$

$$\psi(t) \in \mathbf{W}_0 \subset \mathbf{A}_1, \tag{5.25}$$

we should be able to write $\phi(t)$ and $\psi(t)$ in terms of the bases that generate $\mathbf{A}_1$. In other words, there exist two sequences $\{g_0[k]\}, \{g_1[k]\} \in \ell^2$ such that

$$\phi(t) = \sum_k g_0[k]\phi(2t - k) \tag{5.26}$$

$$\psi(t) = \sum_k g_1[k]\phi(2t - k). \tag{5.27}$$

Equations (5.26) and (5.27) are referred to as *two-scale relations*. In general, for any $j \in \mathbb{Z}$, the relationship between $\mathbf{A}_j$ and $\mathbf{W}_j$ with $\mathbf{A}_{j+1}$ is governed by

$$\phi(2^j t) = \sum_k g_0[k]\phi(2^{j+1}t - k)$$

$$\psi(2^j t) = \sum_k g_1[k]\phi(2^{j+1}t - k).$$

By taking the Fourier transform of the two-scale relations, we have

$$\hat{\phi}(\omega) = G_0(z)\hat{\phi}\left(\frac{\omega}{2}\right) \tag{5.28}$$

$$\hat{\psi}(\omega) = G_1(z)\hat{\phi}\left(\frac{\omega}{2}\right), \tag{5.29}$$

where

$$G_0(z) := \frac{1}{2}\sum_k g_0[k]z^k \tag{5.30}$$

$$G_1(z) := \frac{1}{2}\sum_k g_1[k]z^k, \tag{5.31}$$

with $z = e^{-j\omega/2}$. Observe that the definitions in (5.30) and (5.31) differ slightly from those used in Chapter 2 for z-transform. An example of a two-scale relation for the Haar case (H) is shown in Figure 5.3. Expansions of (5.26) and (5.27) lead to

$$\hat{\phi}(\omega) = \prod_{\ell=1}^{\infty} G_0 \exp\left(-j\frac{\omega}{2^\ell}\right) \tag{5.32}$$

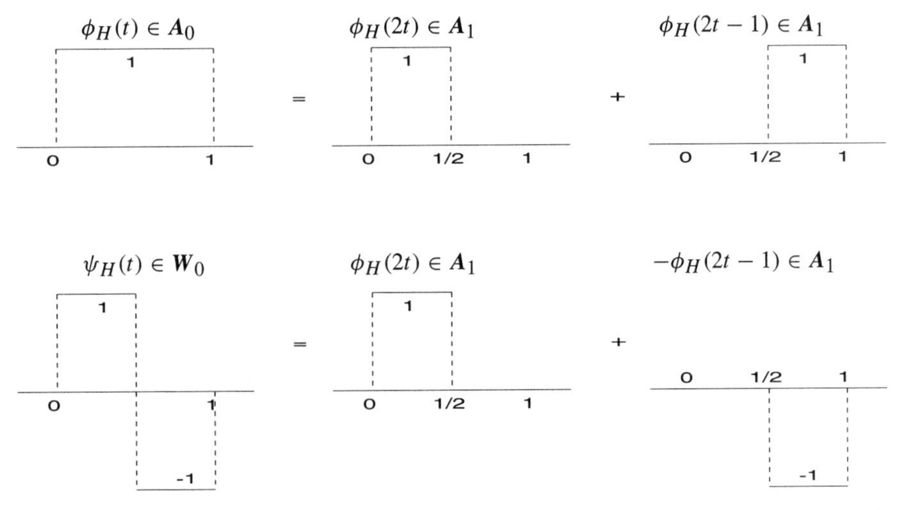

FIGURE 5.3 Two-scale relation for Haar case ($g_0[0] = g_0[1] = 1;\ g_1[0] = -g_1[1] = 1;$ $g_0[k] = g_1[k] = 0$ for all other k).

$$\hat{\psi}(\omega) = G_1 \exp\left(-j\frac{\omega}{2}\right) \prod_{\ell=2}^{\infty} G_0 \exp\left(-j\frac{\omega}{2^{\ell}}\right). \tag{5.33}$$

Since the scaling functions exhibit the lowpass filter characteristic [$\hat{\phi}(0) = 1$], all the coefficients $\{g_0[k]\}$ add up to 2, whereas because of the bandpass filter characteristic of the wavelets ($\hat{\psi}(0) = 0$), the coefficients $\{g_1[k]\}$ add up to 0.

5.4 DECOMPOSITION RELATION

Decomposition relations give the scaling function at any scale in terms of the scaling function and the wavelet at the next-lower scale. Since $\mathbf{A}_1 = \mathbf{A}_0 + \mathbf{W}_0$ and $\phi(2t)$, $\phi(2t-1) \in \mathbf{A}_1$, there exist two sequences $(\{h_0[k]\}, \{h_1[k]\})$ in ℓ^2 such that

$$\phi(2t) = \sum_k \{h_0[2k]\phi(t-k) + h_1[2k]\psi(t-k)\}$$

$$\phi(2t-1) = \sum_k \{h_0[2k-1]\phi(t-k) + h_1[2k-1]\psi(t-k)\}.$$

Combining these two relations, we have

$$\phi(2t-\ell) = \sum_k \{h_0[2k-\ell]\phi(t-k) + h_1[2k-\ell]\psi(t-k)\} \tag{5.34}$$

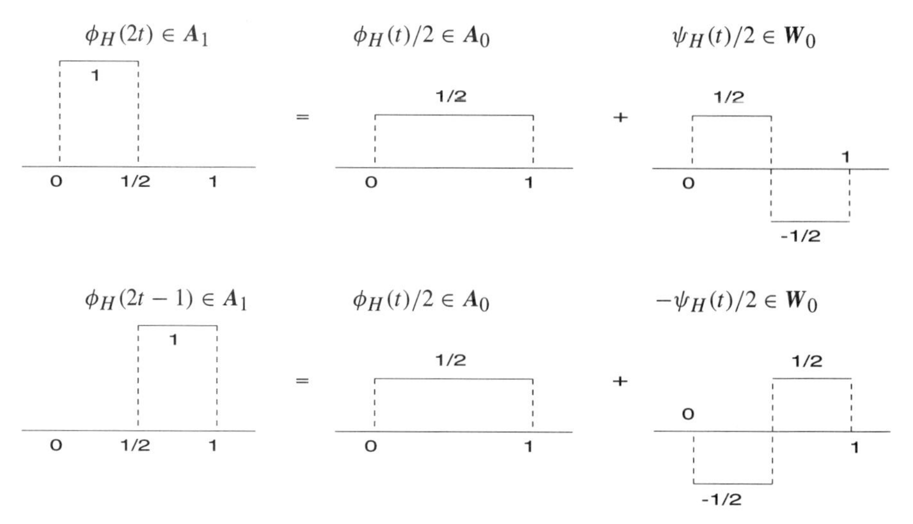

FIGURE 5.4 Decomposition relation for Haar case ($h_0[0] = h_0[-1] = 1/2$; $h_1[0] = -h_1[-1] = 1/2$; $h_0[k] = h_1[k] = 0$ for all other k).

for all $\ell \in \mathbb{Z}$. In general, we have

$$\phi(2^{j+1}t - \ell) = \sum_k \left\{ h_0[2k - \ell]\phi(2^j t - k) + h_1[2k - \ell]\psi(2^j t - k) \right\}. \quad (5.35)$$

Figure 5.4 shows an example of decomposition relation for the Haar case (H).

5.5 SPLINE FUNCTIONS

One of the most basic building blocks for wavelet construction involves cardinal B-splines. Complete coverage of spline theory is beyond the scope of this book. In this section we describe briefly spline functions and their properties that are required to understand the topics discussed in this book. For further details one may refer to many excellent books (e.g., [4–8]).

Spline functions consist of piecewise polynomials (see Figure 5.5) joined together smoothly at the break points (knots: $t_0, t_1, \ldots$), where the degree of smoothness depends on the order of splines. For cardinal B-splines, these break points are equally spaced. Unlike polynomials, these form local bases and have many useful properties that can be applied to function approximation.

The mth-order cardinal B-spline $N_m(t)$ has the knot sequence $\{\ldots, -1, 0, 1, \ldots\}$ and consists of polynomials of order m (degree $m - 1$) between the knots. Let $N_1(t) = \chi_{[0,1)}(t)$ be the characteristic function of $[0, 1)$. Then for each integer $m \geq 2$, the mth-order cardinal B-spline is defined, inductively, by

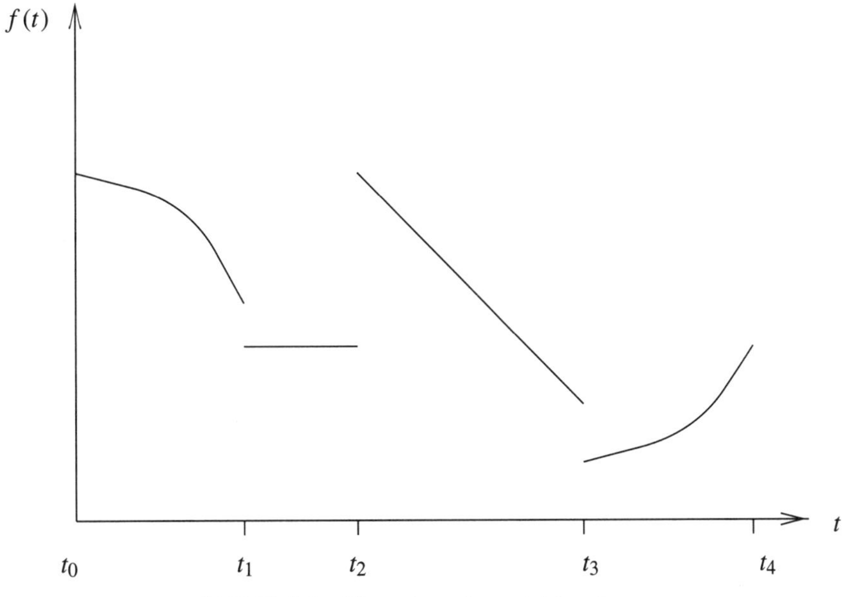

FIGURE 5.5 Piecewise polynomial functions.

$$N_m(t) := (N_{m-1} * N_1)(t) \tag{5.36}$$

$$:= \int_{-\infty}^{\infty} N_{m-1}(t-x) N_1(x)\, dx$$

$$= \int_0^1 N_{m-1}(t-x)\, dx. \tag{5.37}$$

A fast computation of $N_m(t)$ for $m \geq 2$ can be achieved by using the formula [7, p. 131]

$$N_m(t) = \frac{t}{m-1} N_{m-1}(t) + \frac{m-t}{m-1} N_{m-1}(t-1)$$

recursively until we arrive at the first-order B-spline N_1 (see Figure 5.6). Splines of orders 2 to 6, along with their magnitude spectra, are shown in Figure 5.7. The most commonly used splines are linear ($m = 2$) and cubic ($m = 4$) splines. Their explicit expressions are

$$N_2(t) = \begin{cases} t, & t \in [0, 1] \\ 2 - t, & t \in [1, 2] \\ 0, & \text{elsewhere} \end{cases} \tag{5.38}$$

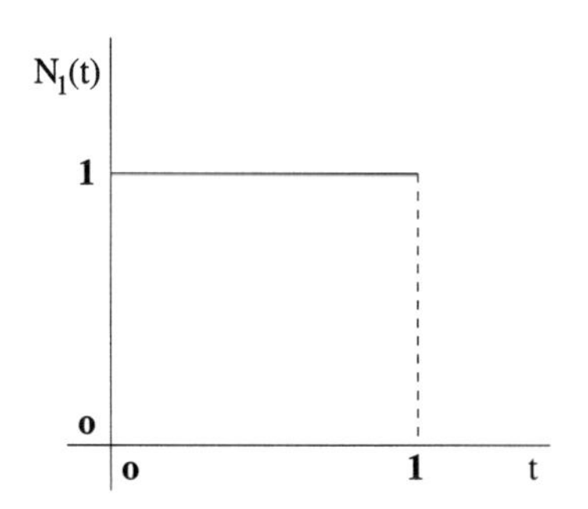

FIGURE 5.6 N_1, the spline of order 1.

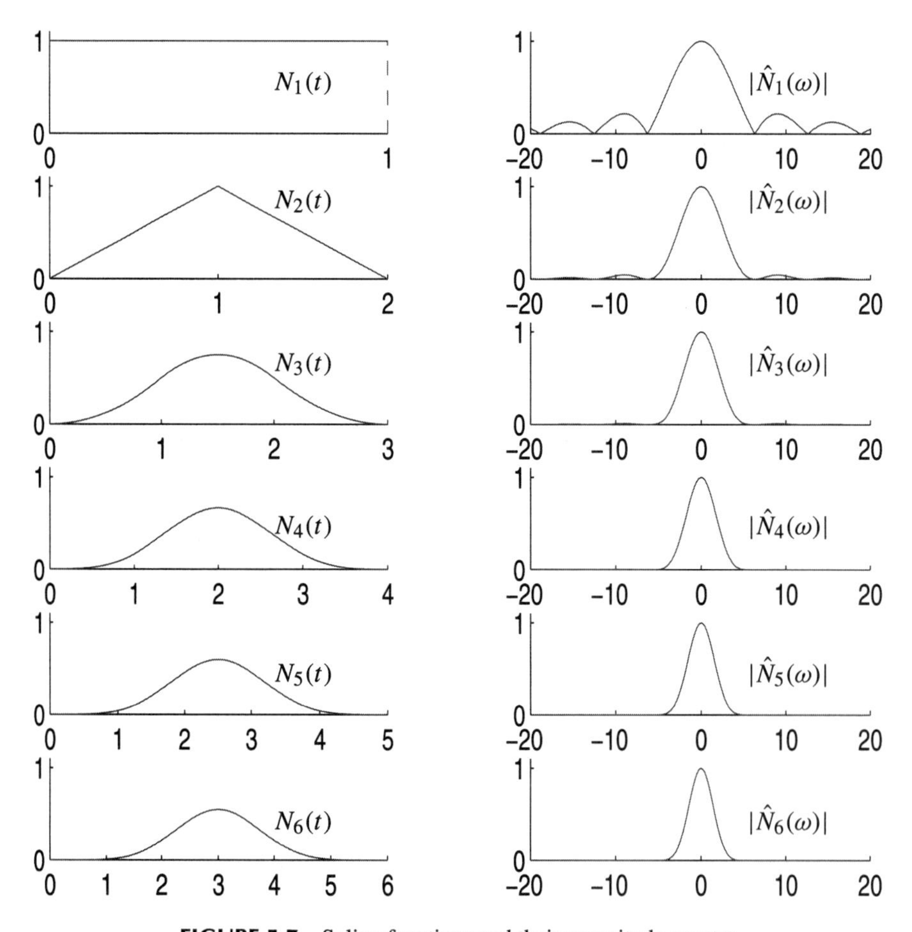

FIGURE 5.7 Spline functions and their magnitude spectra.

$$N_4(t) = \begin{cases} t^3, & t \in [0, 1] \\ 4 - 12t + 12t^2 - 3t^3, & t \in [1, 2] \\ -44 + 60t - 24t^2 + 3t^3, & t \in [2, 3] \\ 64 - 48t + 12t^2 - t^3, & t \in [3, 4] \\ 0, & \text{elsewhere.} \end{cases} \tag{5.39}$$

In many applications we need to compute splines at integer points. Table 5.1 gives spline values at integer locations. The symmetry property can be used to get values at other points.

To obtain the Fourier transform of $N_m(t)$, observe that (5.36) can be written as

$$N_m(t) = \underbrace{(N_1 * \cdots * N_1)}_{m}(t). \tag{5.40}$$

Therefore,

$$\widehat{N}_m(\omega) = \left(\frac{1 - e^{-j\omega}}{j\omega}\right)^m, \tag{5.41}$$

since

$$\widehat{N}_1(\omega) = \int_0^1 e^{-j\omega t}\, dt = \frac{1 - e^{-j\omega}}{j\omega}. \tag{5.42}$$

TABLE 5.1 Cardinal B-Splines at Integer Points

k	$(m-1)!N_m(k)$	k	$(m-1)!N_m(k)$	k	$(m-1)!N_m(k)$
	$m = 3$		$m = 8$		$m = 11$
1	1	1	1	1	1
	$m = 4$	2	120	2	1,013
1	1	3	1,191	3	47,840
2	4	4	2,416	4	455,192
	$m = 5$		$m = 9$	5	1,310,354
1	1	1	1		$m = 12$
2	11	2	247	1	1
	$m = 6$	3	4,293	2	2,036
1	1	4	15,619	3	152,637
2	26		$m = 10$	4	2,203,488
3	66	1	1	5	9,738,114
	$m = 7$	2	502	6	15,724,248
1	1	3	14,608		
2	57	4	88,234		
3	302	5	156,190		

The important property of splines for our purposes is the fact that they are scaling functions. That is, there exists a sequence $\{g_0[m,k]\} \in \ell^2$ such that

$$N_m(t) = \sum_k g_0[m,k]\, N_m(2t-k). \tag{5.43}$$

In Chapter 6 we derive an expression for $g_0[m,k]$.

5.5.1 Properties of Splines

Some important properties of splines, relevant to the topics discussed in this book, are discussed in this section without giving any proof. Proofs of some of the properties are left as exercises.

1. Supp $N_m = [0, m]$ with $N_m(0) = N_m(m) = 0$.
2. $N_m(t) \in C^{m-2}$; C^k is the space of functions that are k times continuously differentiable.
3. $N_m|_{[k-1,k]} \in \pi_{m-1}$, $k \in \mathbb{Z}$; π_k is the polynomial space of degree k (order $k+1$).
4. $\int_{-\infty}^{\infty} N_m(t)\, dt = 1$.
5. $N_m'(t) = N_{m-1}(t) - N_{m-1}(t-1)$.
6. $N_m(t)$ is symmetric with respect to the center $t^* = m/2$, that is,

$$N_m\left(\frac{m}{2}+t\right) = N_m\left(\frac{m}{2}-t\right), \qquad t \in \mathbb{R}. \tag{5.44}$$

7. $N_m(t)$ behaves as a lowpass filter [$\widehat{N}_m(0) = 1$; see Figure 5.7].
8. $N_m(t)$ has mth order of approximation in the sense that $\hat{N}_m(\omega)$ satisfies the Strang–Fix condition

$$\begin{cases} \widehat{N}_m(0) = 1 \\ D^j \widehat{N}_m(2\pi k) = 0, \qquad k \in \mathbb{Z} \setminus \{0\} \quad \text{and} \quad j = 1, \ldots, m-1, \end{cases} \tag{5.45}$$

   where D^j denotes the jth-order derivative. Consequently, $N_m(t)$ locally reproduces all polynomials of order m (see [8, pp. 114–121]).
9. $\sum_k N_m(t-k) \equiv 1$ for all t. This property is referred to as the *partition of unity* property.
10. *Total positivity*: $N_m(t) \geq 0$, for $t \in [0, m]$. By virtue of the total positivity [6, p. 7] property of B-splines, coefficients of a B-spline series follow the shape of the data. For instance, if $g(t) = \sum_j \alpha_j N_m(t-j)$, then

$$\alpha_j \geq 0 \;\; \forall j \Rightarrow g(t) \geq 0$$

$$\alpha_j \uparrow \text{(increasing)} \Rightarrow g(t) \uparrow$$

$$\alpha_j \text{ (convex)} \Rightarrow g(t) \text{ convex.}$$

Furthermore, the number of sign changes of $g(t)$ does not exceed that of the coefficient sequence $\{\alpha_j\}$. The latter property can be used to identify the zero crossing of a signal.

11. As the order m increases, $N_m(t)$ approaches a Gaussian function ($\Delta_{N_m}\Delta_{\widehat{N}_m} \to 0.5$). For instance, in the case of a cubic spline ($m = 4$), the RMS time–frequency window product is 0.501.

5.6 MAPPING A FUNCTION INTO MRA SPACE

As discussed in Section 5.2, before a signal $x(t)$ can be decomposed, it must be mapped into an MRA subspace $\mathbf{A}_M$ for some appropriate scale M, that is,

$$x(t) \longmapsto x_M(t) = \sum_k a_{k,M}\phi(2^M t - k). \tag{5.46}$$

Once we know $\{a_{k,M}\}$ we can use fast algorithms to compute $\{a_{k,s}\}$ for $s < M$. Fast algorithms are discussed in later chapters. Here we are concerned with evaluation of the coefficients $\{a_{k,M}\}$.

If $x(t)$ is known at every t, we can obtain $\{a_{k,M}\}$ by the orthogonal projection (L^2 projection) of the signal, that is,

$$a_{k,M} = 2^M \int_{-\infty}^{\infty} x(t)\widetilde{\phi}(2^M t - k)\,dt. \tag{5.47}$$

However, in practice the signal $x(t)$ is known at some discrete points. The given time step determines the scale M to which the function can be mapped. For a representation such as (5.46), we want it to satisfy two important conditions: (1) interpolatory and (2) polynomial reproducibility. By *interpolatory representation* we mean that the series should be exact, at least at the points at which the function is given, meaning that $x(k/2^M) = x_M(k/2^M)$. As pointed out before, *polynomial reproducibility* means that the representation is exact at every point for polynomials of order m if the basis $\phi(t)$ has the approximation order m. In other words, $x(t) \equiv x_M(t)$ for $x(t) \in \pi_{m-1}$. Cardinal B-splines have m order of approximation. In addition, since they are a local basis, the representation (5.46) is also local. By *local* we mean that to obtain the coefficient $a_{k,M}$ for some k, we do not need all the function values; only a few, determined by the support of the splines, will suffice. The coefficients when $\phi(t) = N_2(t)$ and $\phi(t) = N_4(t)$ are derived below.

***Linear Splines* ($m = 2$)** Suppose that a function $x(t)$ is given at $t = \ell/2^M : \ell \in \mathbb{Z}$. Then to obtain the spline coefficients $\{a_{k,M}\}$ for the representation

$$x(t) \longmapsto x_M(t) = \sum_k a_{k,M}\, N_2(2^M t - k), \tag{5.48}$$

we apply the interpolation condition, namely

$$x\left(\frac{\ell}{2^M}\right) = x_M\left(\frac{\ell}{2^M}\right). \tag{5.49}$$

By using (5.49) along with the fact that

$$N_2(1) = 1 \quad \text{and} \quad N_2(k) = 0, \qquad k \in \mathbb{Z} \setminus \{0\}, \tag{5.50}$$

wc gct

$$a_{k,M} = x\left(\frac{k+1}{2^M}\right). \tag{5.51}$$

The representation (5.48) preserves all polynomials of degree at most 1.

***Cubic Splines* (*m* = 4)** In this case

$$x(t) \longmapsto x_M(t) = \sum_k a_{k,M} N_4(2^M t - k), \tag{5.52}$$

where [4, p. 117]

$$a_{k,M} = \sum_{n=k-2}^{k+6} v_{k+2-2n}\, x\left(\frac{n}{2^{M-1}}\right)$$

and

$$v_n = \begin{cases} \frac{29}{24}, & n = 0 \\ \frac{7}{12}, & n = \pm 1 \\ -\frac{1}{8}, & n = \pm 2 \\ -\frac{1}{12}, & n = \pm 3 \\ \frac{1}{48}, & n = \pm 4 \\ 0, & \text{otherwise.} \end{cases}$$

The representation (5.52) preserves all polynomials of degree at most 3.

5.7 EXERCISES

1. For a given $j \in \mathbb{Z}$, a projection $P_{2,j} f(t)$ of any given function $f(t) \in L^2(-\infty, \infty)$ onto the hat function space

$$V_j = \left\{ \sum_{k=-\infty}^{\infty} c_k N_2(2^j t - k) : \{c_k\}_{k \in Z} \in \ell_2 \right\}$$

can be determined by the interpolation conditions $P_{2,j} f(k/2^j) = f(k/2^j)$ for all $k \in Z$. Find the formulas for the coefficients $\{a_n\}$ if $P_{2,j} f$ is written as

$$P_{2,j} f(t) = \sum_{n=-\infty}^{\infty} a_n N_2(2^j t - n).$$

2. For the Haar wavelet

$$\psi_H(t) := N_1(t) = \begin{cases} 1 & \text{for} \quad t \in [0, \frac{1}{2}) \\ -1 & \text{for} \quad t \in [\frac{1}{2}, 1) \\ 0, & \text{otherwise,} \end{cases}$$

define

$$\psi_{H,k,s}(t) = 2^{s/2}\psi_H(2^s t - k), \qquad k, s \in Z.$$

Show the orthogonality relations

$$\int_{-\infty}^{\infty} \psi_{H,k,s}(t)\psi_{H,m,p}(t)\, dt = \delta_{m,k}\delta_{s,p}, \qquad m, k, p, s \in Z.$$

Due to these relations, we say that the set $\{\psi_{H,k,s}\}_{k,s \in Z}$ forms an orthonormal family in $L^2(-\infty, \infty)$.

3. Show that the Gaussian function $\phi(t) = e^{-t^2}$ cannot be the scaling function of a multiresolution analysis. (Hint: Assume that e^{-t^2} can be written as $e^{-t^2} = \sum_{k=-\infty}^{\infty} a_k e^{-(2t-k)^2}$ for a sequence $\{a_k\}_{k \in Z}$ in ℓ_2, which has to be true if $e^{-t^2} \in V_0 \subset V_1$. Then show that this leads to a contradiction by taking Fourier transforms on both sides of the equation and comparing the results.)

4. Show that the mth-order B-spline $N_m(t)$ and its integer translates form a partition of unity, that is,

$$\sum_{k=-\infty}^{\infty} N_m(t-k) = 1 \qquad \text{for all } x \in \mathbb{R}.$$

(Hint: Use Poisson's sum formula.)

5. Show the following symmetry property of $N_m(t)$:

$$N_m\left(\frac{m}{2} + t\right) = N_m\left(\frac{m}{2} - t\right), \qquad x \in R.$$

6. Use Exercise 5 to show that

$$\int_{-\infty}^{\infty} N_m(t+k)N_m(t)\, dt = N_{2m}(m+k) \qquad \text{for any } k \in \mathbb{Z}.$$

7. Show that the hat function

$$N_2(t) = \begin{cases} t & \text{for} \quad t \in [0, 1] \\ 2-t & \text{for} \quad t \in [1, 2] \\ 0, & \text{otherwise} \end{cases}$$

and the function $N_1(t)$ are related by convolution: $N_2(t) = N_1(t) * N_1(t)$. Find the defining equations (the polynomial expression) for the functions given by

$$N_3(t) := N_2(t) * N_1(t)$$
$$N_4(t) := N_3(t) * N_1(t), \qquad t \in \mathbb{R}.$$

5.8 COMPUTER PROGRAMS

5.8.1 B-Splines

```
%
% PROGRAM Bspline.m
%
% Computes uniform Bsplines

function y = Bspline(m,x)

y = 0;

% Characteristic function

if m == 1
  if x >= 0.0 & x < 1.0
    y = 1.0;
  else
    y = 0.0;
  end
end

% Higher order

a = zeros(1,500);

if m >= 2 & m < 100
  for k = 1:m-1
    a(k) = 0.0;
    x1 = x - k + 1;
    if x1 >= 0.0 & x1 < 1.0
      a(k) = x1;
    end
    if x1 >= 1.0 & x1 < 2.0
      a(k) = 2 - x1;
    end
  end

  for p=1:m-2
    for q=1:m-1-p
```

```
      a(q)  = ((x-q+1) * a(q)+(p+q+1-x)*a(q+1)) / (p+1);
    end
  end

  y = a(1);
end
```

REFERENCES

1. Y. Meyer, *Wavelets: Algorithms and Applications.* Philadelphia: SIAM, 1993.
2. S. Mallat, "A theory of multiresolution signal decomposition: the wavelet representation," *IEEE Trans. Pattern Anal. Machine Intell.*, **11**, pp. 674–693, 1989.
3. S. Mallat, Multiresolution representation and wavelets, Ph.D. thesis, University of Pennsylvania, Philadelphia, 1988.
4. C. K. Chui, *An Introduction to Wavelets.* San Diego, Calif.: Academic Press, 1992.
5. I. J. Schoenberg, *Cardinal Spline Interpolation*, CBMS Ser. 12. Philadelphia: SIAM, 1973.
6. L. L. Schumaker, *Spline Functions: Basic Theory.* New York: Wiley-Interscience, 1981.
7. C. de Boor, *A Practical Guide to Splines.* New York: Springer-Verlag, 1978.
8. C. K. Chui, *Multivariate Splines.* CBMS-NSF Ser. Appl. Math. 54. Philadelphia: SIAM, 1988.

CHAPTER SIX

Construction of Wavelets

In this chapter we are concerned with the construction of orthonormal, semiorthogonal, and biorthogonal wavelets. The construction problem is tantamount to finding suitable two-scale and decomposition sequences as introduced in Chapter 5. It turns out that these coefficients for orthonormal wavelets can easily be derived from those of semiorthogonal wavelets. Therefore, we first discuss the semiorthogonal wavelet followed by orthonormal and biorthogonal wavelets.

Recall that for the semiorthogonal wavelet, both $\phi(t)$ and $\tilde{\phi}(t)$ are in $\mathbf{A}_0$, and $\psi(t)$ and $\tilde{\psi}(t)$ are in $\mathbf{W}_0$. Consequently, we can write $\phi(t)$ in terms of $\tilde{\phi}(t)$; similarly for $\psi(t)$. These relations as given by (5.21) and (5.22) are

$$\hat{\tilde{\phi}}(\omega) = \frac{\hat{\phi}(\omega)}{E_{\phi}(e^{j\omega})} \tag{6.1}$$

and

$$\hat{\tilde{\psi}}(\omega) = \frac{\hat{\psi}(\omega)}{E_{\psi}(e^{j\omega})}, \tag{6.2}$$

with the Euler–Frobenius–Laurent polynomial $E_f(e^{j\omega})$ given by

$$E_f(e^{j\omega}) := \sum_{k=-\infty}^{\infty} |\hat{f}(\omega + 2\pi k|^2 = \sum_{k=-\infty}^{\infty} A_f(e^{j\omega k}). \tag{6.3}$$

We will, therefore, concentrate on the construction of ϕ and ψ only.

As the first step toward constructing wavelets, we express $\{h_0[k]\}$, $\{h_1[k]\}$, and $\{g_1[k]\}$ in terms of $\{g_0[k]\}$ so that only $\{g_0[k]\}$ and hence the scaling functions need to be constructed. In semiorthogonal cases, all these sequences have different lengths, in general. Later we will show that for orthonormal cases, all of these sequences have the same length and that there is a very simple relation among them which can easily be derived as a special case of the relationship for semiorthogonal cases. The construction of a semiorthogonal wavelet is followed by the construction of several popular orthonormal wavelets: the Shannon, Meyer, Battle–Lemarié, and Daubechies wavelets. Finally, we construct a biorthogonal wavelet.

6.1 NECESSARY INGREDIENTS FOR WAVELET CONSTRUCTION

As pointed out before, we need to obtain the coefficient sequences $\{g_0[k]\}$ and $\{g_1[k]\}$ to be able to construct wavelets. In this section our goal is to find a relationship among various sequences. This will help us in reducing our task. Here we consider the case of semiorthogonal decomposition of a multiresolution space.

6.1.1 Relationship Between Two-Scale Sequences

Recall from Chapter 5 that as a result of the multiresolution properties, the scaling functions and wavelets at one scale (coarser) are related to the scaling functions at the next-higher scale (finer) by the two-scale relations, namely

$$\phi(t) = \sum_k g_0[k]\phi(2t-k) \tag{6.4}$$

$$\psi(t) = \sum_k g_1[k]\phi(2t-k). \tag{6.5}$$

By taking the Fourier transform of the relation above, we have

$$\hat{\phi}(\omega) = G_0(z)\hat{\phi}\left(\frac{\omega}{2}\right) \tag{6.6}$$

$$\hat{\psi}(\omega) = G_1(z)\hat{\phi}\left(\frac{\omega}{2}\right), \tag{6.7}$$

where $z = e^{-j\omega/2}$ and

$$G_0(z) := \frac{1}{2}\sum_k g_0[k]z^k \tag{6.8}$$

$$G_1(z) := \frac{1}{2}\sum_k g_1[k]z^k. \tag{6.9}$$

Observe that $\phi(t) \in \mathbf{A}_0, \phi(2t) \in \mathbf{A}_1$, and $\psi(t) \in \mathbf{W}_0$. From the nested property of MRA we know that $\mathbf{A}_0 \subset \mathbf{A}_1$ and $\mathbf{A}_0 \perp \mathbf{W}_0$ such that $\mathbf{A}_0 \oplus \mathbf{W}_0 = \mathbf{A}_1$. Orthogonality of the subspaces $\mathbf{A}_0$ and $\mathbf{W}_0$ implies that for any $\ell \in \mathbb{Z}$,

$$\langle \phi(t-\ell), \psi(t)\rangle = 0. \tag{6.10}$$

Equation (6.10) can be rewritten using Parseval's identity as

$$\begin{aligned}
0 &= \frac{1}{2\pi}\int_{-\infty}^{\infty} \widehat{\phi}(\omega)e^{-j\ell\omega}\overline{\widehat{\psi}(\omega)}\,d\omega \\
&= \frac{1}{2\pi}\int_{-\infty}^{\infty} G_0(z)\overline{G_1(z)}\left|\widehat{\phi}\left(\frac{\omega}{2}\right)\right|^2 e^{-j\ell\omega}\,d\omega \\
&= \frac{1}{2\pi}\sum_k \int_{4\pi k}^{4\pi(k+1)} G_0(z)\overline{G_1(z)}\left|\widehat{\phi}\left(\frac{\omega}{2}\right)\right|^2 e^{-j\ell\omega}\,d\omega
\end{aligned}$$

$$= \frac{1}{2\pi} \sum_k \int_0^{4\pi} G_0(z)\overline{G_1(z)} \left|\widehat{\phi}\left(\frac{\omega}{2} + 2\pi k\right)\right|^2 e^{-j\ell\omega}\, d\omega$$

$$= \frac{1}{2\pi} \int_0^{4\pi} G_0(z)\overline{G_1(z)} E_\phi(z) e^{-j\ell\omega}\, d\omega, \tag{6.11}$$

where $z = e^{-j\omega/2}$. By partitioning the integration limit $[0, 4\pi]$ into $[0, 2\pi]$ and $[2\pi, 4\pi]$, and with a simple change of variable, it is easy to verify that (6.11) is the same as

$$\frac{1}{2\pi} \int_0^{2\pi} \left[G_0(z)\overline{G_1(z)} E_\phi(z) + G_0(-z)\overline{G_1(-z)} E_\phi(-z)\right] e^{-j\ell\omega}\, d\omega = 0. \tag{6.12}$$

The expression (6.12) holds for all $\ell \in \mathbb{Z}$. What does it mean? To understand this, let us recall that an integrable 2π-periodic function $f(t)$ has the Fourier series representation

$$f(\omega) = \sum_\ell c_\ell e^{j\ell\omega}, \tag{6.13}$$

where

$$c_\ell = \frac{1}{2\pi} \int_0^{2\pi} f(\omega) e^{-j\ell\omega}\, d\omega. \tag{6.14}$$

From the above it is clear that the quantity on the left of (6.12) represents the ℓth Fourier coefficient of a periodic function $G_0(z)\overline{G_1(z)} E_\phi(z) + G_0(-z)\overline{G_1(-z)} E_\phi(-z)$. Since all these coefficients are zero, it implies that

$$G_0(z)\overline{G_1(z)} E_\phi(z) + G_0(-z)\overline{G_1(-z)} E_\phi(-z) \equiv 0 \tag{6.15}$$

for $|z| = 1$.

The solution of (6.15) gives the relationship between $G_1(z)$ and $G_0(z)$. By direct substitution, we can verify that

$$G_1(z) = -cz^{2m+1}\overline{G_0(-z)} E_\phi(-z) \tag{6.16}$$

for any integer m, and a constant $c > 0$ is a solution (6.15). Without any loss of generality we can set $c = 1$. The effect of m is to shift the index of the sequence $\{g_1[k]\}$. Usually, m is chosen such that the index begins with 0.

6.1.2 Relationship Between Reconstruction and Decomposition Sequences

Recall from Chapter 5 that the scaling function at a certain scale (finer) can be obtained from the scaling functions and wavelets at the next lower (coarse) scale. In mathematical terms, it means that there exist finite-energy sequences $\{h_0[k]\}$, $\{h_1[k]\}$ such that

$$\phi(2t - \ell) = \sum_k \{h_0[2k - \ell]\phi(t - k) + h_1[2k - \ell]\psi(t - k)\}, \tag{6.17}$$

where, as discussed in Chapter 5, $\{h_0[k]\}$ and $\{h_1[k]\}$ are the decomposition sequences.

By taking the Fourier transform of the decomposition relation, we get

$$\frac{1}{2}\hat{\phi}\left(\frac{\omega}{2}\right)e^{-j\omega\ell/2} = \sum_k h_0[2k-\ell]e^{-jk\omega}\hat{\phi}(\omega) + \sum_k h_1[2k-\ell]e^{-jk\omega}\hat{\psi}(\omega)$$

$$= \left\{G_0(z)\sum_k h_0[2k-\ell]e^{-jk\omega} + G_1(z)\sum_k h_1[2k-\ell]e^{-jk\omega}\right\}\hat{\phi}\left(\frac{\omega}{2}\right).$$

The equation above reduces to

$$\left(\sum_k h_0[2k-\ell]e^{-j(2k-\ell)\omega/2}\right)G_0(z) + \left(\sum_k h_1[2k-\ell]e^{-j(2k-\ell)\omega/2}\right)G_1(z) \equiv \frac{1}{2} \qquad \forall \ell \in \mathbb{Z}. \tag{6.18}$$

Combining the Fourier transforms of the decomposition and two-scale relations, we get

$$[H_0(z)+H_0(-z)]G_0(z) + [H_1(z)+H_1(-z)]G_1(z) = \frac{1}{2} \qquad \text{for even } \ell; \tag{6.19}$$

$$[H_0(z)-H_0(-z)]G_0(z) + [H_1(z)-H_1(-z)]G_1(z) = 0 \qquad \text{for odd } \ell, \tag{6.20}$$

where $z = e^{-j\omega/2}$ and

$$H_0(z) := \frac{1}{2}\sum_k h_0[k]z^k$$

$$H_1(z) := \frac{1}{2}\sum_k h_1[k]z^k.$$

These equations lead to

$$H_0(z)G_0(z) + H_1(z)G_1(z) = \frac{1}{2}$$

$$H_0(-z)G_0(z) + H_1(-z)G_1(z) = 0.$$

The last equation can also be written as

$$H_0(z)G_0(-z) + H_1(z)G_1(-z) = 0. \tag{6.21}$$

In matrix form we have

$$\begin{bmatrix} G_0(z) & G_1(z) \\ G_0(-z) & G_1(-z) \end{bmatrix} \begin{bmatrix} H_0(z) \\ H_1(z) \end{bmatrix} = \begin{bmatrix} \frac{1}{2} \\ 0 \end{bmatrix}, \tag{6.22}$$

the solution of which gives

$$H_0(z) = \frac{1}{2} \times \frac{G_1(-z)}{\Delta_{G_0G_1}(z)} \tag{6.23}$$

$$H_1(z) = -\frac{1}{2} \times \frac{G_0(-z)}{\Delta_{G_0G_1}(z)} \tag{6.24}$$

with

$$\Delta_{G_0G_1}(z) = G_0(z)G_1(-z) - G_0(-z)G_1(z). \tag{6.25}$$

It can be shown that

$$\Delta_{G_0G_1}(z) = cz^m E_\phi(z^2), \tag{6.26}$$

where $c > 0$ and m is an integer. Since ϕ generates a Riesz or stable basis, $E_\phi(z)$ and hence $\Delta_{G_0G_1}(z) \neq 0$.

6.2 CONSTRUCTION OF SEMIORTHOGONAL SPLINE WAVELETS

The significance of the results obtained in Section 6.1 is that we need to construct only the scaling functions (i.e., we need to find only the sequence $\{g_0[k]\}$). In this section we obtain these sequences for the semiorthogonal spline wavelets introduced by Chui and Wang [1]. Here the cardinal B-splines N_m are chosen to be the scaling functions. We will show that a finite-energy sequence $\{g_0[m, k]\}$ exists such that the scaling relation

$$N_m(t) = \sum_k g_0[m, k] N_m(2t - k) \tag{6.27}$$

is satisfied and therefore $N_m(t)$ is a scaling function. For $m = 1$, $\{N_1(t-k) : k \in \mathbb{Z}\}$ form an orthonormal basis of $\mathbf{A}_0$. For this case we have already seen that $g_0[0] = g_0[1] = 1$ (see Figure 5.3). In this section we consider cases for which $m \geq 2$.

For $m \geq 2$, the scaling functions $\{N_m(t-k) : k \in \mathbb{Z}\}$ are no longer orthogonal; that is,

$$\int_{-\infty}^{\infty} N_m(t) N_m(t - \ell)\, dt \neq \delta_{0,\ell}, \tag{6.28}$$

for all $\ell \in \mathbb{Z}$ and $m \geq 2$. An example of nonorthogonality of $N_2(t)$ is shown in Figure 6.1. The $\int_{-\infty}^{\infty} N_2(t)N_2(t-\ell)\, dt$ is shown by the shaded area, which is nonzero.

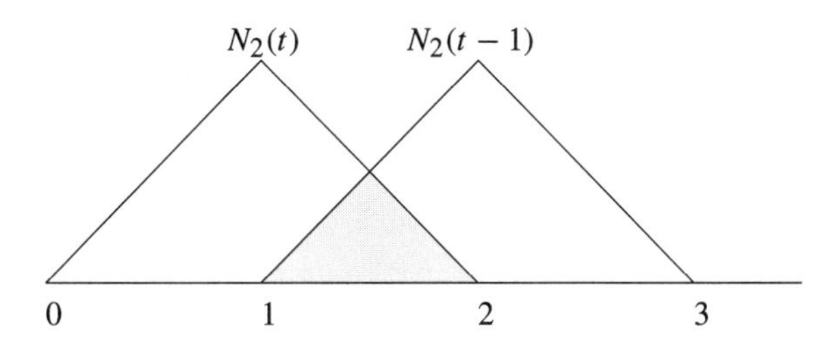

FIGURE 6.1 Nonorthogonality of linear spline shown by the shaded area.

6.2.1 Expression for $\{g_0[k]\}$

Recall from the definition of $N_m(t)$ in Chapter 5 that

$$N_m(t) = \underbrace{(N_1 * \cdots * N_1)}_{m}(t)$$

and that

$$\widehat{N}_m(\omega) = \left(\frac{1-e^{-j\omega}}{j\omega}\right)^m. \tag{6.29}$$

From the Fourier transform of the two-scale relation, we have

$$G_0(z) = \frac{1}{2}\sum_k g_0[m,k]z^k = \frac{\widehat{N}_m(\omega)}{\widehat{N}_m(\omega/2)} \tag{6.30}$$

$$= \left(\frac{1+e^{-j(\omega/2)}}{2}\right)^m$$

$$= 2^{-m}(1+z)^m \qquad z = e^{-j\omega/2}$$

$$= 2^{-m}\sum_{k=0}^{m}\binom{m}{k}z^k. \tag{6.31}$$

By comparing the coefficient of powers of z, we get

$$g_0[k] := g_0[m,k] = \begin{cases} 2^{-m+1}\binom{m}{k}, & 0 \le k \le m \\ 0, & \text{otherwise.} \end{cases} \tag{6.32}$$

Once we have $\{g_0[k]\}$, the rest of the sequences $\{g_1[k]\}$, $\{h_0[k]\}$, and $\{h_1[k]\}$ can be found by using the relations derived in Section 6.1. The expression of $\{g_1[k]\}$ is derived below.

For $N_m(t)$, the Euler–Frobenius–Laurent polynomial $E_{N_m}(z)$ takes the form

$$\begin{aligned} E_{N_m}(z) &= \sum_{k=-\infty}^{\infty} \left|\widehat{N}_m\left(\frac{\omega}{2}+2\pi k\right)\right|^2 \\ &= \sum_{k=-\infty}^{\infty} A_{N_m}(k) z^k \\ &= \sum_{k=-m+1}^{m-1} N_{2m}(m+k) z^k \end{aligned} \tag{6.33}$$

with $z := e^{-j\omega/2}$ and the autocorrelation function

$$A_{N_m}(k) = \int_{-\infty}^{\infty} N_m(x) N_m(k+x)\,dx = N_{2m}(m+k). \tag{6.34}$$

Finally, by using the relation (6.16), we have

$$g_1[k] := g_1[m,k] = (-1)^k \cdot 2^{-m+1} \sum_{\ell=0}^{m} \binom{m}{\ell} N_{2m}(k+1-\ell), \tag{6.35}$$

$$0 \le k \le 3m-2. \tag{6.36}$$

Remarks: Recall that in the expressions for $H_0(z)$ and $H_1(z)$, in terms of $G_0(z)$ and $G_1(z)$, there is a term $\Delta_{G_0G_1}(z) = z^m E_{N_m}(z)$ in the denominator. Consequently, the sequences $\{h_0[k]\}$ and $\{h_1[k]\}$ are infinitely long, although their magnitude decays exponentially. These are the sequences that will be used in the development of decomposition and reconstruction algorithms in Chapters 7 and 8. It is clear that while G_0 and G_1 form FIR filters, H_0 and H_1 are always IIR. We will, however, prove in Chapter 7 that we can use G_0 and G_1 for both reconstruction as well as decomposition purposes. This is a consequence of the duality principle that we mentioned briefly in Chapter 5.

The commonly used cubic spline and the corresponding semiorthogonal wavelet with their duals and magnitude spectra are shown in Figures 6.2 and 6.3. See Chapter 10 for the expressions of commonly used semiorthogonal scaling functions and wavelets. Table 6.1 gives the coefficients $\{g_1[m,k]\}$ for $m = 2$ through 6.

6.3 CONSTRUCTION OF ORTHONORMAL WAVELETS

Recall from Chapter 2 that the Riesz bounds for orthonormal bases are 1. Therefore, for orthonormal scaling functions ϕ and the corresponding wavelets ψ, we have

$$E_\phi(e^{jw}) = \sum_k |\hat{\phi}(\omega+2\pi k)|^2 \equiv 1,$$

$$E_\psi(e^{jw}) = \sum_k |\hat{\psi}(\omega+2\pi k)|^2 \equiv 1,$$

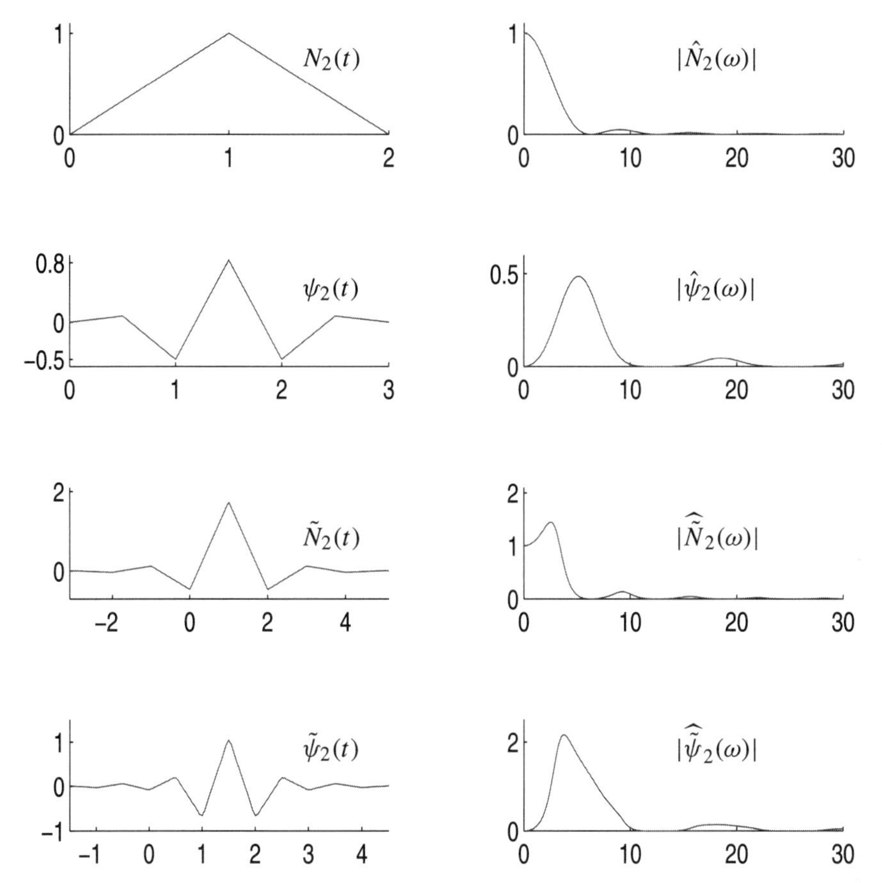

FIGURE 6.2 Linear spline, dual linear spline, the corresponding wavelets, and their magnitude spectra.

for almost all ω. Consequently,

$$\tilde{\phi}(t) = \phi(t) \quad \text{and} \quad \tilde{\psi}(t) = \psi(t);$$

that is, they are self-duals. Remember from our discussion in Chapter 5 that because of the nested nature of MRA subspaces, the scaling functions are not orthogonal with respect to scales. Orthonormal scaling functions imply that these are orthogonal with respect to translation on a given scale. Orthonormal wavelets, on the other hand, are orthonormal with respect to scale as well as the translation.

By starting with

$$\langle \phi(t-\ell), \phi(t) \rangle = \delta_{0,\ell} \tag{6.37}$$

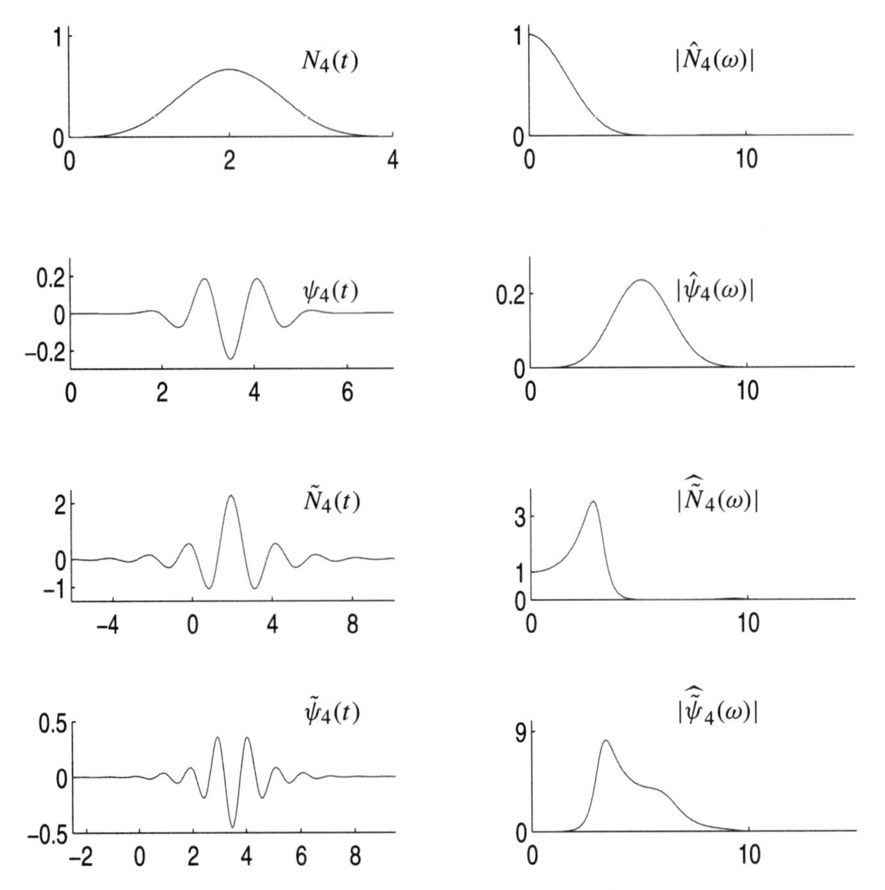

FIGURE 6.3 Cubic spline, dual cubic spline, the corresponding wavelets and their magnitude spectra.

and following the derivation of (6.12), we arrive at the following results:

$$|G_0(z)|^2 + |G_0(-z)|^2 \equiv 1, \qquad |z| = 1. \tag{6.38}$$

For orthonormal scaling functions and wavelets, the relationships among the various sequences $\{g_0[k]\}$, $\{g_1[k]\}$, $\{h_0[k]\}$, and $\{h_1[k]\}$ can be obtained from the results of Section 6.1 by setting $m = 0$. These results are summarized below.

$$G_1(z) = -z\overline{G_0(-z)}$$
$$\Rightarrow g_1[k] = (-1)^k g_0[1-k]; \tag{6.39}$$
$$\Delta_{G_0,G_1}(z) = G_0(z)G_1(-z) - G_0(-z)G_1(z) = z \tag{6.40}$$

TABLE 6.1 Coefficients $u_{m,k} := 2^{m-1}(2m-1)!\, g_1[m,k]$ for the Semiorthogonal Wavelet ($g_1[m,k] = (-1)^m g_1[m, 3m-2-k]$)

k	$u_{m,k}$	k	$u_{m,k}$	k	$u_{m,k}$
	$m=2$		$m=5$		$m=6$
0	1	0	1	0	1
1	−6	1	−507	1	−2,042
2	10	2	17,128	2	164,868
	$m=3$	3	−166,304	3	−3,149,870
0	1	4	748,465	4	25,289,334
1	−29	5	−1,900,115	5	−110,288,536
2	147	6	2,973,560	6	296,526,880
3	−303			7	−525,228,384
	$m=4$			8	633,375,552
0	1				
1	−124				
2	1,677				
3	−7,904				
4	18,482				
5	−24,264				

$$H_0(z) = \frac{1}{2} \times \frac{G_1(-z)}{z} = \frac{1}{2} \times \overline{G_0(z)}$$

$$\Rightarrow h_0[k] = \frac{1}{2} g_0[-k]; \tag{6.41}$$

$$H_1(z) = -\frac{1}{2} \times \frac{G_0(-z)}{z} = \frac{1}{2} \times \overline{G_1(z)}$$

$$\Rightarrow h_1[k] = \frac{1}{2}(-1)^k g_0[k+1] \tag{6.42}$$

As an example, for a Haar scaling function and wavelet (see Figures 5.3 and 5.4) ,

$$g_0[0] = g_0[1] = 1$$

$$g_1[0] = 1,\ g_1[1] = -1$$

$$h_0[0] = h_0[1] = \frac{1}{2}$$

$$h_1[0] = \frac{1}{2}, \qquad h_1[1] = -\frac{1}{2}$$

$$g_0[k] = g_1[k] = h_0[k] = h_1[k] = 0 \qquad \forall\, k \in \mathbb{Z} \setminus \{0, 1\}.$$

Remarks: One of the most important features of orthonormal bases is that all of the decomposition and reconstruction filters are FIR and have the same length. This helps tremendously in the decomposition and reconstruction algorithm discussed in

Chapter 7. One of the disadvantages of orthonormal wavelets is that they generally do not have closed-form expressions, nor does a compactly supported orthonormal wavelet have a linear phase (no symmetry). The importance of a linear phase in signal reconstruction is discussed in Chapter 7. It has also been shown [2] that the higher-order orthonormal scaling functions and wavelets have poor time–frequency localization.

6.4 ORTHONORMAL SCALING FUNCTIONS

In this section we discuss the commonly used orthonormal wavelets of Shannon, Meyer, Battle–Lemarié, and Daubechies. We will derive only the expressions for the sequence $\{g_0[k]\}$ since other sequences can be obtained from the relationships of Section 6.1.

6.4.1 Shannon Scaling Function

The Shannon sampling function

$$\phi_{SH}(t) := \frac{\sin \pi t}{\pi t} \tag{6.43}$$

is an orthonormal scaling function with $\hat{\phi}_{SH}(\omega) = \chi_{(-\pi,\pi)}(\omega)$. Proving the orthogonality of (6.43) in the time domain by the relation

$$\langle \phi_{SH}(t-\ell), \phi_{SH}(t) \rangle = \delta_{0,\ell} \tag{6.44}$$

is cumbersome. Here it is rather easy to show that the Riesz bounds are 1; that is,

$$\sum_k |\hat{\phi}_{SH}(\omega + 2\pi k)|^2 \equiv 1. \tag{6.45}$$

The sequence $\{g_0[k]\}$ can be obtained from the two-scale relation

$$\frac{1}{2}\sum_k g_0[k] e^{-jk\omega/2} = \frac{\widehat{\phi_{SH}}(\omega)}{\widehat{\phi_{SH}}(\omega/2)}. \tag{6.46}$$

Since the left-hand side of the expression is a 4π-periodic function, we need a 4π-periodic extension of the right-hand side. In other words, $G_0(z)$ is nothing but a 4π-periodic extension of $\hat{\phi}_{SH}(\omega)$ (see Figure 6.4):

$$G_0(z) = \frac{1}{2}\sum_\ell g_0[\ell] \exp\left(-j\frac{\omega\ell}{2}\right) = \sum_\ell \hat{\phi}_{SH}(\omega + 4\pi\ell). \tag{6.47}$$

From (6.47) we can get the expression for the coefficients $\{g_0[k]\}$:

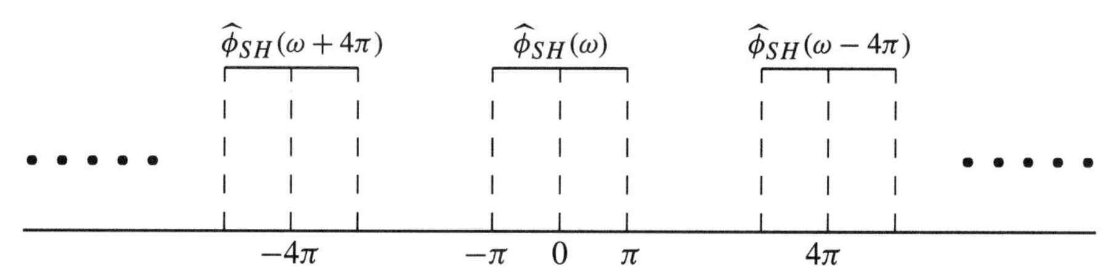

FIGURE 6.4 4π-periodic extension of $\hat{\phi}_{\mathrm{SH}}(\omega)$.

$$
\begin{aligned}
g_0[k] &= \frac{1}{2\pi}\int_0^{4\pi}\sum_{\ell}\hat{\phi}_{\mathrm{SH}}(\omega+4\pi\ell)\exp\left(j\frac{\omega k}{2}\right)d\omega \\
&= \frac{1}{2\pi}\sum_{\ell=-\infty}^{\infty}\int_{4\pi\ell}^{4\pi(\ell+1)}\hat{\phi}_{\mathrm{SH}}(\omega)\exp\left(j\frac{\omega k}{2}\right)d\omega \\
&= \frac{1}{2\pi}\int_{-\infty}^{\infty}\hat{\phi}_{\mathrm{SH}}(\omega)\exp\left(j\frac{\omega k}{2}\right)d\omega \\
&= \phi_{\mathrm{SH}}\left(\frac{k}{2}\right).
\end{aligned}
\tag{6.48}
$$

By using (6.43), we get

$$
g_0[k] = \begin{cases} 1 & \text{for } k = 0 \\ (-1)^{(k-1)/2}\dfrac{2}{k\pi} & \text{for odd } k \\ 0 & \text{for even } k \neq 0. \end{cases}
$$

Figure 6.5 shows the Shannon scaling function and the wavelet.

6.4.2 Meyer Scaling Function

The Shannon scaling function $\phi_{\mathrm{SH}}(t)$ has poor time localization $\left(\Delta_{\phi_{\mathrm{SH}}} = \infty\right)$. The reason for this is that in the frequency domain, $\hat{\phi}_{\mathrm{SH}}(\omega)$ has a discontinuity at $-\pi$ and π. Consequently, in the time domain, as given by (6.43), the function decays as $1/t$ and hence its RMS time window width (4.3) is ∞. To improve it, Meyer [3, 4] obtained the scaling function $\hat{\phi}_{M,m}(\omega)$ by applying a smoothing function near the discontinuities of $\widehat{\phi}_{\mathrm{SH}}(\omega)$ in such a way that the orthogonality condition

$$
\sum_k \left|\widehat{\phi}_{M,m}(\omega+2\pi k)\right|^2 = 1 \tag{6.49}
$$

is satisfied. In (6.49) the index m indicates the degree of smoothness [i.e., the mth-order corner smoothing function $S_m(\omega)$ is m times continuously differentiable].

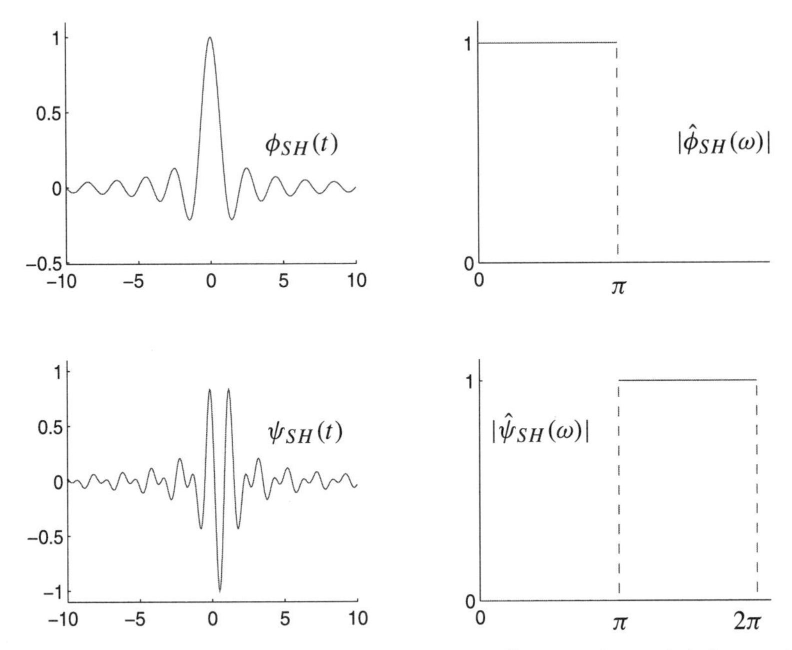

FIGURE 6.5 Shannon scaling function, the corresponding wavelet, and their magnitude spectra.

To satisfy the orthogonality requirement (6.38), these corner smoothing functions should have the following properties:

$$
\begin{aligned}
&S_m(y) + S_m(1-y) = 1, && 0 \le y \le 1 \\
&S_m(y) = 0, && y < 0 \\
&S_m(y) = 1, && y > 1.
\end{aligned}
$$

Examples of corner smoothing functions are

$$
S_0(y) = \begin{cases} 0, & y < 0 \\ y, & 0 \le y \le 1 \\ 1, & y > 1 \end{cases}
$$

$$
S_1(y) = \begin{cases} 0, & y < 0 \\ 2y^2, & 0 \le y < 0.5 \\ -2y^2 + 4y - 1, & 0.5 \le y \le 1 \\ 1, & y > 1. \end{cases}
$$

Let $S_m(y)$ be a desirable corner smoothing function. Then

$$\widehat{\phi}_{M,m}(\omega) = \begin{cases} \cos\left[\frac{\pi}{2} S_m\left(\frac{3}{2\pi}|\omega| - 1\right)\right], & \frac{2\pi}{3} \le |\omega| \le \frac{4\pi}{3} \\ 0, & |\omega| \ge \frac{4\pi}{3} \\ 1, & |\omega| \le \frac{2\pi}{3}. \end{cases}$$

The scaling function in the time domain then becomes

$$\phi_{M,m}(t) = \frac{2}{3}\frac{\sin(2\pi t/3)}{2\pi t/3} + \frac{2}{3}\int_0^1 \cos\left[\frac{\pi}{2} S_m(\xi)\right]\cos\left[\frac{2\pi}{3}(1+\xi)t\right]d\xi. \quad (6.50)$$

For a linear smoothing function $m = 0$, the integral above can easily be evaluated. The result is

$$\phi_{M,0}(t) = \frac{2}{3}\frac{\sin(2\pi t/3)}{2\pi t/3} + \frac{4}{\pi}\frac{4t\sin(2\pi t/3) + \cos(4\pi t/3)}{9 - 16t^2}. \quad (6.51)$$

For higher values of m, the integral in (6.50) needs to be evaluated numerically. In Figure 6.6 we show the scaling function and wavelet of Meyer for $m = 0$ and 1.

As done before, the two-scale coefficients $\{g_0[k]\}$ can be obtained by a 4π-periodic extension of $\widehat{\phi}_{M,m}(\omega)$. An example of such an extension is shown in Figure 6.7.

$$\begin{aligned} \frac{1}{2}\sum_k g_0[k]e^{-jk\omega/2} &= \frac{\widehat{\phi}_{M,m}(\omega)}{\widehat{\phi}_{M,m}(\omega/2)} \\ &= \sum_k \widehat{\phi}_{M,m}(\omega + 4\pi k). \end{aligned}$$

Similar to the case of the Shannon scaling function, here too we get

$$g_0[k] = \phi_{M,N}\left(\frac{k}{2}\right). \quad (6.52)$$

Therefore, for $m = 0$, we can obtain $g_0[k]$ simply by substituting $k/2$ for t in (6.51). Table 6.2 gives the two-scale coefficients for $m = 1$.

Meyer's wavelets can be obtained by using the two-scale relations. Since the scaling functions have compact support in the frequency domain, Meyer wavelets are related to the scaling function in a more direct way; that is,

$$\begin{aligned} \widehat{\psi}_{M,m}(\omega) &= G_1(z)\widehat{\phi}_{M,m}\left(\frac{\omega}{2}\right) \qquad z = \exp\left(-j\frac{\omega}{2}\right) \\ &= -z\overline{G_0(-z)}\widehat{\phi}_{M,m}\left(\frac{\omega}{2}\right) \end{aligned}$$

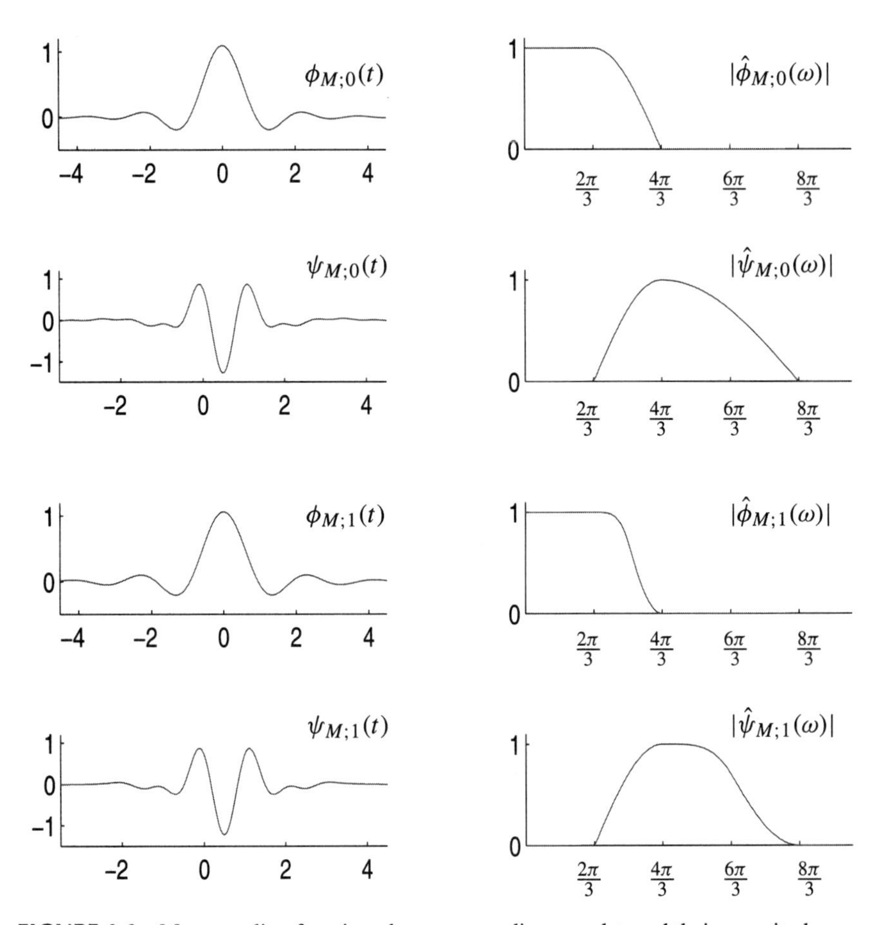

FIGURE 6.6 Meyer scaling function, the corresponding wavelet, and their magnitude spectra.

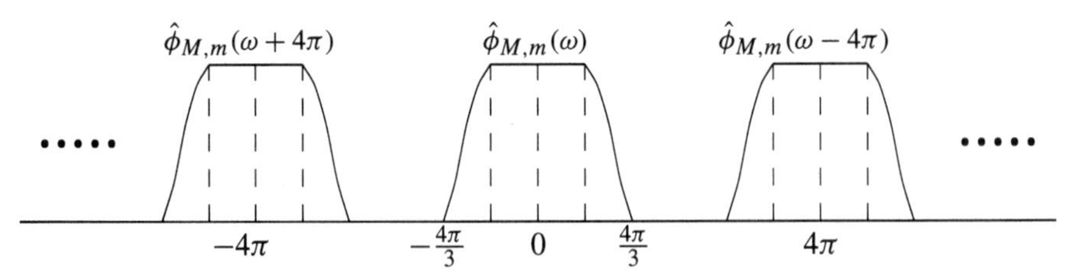

FIGURE 6.7 4π-periodic extension of $\hat{\phi}_{M,m}(\omega)$.

TABLE 6.2 Two-Scale Sequence for First-Order Meyer Scaling Function $\phi_{M;1}$

n	$g_0[n] = g_0[-n]$	n	$g_0[n] = g_0[-n]$	n	$g_0[n] = g_0[-n]$
0	1.0635133307325022	13	0.0018225696961070	26	0.0000781583234904
1	0.6237929148320031	14	−0.0001225788843060	27	−0.0002817686403039
2	−0.0594319217681172	15	−0.0019003177368828	28	−0.0000686017777485
3	−0.1762971983704155	16	−0.0000361315305005	29	0.0003520515347881
4	0.0484777578300750	17	0.0018514320187282	30	−0.0000591760677635
5	0.0751184531725782	18	−0.0004792529715153	31	−0.0002870818672708
6	−0.0339527984193033	19	−0.0013039128005108	32	0.0001435155716864
7	−0.0311015336438103	20	0.0007208498373768	33	0.0001507339706291
8	0.0197659340813598	21	0.0006265171401084	34	−0.0001171599560112
9	0.0110906323385240	22	−0.0005163028169833	35	−0.0000530482980227
10	−0.0089132072379117	23	−0.0002172396357380	36	0.0000282695514764
11	−0.0035390831203475	24	0.0001468883466883	37	0.0000443263271494
12	0.0025690718118815	25	0.0001627491841323	38	0.0000355188445237

$$= -z\left[\sum_{k=-\infty}^{\infty} \widehat{\phi}_{M,m}\left(\omega + 2\pi + 4\pi k\right)\right] \widehat{\phi}_{M,m}\left(\frac{\omega}{2}\right)$$

$$= -z\left[\widehat{\phi}_{M,m}(\omega + 2\pi) + \widehat{\phi}_{M_m}(\omega - 2\pi)\right] \widehat{\phi}_{M,m}\left(\frac{\omega}{2}\right). \tag{6.53}$$

6.4.3 Battle–Lemarié Scaling Function

Battle–Lemarié [5,6] scaling functions are constructed by orthonormalizing the mth-order cardinal B-spline $N_m(t)$ for $m \geq 2$. As pointed out before, the set of basis functions $\{N_m(t-k) : k \in \mathbb{Z}\}$ is not orthogonal for $m \geq 2$. The corresponding orthonormal scaling function $N_m^{\perp}(t)$ can be obtained as

$$\widehat{N}_m^{\perp}(\omega) = \frac{\widehat{N}_m(\omega)}{\left[E_{N_m}(e^{-j\omega})\right]^{1/2}}. \tag{6.54}$$

The Battle–Lemarié scaling function $\phi_{BL,m}(t)$ is, then,

$$\phi_{BL,m}(t) = N_m^{\perp}(t),$$

and the coefficients $\{g_0[k]\}$ can be found from

$$\frac{1}{2}\sum_k g_0[k]z^k = G_0 e^{-j(\omega/2)} = \frac{\widehat{N}_m^{\perp}(\omega)}{\widehat{N}_m^{\perp}(\omega/2)}, \tag{6.55}$$

where $z = e^{-j(\omega/2)}$. By combining (6.54) and (6.55), we have

$$G_0(z) = \left(\frac{1+z}{2}\right)^m \left[\frac{\sum_{k=-m+1}^{m-1} N_{2m}(m+k)z^k}{\sum_{k=-m+1}^{m-1} N_{2m}(m+k)z^{2k}}\right]^{1/2}. \tag{6.56}$$

As an example, consider the linear Battle–Lemarié scaling function, for which $m = 2$. For this case we have

$$\frac{1}{2}\sum_k g_0[k]z^k = G_0(z) = \frac{(1+z)^2}{4} \times \frac{z^2+4z+1}{z^4+4z^2+1}. \tag{6.57}$$

The coefficients $\{g_0[k]\}$ can be found by expanding the expression on the right-hand side as a polynomial in z and then comparing the coefficients of the like powers of z. These coefficients can also be found by computing the Fourier coefficients of the right-hand side expression. Observe that $G_0(1) = 1$ is satisfied, thus giving the sum of all $\{g_0[k]\}$ to be 2. In Tables 6.3 and 6.4 we provide the coefficients of the

TABLE 6.3 Two-Scale Sequence for Linear Battle–Lemarié Scaling Function $\phi_{BL;2}$

n	$g_0[n] = g_0[2-n]$	n	$g_0[n] = g_0[2-n]$	n	$g_0[n] = g_0[2-n]$
1	1.1563266304457929	14	0.0000424422257478	27	−0.0000000053986543
2	0.5618629285876487	15	−0.0000195427343909	28	−0.0000000028565276
3	−0.0977235484799832	16	−0.0000105279065482	29	0.0000000013958989
4	−0.0734618133554703	17	0.0000049211790530	30	0.0000000007374693
5	0.0240006843916324	18	0.0000026383701627	31	−0.0000000003617852
6	0.0141288346913845	19	−0.0000012477015924	32	−0.0000000001908819
7	−0.0054917615831284	20	−0.0000006664097922	33	0.0000000000939609
8	−0.0031140290154640	21	0.0000003180755856	34	0.0000000000495170
9	0.0013058436261069	22	0.0000001693729269	35	−0.0000000000244478
10	0.0007235625130098	23	−0.0000000814519590	36	−0.0000000000128703
11	−0.0003172028555467	24	−0.0000000432645262	37	0.0000000000063709
12	−0.0001735046359701	25	0.0000000209364375	38	0.0000000000033504
13	0.0000782856648652	26	0.0000000110975272	39	−0.0000000000016637

TABLE 6.4 Two-Scale Sequence for Cubic Battle–Lemarié Scaling Function $\phi_{BL;4}$

n	$g_0[n] = g_0[4-n]$	n	$g_0[n] = g_0[4-n]$	n	$g_0[n] = g_0[4-n]$
2	1.0834715125686560	15	0.0026617387556783	28	−0.0000282171646500
3	0.6136592734426418	16	−0.0015609238233188	29	−0.0000222283943141
4	−0.0709959598848591	17	−0.0013112570210398	30	0.0000146073867894
5	−0.1556158437675466	18	0.0007918699951128	31	0.0000114467590896
6	0.0453692402954247	19	0.0006535296221413	32	−0.0000075774407788
7	0.0594936331541212	20	−0.0004035935254263	33	−0.0000059109049365
8	−0.0242909783203567	21	−0.0003285886943928	34	0.0000039378865616
9	−0.0254308422142201	22	0.0002065343929212	35	0.0000030595965005
10	0.0122828617178522	23	0.0001663505502899	36	−0.0000020497919302
11	0.0115986402962103	24	−0.0001060637892378	37	−0.0000015870262674
12	−0.0061572588095633	25	−0.0000846821755363	38	0.0000010685382577
13	−0.0054905784655009	26	0.0000546341264354	39	0.0000008247217560
14	0.0030924782908629	27	0.0000433039957782	40	−0.0000005577533684

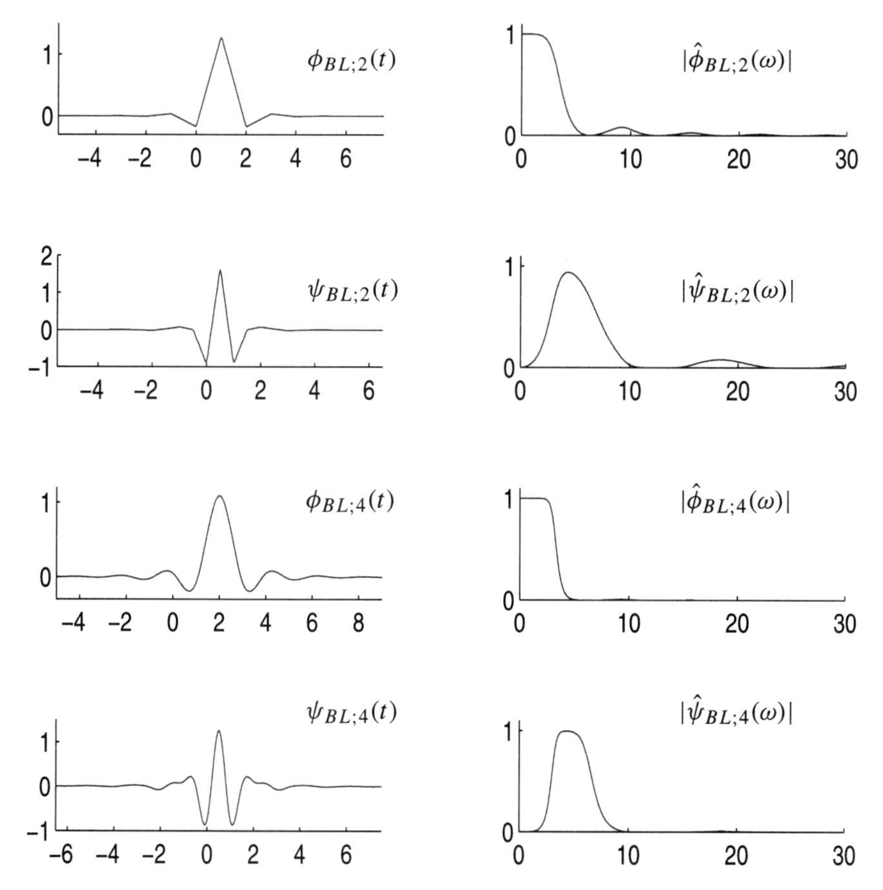

FIGURE 6.8 Battle–Lemarié scaling function, the corresponding wavelet, and their magnitude spectra.

linear and cubic Battle–Lemarié scaling functions. The linear Battle–Lemarié scaling function and the wavelet are shown in Figure 6.8.

6.4.4 Daubechies Scaling Function

Battle–Lemarié obtained orthonormal scaling functions by orthonormalizing mth-order cardinal B-splines $N_m(t)$ for $m \geq 2$. However, because of the presence of $E_{N_m}(z)$ in the denominator for the orthonormalization process, the sequence $\{g_0[k]\}$ becomes infinitely long.

To obtain orthonormality but preserve the finite degree of the (Laurent) polynomial, Daubechies [7,8] considered the two-scale symbol for the scaling function $\phi_{D,m}$:

$$G_0(z) = \left(\frac{1+z}{2}\right)^m S(z), \tag{6.58}$$

where $S(z) \in \pi_{m-1}$. So our objective is to find $S(z)$. First, observe that since $G_0(1) = 1$, we must have $S(1) = 1$. Furthermore, we also want $S(-1) \neq 0$, because if $S(-1) = 0$, then $z+1$ is a factor of $S(z)$ and hence can be taken out. Now $G_0(z)$ given by (6.58) must satisfy the orthogonality condition, namely

$$|G_0(z)|^2 + |G_0(-z)|^2 = 1, \qquad z = e^{-j\omega/2} \tag{6.59}$$

$$\Rightarrow \cos^{2m}\frac{\omega}{4}|S(z)|^2 + \sin^{2m}\frac{\omega}{4}|S(-z)|^2 = 1. \tag{6.60}$$

By defining

$$x := \sin^2\frac{\omega}{4}$$

and

$$f(x) := |S(z)|^2,$$

(6.60) can be rewritten as (Exercise 11)

$$(1-x)^m f(x) + x^m f(1-x) = 1$$

$$\Rightarrow f(x) = (1-x)^{-m}\left[1 - x^m f(1-x)\right]$$

$$= \sum_{k=0}^{m-1}\binom{m+k-1}{k}x^k + R_m(x), \tag{6.61}$$

where the remainder $R_m(x)$ is

$$R_m(x) := \sum_{k=m}^{\infty}\binom{m+k-1}{k}x^k + (-x)^m f(1-x)\sum_{k=0}^{\infty}\binom{m+k-1}{k}x^k. \tag{6.62}$$

Since $f(x)$ is a polynomial of order m, $R_m(x) \equiv 0$. Therefore, we have

$$|S(z)|^2 = \sum_{k=0}^{m-1}\binom{m+k-1}{k}\sin^{2k}\frac{\omega}{4}, \qquad z = e^{-j\omega/2}. \tag{6.63}$$

The polynomial above can be converted to

$$|S(z)|^2 = \frac{a_0}{2} + \sum_{k=1}^{m-1} a_k \cos\frac{k\omega}{2}, \tag{6.64}$$

where [9, p. 80]

$$a_k = (-1)^k \sum_{n=0}^{m-k-1} \frac{1}{2^{2(k+n)-1}} \binom{2(k+n)}{n} \binom{m+k+n-1}{k+n}. \tag{6.65}$$

Our next task is to retrieve $S(z)$ from $|S(z)|^2$.

According to Riesz's lemma [10, p. 172], corresponding to a cosine series

$$\widehat{f}(\omega) = \frac{a_0}{2} + \sum_{k=1}^{N} a_k \cos k\omega \tag{6.66}$$

with $a_0, \ldots, a_N \in \mathbb{R}$ and $a_N \neq 0$, there exists a poynomial

$$g(z) = \sum_{k=0}^{N} b_k z^k \tag{6.67}$$

with $a_0, \ldots, a_N \in \mathbb{R}$, such that

$$|g(z)|^2 = \widehat{f}(\omega), \qquad z = e^{-j\omega}. \tag{6.68}$$

By applying Riesz's lemma to (6.64) it is easy to verify [9] that $S(z)$ has the following form:

$$S(z) = C \prod_{k=1}^{K} (z - r_k) \prod_{\ell=1}^{L} (z - z_\ell)(z - \overline{z_\ell}), \qquad K + 2L = m - 1, \tag{6.69}$$

where $\{r_k\}$ are the nonzero real roots and $\{z_\ell\}$ are the complex roots of $z^{m-1}|S(z)|^2$ inside a unit circle and C is a constant such that $S(1) = 1$.

Once we have $S(z)$, we can substitute this into (6.58) and compare the coefficients of powers of z to get the sequence $\{g_0[k]\}$. We will show the steps to get these sequences with an example.

Consider $m = 2$. For this, we have $a_0 = 4$ and $a_1 = -1$, which gives

$$\begin{aligned} |S(z)|^2 &= 2 - \cos\frac{\omega}{2} \\ &= 2 - \frac{1}{2}(z + z^{-1}) \end{aligned}$$

$$\Rightarrow z|S(z)|^2 = \frac{1}{2}(-1 + 4z - z^2) = -\frac{1}{2}(z - r_1)\left(z - \frac{1}{r_1}\right), \tag{6.70}$$

where $r_1 = 2 - \sqrt{3}$. From (6.69), we have

$$S(z) = \frac{1}{1 - r_1}(z - r_1) = \frac{1}{-1 + \sqrt{3}}(z - 2 + \sqrt{3}). \tag{6.71}$$

So, for $m = 2$, we get

$$G_0(z) = \frac{1}{2}(g_0[0] + g_0[1]z + g_0[2]z^2 + g_0[3]z^3)$$

$$= \left(\frac{1+z}{2}\right)^2 \times \frac{1}{2}\left[(1+\sqrt{3})z + (1-\sqrt{3})\right]$$

$$= \frac{1}{2}\left(\frac{1-\sqrt{3}}{4} + \frac{1-\sqrt{3}}{4}z + \frac{1-\sqrt{3}}{4}z^2 + \frac{1-\sqrt{3}}{4}z^3\right). \tag{6.72}$$

Since $S(z)$ is a polynomial of order m, the length of two-scale sequence for $\phi_{D;m}$ is $2m$.

For $m = 2$ and 7, the scaling functions and wavelets, along with their magnitude spectra, are shown in Figure 6.9. Two-scale sequences for some Daubechies scaling

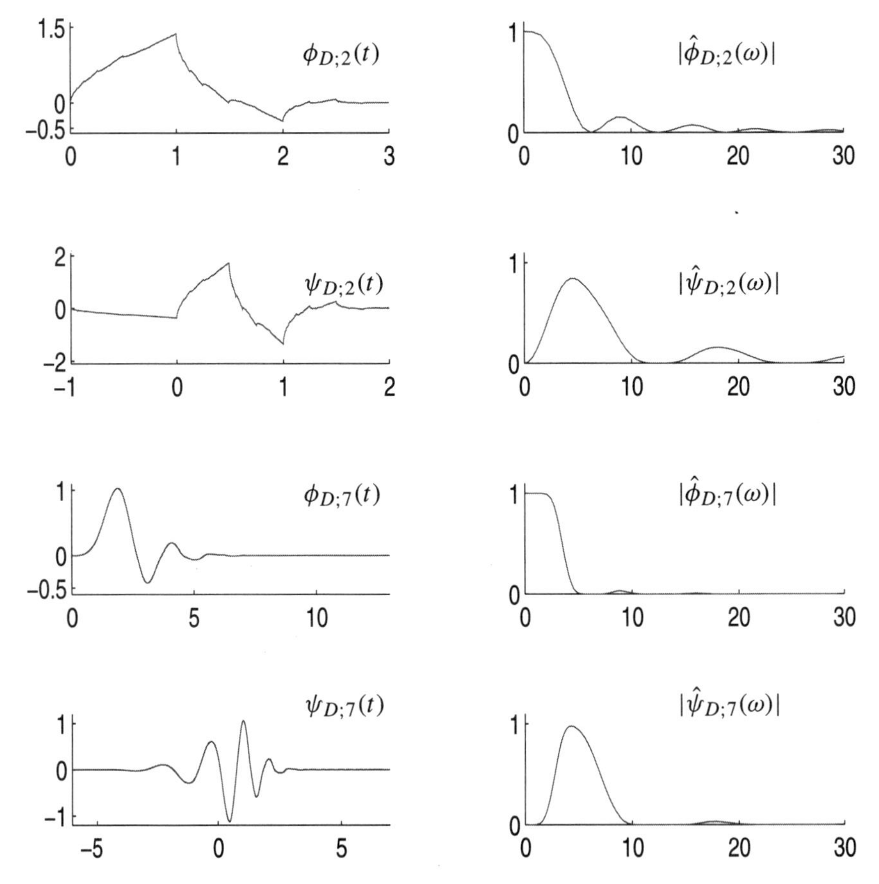

FIGURE 6.9 Daubechies scaling function, the corresponding wavelet, and their magnitude spectra.

TABLE 6.5 Two-Scale Sequence for the Daubechies Scaling Function

n	$g_0[n] = g_0[-n]$	n	$g_0[n] = g_0[-n]$	n	$g_0[n] = g_0[-n]$
	$m = 2$		$m = 5$		$m = 7$
0	0.6830127018922193	0	0.2264189825835584	0	0.1100994307456160
1	1.1830127018922192	1	0.8539435427050283	1	0.5607912836254882
2	0.3169872981077807	2	1.0243269442591967	2	1.0311484916361415
3	−0.1830127018922192	3	0.1957669613478087	3	0.6643724822110735
	$m = 3$	4	−0.3426567153829353	4	−0.2035138224626306
0	0.4704672077841636	5	−0.0456011318835469	5	−0.3168350112806179
1	1.1411169158314436	6	0.1097026586421339	6	0.1008464650093839
2	0.6503650005262323	7	−0.0088268001083583	6	0.1008464650093839
3	−0.1909344155683274	8	−0.0177918701019542	7	0.1140034451597351
4	−0.1208322083103963	9	0.0047174279390679	8	−0.0537824525896852
5	0.0498174997368837		$m = 6$	9	−0.0234399415642046
	$m = 4$	0	0.1577424320027466	10	0.0177497923793598
0	0.3258034280512982	1	0.6995038140774233	10	0.0177497923793598
1	1.0109457150918286	2	1.0622637598801890	11	0.0006075149954022
2	0.8922001382609015	3	0.4458313229311702	12	−0.0025479047181871
3	−0.0395750262356447	4	−0.3199865989409983	13	0.0005002268531225
4	−0.2645071673690398	5	−0.1835180641065938		
5	0.0436163004741772	6	0.1378880929785304		
6	0.0465036010709818	7	0.0389232097078970		
7	−0.0149869893303614	8	−0.0446637483054601		
		9	0.0007832511506546		
		10	0.0067560623615907		
		11	−0.0015235338263795		

functions are given in Table 6.5. Readers should keep in mind that in some books (e.g. [8]), there is a factor of $\sqrt{2}$ in the two-scale sequences.

6.5 CONSTRUCTION OF BIORTHOGONAL WAVELETS

In previous sections we discussed semiorthogonal and orthonormal wavelets. We developed orthogonal wavelets as a special case of semiorthogonal wavelets by using

$$\begin{aligned} \widetilde{\phi} &= \phi \\ \widetilde{\psi} &= \psi. \end{aligned} \tag{6.73}$$

One of the major difficulties with compactly supported orthonormal wavelets is that they lack spatial symmetry. This means that the processing filters are nonsymmetric and do not possess a linear phase property. Lacking this property results in severe undesirable phase distortions in signal processing. This topic is dealt with in more detail in Chapter 7. Semiorthogonal wavelets, on the other hand, are symmetric but suffer from the drawback that their duals do not have compact support. This is also undesirable since truncation of the filter coefficients is necessary for real-time processing. Biorthogonal wavelets may have both symmetry and compact support.

Cohen et al. [11] extended the framework of the theory of orthonormal wavelets to the case of biorthogonal wavelets by modification of the approximation space structure. Let us recall that in both the semiorthogonal and orthonomal cases, there exists only one sequence of nested approximation subspaces,

$$\{0\} \longleftarrow \cdots \subset \mathbf{A}_{-2} \subset \mathbf{A}_{-1} \subset \mathbf{A}_0 \subset \mathbf{A}_1 \subset \mathbf{A}_2 \subset \cdots \longrightarrow L^2. \tag{6.74}$$

The wavelet subspace, W_s, is the orthogonal complement to $\mathbf{A}_s$ within $\mathbf{A}_{s+1}$ such that

$$\mathbf{A}_s \cap \mathbf{W}_s = \{0\}, \qquad s \in \mathbb{Z},$$

and

$$\mathbf{A}_s + \mathbf{W}_s = \mathbf{A}_{s+1}. \tag{6.75}$$

This framework implies that the approximation space is orthogonal to the wavelet space at any given scale s, and the wavelet spaces are orthogonal across scales:

$$\mathbf{W}_s \perp \mathbf{W}_p \qquad \text{for } s \neq p. \tag{6.76}$$

In the orthonomal case, the scaling functions and wavelets are orthogonal to their translates at any given scale s:

$$\begin{aligned} \langle \phi_{k,s}(t), \phi_{k,m}(t) = \delta_{k,m} \rangle \\ \langle \psi_{k,s}(t), \psi_{k,m}(t) = \delta_{k,m} \rangle. \end{aligned} \tag{6.77}$$

In the semiorthogonal case, (6.77) no longer holds for ϕ and ψ. Instead, they are orthogonal to their respective duals,

$$\begin{aligned} \langle \phi_{k,s}(t), \widetilde{\phi}_{m,s}(t) \rangle = \delta_{k,m} \\ \langle \psi_{k,s}(t), \widetilde{\psi}_{m,s}(t) \rangle = \delta_{k,m}, \end{aligned} \tag{6.78}$$

and the duals span dual spaces in the sense that $\widetilde{A_s} := \operatorname{span}\{\widetilde{\phi}_{k,s}(t)(2^s t - m), s, m, \in \mathbb{Z}\}$ and $\widetilde{\mathbf{W}_s} := \operatorname{span}\{\widetilde{\psi}(2^s t - m), s, m, \in \mathbb{Z}\}$. As described in Chapter 5, semiorthogonality implies that $\widetilde{\mathbf{A}_s} = \mathbf{A}_s$ and $\widetilde{\mathbf{W}_s} = \mathbf{W}_s$.

In biorthogonal systems, there exists an additional dual nested space:

$$\{0\} \longleftarrow \cdots \subset \widetilde{\mathbf{A}}_{-2} \subset \widetilde{\mathbf{A}}_{-1} \subset \widetilde{\mathbf{A}}_0 \subset \widetilde{\mathbf{A}}_1 \subset \widetilde{\mathbf{A}}_2 \subset \cdots \longrightarrow L^2. \tag{6.79}$$

In association with this nested sequence of spaces is a set of dual wavelet subspaces (not nested) $\widetilde{W}_s$, $s \in \mathbb{Z}$, that complements the nested subspaces $\mathbf{A}_s$, $s \in \mathbb{Z}$. To be more specific, the relations of these subspaces are

$$\mathbf{A}_s + \widetilde{\mathbf{W}}_s = \mathbf{A}_{s+1} \tag{6.80}$$

$$\widetilde{\mathbf{A}}_s + \mathbf{W}_s = \widetilde{\mathbf{A}}_{s+1}. \tag{6.81}$$

The orthogonality conditions then become

$$\mathbf{A}_s \perp \widetilde{\mathbf{W}}_s \tag{6.82}$$

$$\widetilde{\mathbf{A}}_s \perp \mathbf{W}_s, \tag{6.83}$$

giving us

$$\left\langle \phi_{k,s}(t), \widetilde{\psi}_{m,s}(t) \right\rangle = 0 \tag{6.84}$$

$$\left\langle \widetilde{\phi}_{k,s}(t), \psi_{m,s}(t) \right\rangle = 0. \tag{6.85}$$

In addition, the biorthogonality between the scaling functions and the wavelets in (6.78) still holds. The two-scale relations for these bases are

$$\phi(t) = \sum_k g_0[k]\phi(2t - k) \tag{6.86}$$

$$\widetilde{\phi}(t) = \sum_k \widetilde{h}_0[k]\widetilde{\phi}(2t - k) \tag{6.87}$$

$$\psi(t) = \sum_k g_1[k]\phi(2t - k) \tag{6.88}$$

$$\widetilde{\psi}(t) = \sum_k \widetilde{h}_1[k]\widetilde{\phi}(2t - k). \tag{6.89}$$

The orthogonality and biorthogonality between these bases give the following four conditions on the filtering sequences:

$$\left\langle g_0[k - 2m], \widetilde{h}_1[k - 2n] \right\rangle = 0 \tag{6.90}$$

$$\left\langle g_1[k - 2m], \widetilde{h}_0[k - 2n] \right\rangle = 0 \tag{6.91}$$

$$\left\langle g_0[k - 2m], \widetilde{h}_0[k] \right\rangle = \delta_{m,0} \tag{6.92}$$

$$\left\langle g_1[k - 2m], \widetilde{h}_1[k] \right\rangle = \delta_{m,0}. \tag{6.93}$$

Biorthogonal wavelet design consists of finding the filter sequences that satisfy (6.90) through (6.93). Because there is quite a bit of freedom in designing the biorthogonal wavelets, there are no set steps in the design procedure. For example, one may begin with $g_0[k]$ being the two-scale sequence of a B-spline and proceed to determine the rest of the sequences. Another way is to design biorthogonal filter banks and then iterate the sequences to obtain the scaling functions and the wavelet (discussed in Section 6.6). Unlike the orthonormal wavelet, where the analysis filter is a simple time-reversed version of the synthesis filter, one must iterate both the synthesis filter and the analysis filter to get both wavelets and both scaling functions. We will follow this approach and defer our discussion of biorthogonal wavelet design by way of an example in Chapter 7.

6.6 GRAPHICAL DISPLAY OF WAVELETS

Many wavelets are mathematical functions that may not be described analytically. As examples, the Daubechies compactly supported wavelets are given in terms of two-scale sequences, and the spline wavelets are described in terms of infinite polynomials. It is difficult for the user to visualize the scaling function and the wavelet based on parameters and indirect expressions. We describe three methods here to display the graph of the scaling function and the wavelet.

6.6.1 Iteration Method

This method is the simplest to implement. We include a Matlab program with this book for practice. Let us write

$$\phi_{m+1}(t) = \sum g_0[k]\phi_m(2t - k), \qquad m = 0, 1, 2, 3, \ldots \tag{6.94}$$

and compute all values of t. In practice, we may initialize the program by taking

$$\phi_0(t) = \delta(t) \tag{6.95}$$

and setting $\phi_0(n) = \delta(n) = 1$. After upsampling by 2, the sequence is convolved with the $g_0[k]$ sequence to give $\phi_1(n)$. This sequence is upsampled and convolved with $g_0[k]$ again to give $\phi_2(n)$, and so on. In most cases the procedure usually converges within 10 iterations. For biorthogonal wavelets, the convergence time may be longer. Once the scaling function has been obtained, the associated wavelet can be computed and displayed using the two-scale relation for the wavelet:

$$\psi(t) = \sum g_1[k]\phi(2t - k).$$

A display indicating the iterative procedure is given in Figure 6.10. The figure indicates the number of points in each iteration. To get the corresponding position along the time axis, the abscissa needs to be divided by 2^m for each iteration m.

6.6.2 Spectral Method

In this method, the two-scale relation for the scaling function is expressed in the spectral domain

$$\begin{aligned}\widehat{\phi}(\omega) &= G_0(e^{j(\omega/2)})\widehat{\phi}\left(\frac{\omega}{2}\right), \qquad z = e^{j(\omega/2)}, \\ &= G_0(e^{j(\omega/2)})G_0(e^{j(\omega/4)})\widehat{\phi}\left(\frac{\omega}{4}\right) \\ &\;\;\vdots\end{aligned} \tag{6.96}$$

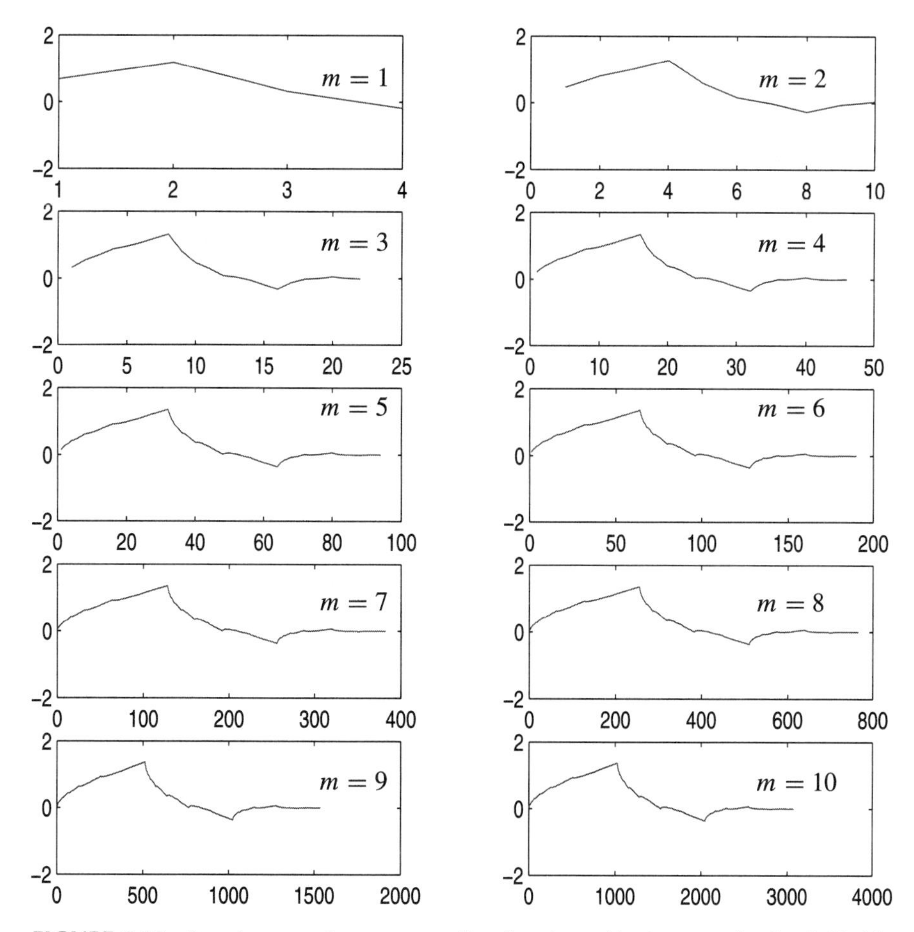

FIGURE 6.10 Iterative procedure to get scaling functions. Abscissas need to be divided by 2^m to get the correct position in time.

$$= \prod_{k=1}^{N} G_0(e^{j(\omega/2^k)})\widehat{\phi}\left(\frac{\omega}{2^N}\right)$$

$$= \prod_{k=1}^{N\to\infty} G_0(e^{j(\omega/2^k)})\widehat{\phi}(0). \tag{6.97}$$

Since $\widehat{\phi}(0) = 1$, we may take the inverse Fourier transform of (6.97) to yield

$$\phi(t) = \frac{1}{2\pi}\int_{-\infty}^{\infty}\left(\prod_{k=1}^{N\to\infty} G_0(e^{j(\omega/2^k)})\right) e^{j\omega t}\, d\omega. \tag{6.98}$$

To compute (6.98), the user has to evaluate the truncated infinite product and then take the fast Fourier transform.

6.6.3 Eigenvalue Method

This method converts the two-scale relation into an eigen equation. Let us consider the two-scale relation by setting $x = n$ to yield the following matrix equation:

$$\begin{aligned}\phi(n) &= \sum_k g_0[k]\phi(2n-k)\\ &= \sum_m g_0[2n-m]\phi(m)\\ &= [g_0(n,m)]\,\phi(m),\end{aligned} \tag{6.99}$$

where the matrix element $g_0(n,m) = g_0(2n-m)$. In matrix form, we write (6.99) as

$$\begin{bmatrix} \cdot & \cdot & \cdot & \cdot & \cdot \\ \cdot & g_0[0] & g_0[-1] & g_0[-2] & \cdot \\ \cdot & g_0[2] & g_0[1] & g_0[0] & \cdot \\ \cdot & g_0[4] & g_0[3] & g_0[2] & \cdot \\ \cdot & \cdot & \cdot & \cdot & \cdot \end{bmatrix} \begin{bmatrix} \cdot \\ \phi(0) \\ \phi(1) \\ \phi(2) \\ \cdot \end{bmatrix} = 1 \begin{bmatrix} \cdot \\ \phi(0) \\ \phi(1) \\ \phi(2) \\ \cdot \end{bmatrix}.$$

The eigenvalue of this eigenmatrix is 1, so we can compute $\phi(n)$ for all integers n. This procedure can be repeated for a twofold increase in resolution. Let $x = n/2$, and the two-scale relation becomes

$$\phi\left(\frac{n}{2}\right) = \sum_k g_0[k]\phi(n-k). \tag{6.100}$$

By repeating this procedure for $x = n/4, n/8, \ldots,$ we compute the discretized $\phi(t)$ to an arbitrarily fine resolution.

6.7 EXERCISES

1. Show that the support of semiorthogonal wavelet, $\psi_m(t) = [0, 2m-1]$.
2. Show that the integer translates of the Shannon wavelet $\psi_s(t-k)$ form an orthonormal basis.
3. Find the cubic polynomial S_4 that satisfies the conditions $S_4(0) = S_4'(0) = 0$, $S_4(1) = 1$, $S_4'(1) = 0$, and $S_4(x) + S_4(1-x) \equiv 1$. Use this polynomial as the smoothing function for Meyer's scaling function and compute the two-scale coefficients.
4. Show that if $\{\phi(t-k), k \in Z\}$ is a Riesz basis of $V_0 = \{\phi(t-k) : k \in Z\}$, then $\left\{\phi_{k,s}\right\}_{k\in Z}$ is a Riesz basis of $V_s = \left\{\phi_{k,s}(t), k \in Z\right\}$ for a fixed $s \in Z$. That is,

$$A \sum_{k=-\infty}^{\infty} |a_k|^2 \le \left\| \sum_{k=-\infty}^{\infty} a_k \phi(t-k) \right\|_2^2 \le B \sum_{k=-\infty}^{\infty} |a_k|^2$$

implies that

$$A \sum_{k=-\infty}^{\infty} |a_k|^2 \le \left\| \sum_{k=-\infty}^{\infty} a_k \phi_{k,s}(t) \right\|_2^2 \le B \sum_{k=-\infty}^{\infty} |a_k|^2$$

with the same constants A and B.

5. Show that the following statements are equivalent: (**a**) $\{\phi(\cdot - k) : k \in Z\}$ is an orthonormal family, and (**b**) $\sum_{k=-\infty}^{\infty} |\widehat{\phi}(\omega + 2k\pi)|^2 = 1$ almost everywhere.

6. Prove that $\{N_1(\cdot - k) : k \in Z\}$ is an orthonormal family by using this theorem, that is, by showing that

$$\sum_{k=-\infty}^{\infty} |\widehat{N}_1(\omega + 2k\pi)|^2 = 1.$$

7. Obtain an algebraic polynomial corresponding to Euler–Frobenius–Laurent polynomial $E_{N_4}(z)$ and find its roots, $\lambda_1 > \cdots > \lambda_6$. Check that these zeros are simple, real, negative, and come in reciprocal pairs, i.e.,

$$\lambda_1 \lambda_6 = \lambda_2 \lambda_5 = \lambda_3 \lambda_4 = 1.$$

8. The autocorrelation function F for a given function $f \in L^2(-\infty, \infty)$ is defined as

$$F(x) = \int_{-\infty}^{\infty} f(t+x)\overline{f(t)}\, dt, \quad x \in R.$$

Compute the autocorrelation function of the hat function N_2 and compare it to the function N_4 introduced in Exercise 7.

9. Construct a linear Battle–Lemarié scaling function to show that for the hat function $N_2(t)$, it holds (let $z = e^{-j(\omega/2)}$) that

$$\sum_{k=-\infty}^{\infty} |\widehat{N}_2(\omega + 2k\pi)|^2 = \frac{1}{6}(z^{-2} + 4 + z^2)$$

and

$$\widehat{N}_2(\omega) = \frac{-1}{\omega^2}(1 - z^2)^2.$$

The Fourier transform of the orthonormalized scaling function $N_2^{\perp}(t)$ is given by

$$\widehat{N_2^{\perp}}(\omega) = \frac{(-1/\omega^2)\,(1-z^2)^2}{\left[(1/6)(z^{-2}+4+z^2)\right]^{1/2}}.$$

We have shown that the symbol

$$G_0(z) = \frac{\widehat{N_2^{\perp}}(\omega)}{\widehat{N_2^{\perp}}(\omega/2)}.$$

Compute the ratio to show that the result is

$$\left(\frac{1+z}{2}\right)^2 (1+\eta)^{1/2},$$

where

$$\eta = \frac{[(z^{-1}+z)/4] - [(z^{-2}+z^2)/4]}{1+(z^{-2}+z^2)/4}.$$

Use the power series expansions

$$(1+\eta)^{1/2} = 1 + \frac{1}{2}\eta + \sum_{n=2}^{\infty}(-1)^{n+1}\frac{1}{2^n}\frac{1}{n!}1\cdot 3\cdots(2n-3)\eta^n$$

and

$$(1+x)^{-n} = \sum_{j=0}^{\infty}(-1)^j \binom{n-1+j}{j} x^j$$

as well as the binomial theorem to expand the expression $[(1+z)/2]^2(1+\eta)^{1/2}$ in powers of z and determine the coefficients $g[k]$ for $k = -5, \ldots, 5$ by comparing the corresponding coefficients of $z^{-5}, \ldots, z^5$ in $G_0(z)$ and $[(1+z)/2]^2(1+\eta)^{1/2}$. You should use symbolic packages such as *Mathematica* or *Maple* for these computations.

10. *Construction of linear B-spline wavelet*: Given the two-scale relation for the hat function

$$N_2(t) = \sum_{k=0}^{2}\frac{1}{2}\binom{2}{k}N_2(2t-k),$$

we want to determine the two-scale relation for a linear wavelet with minimal support,

$$\psi_2(t) = \sum_k g_1[k]N_2(2t-k),$$

using the corresponding Euler–Frobenius–Laurent polynomial $E(z) = z^{-1} + 4 + z$. It was shown that for the corresponding symbols

$$G_0(z) = \frac{1}{2}\sum_k g_0[k]z^k \quad \text{and} \quad G_1(z) = \frac{1}{2}\sum_k g_1[k]z^k, \qquad z = e^{-j(\omega/2)},$$

the orthogonality condition is equivalent to

$$G_0(z)\overline{G_1(z)}E(z) + G_0(-z)\overline{G_1(-z)}E(-z) = 0$$

with $|z| = 1$. We need to determine the polynomial $G_1(z)$ from the equation above. There is no unique solution to this equation.

(a) Show that $G_1(z) = (-1/3!)z^3G_0(-z)E(-z)$ is a solution of the equation above.

(b) Show that $G_0(z) = [(1+z)/2]^2$.

(c) Expand $G_1(z) = (-1/3!)z^3G_0(-z)E(-z)$ in powers of z and thus determine the two-scale relation for the function ψ_{N_2} by comparing coefficients in

$$G_1(z) = \frac{1}{2}\sum_k g_1[k]z^k = \frac{-1}{3!}z^3G_0(-z)E(-z).$$

(d) Graph ψ_2.

11. Complete the missing steps in the derivation of Daubechies wavelet in Section 6.4.4. Note that $|S(z)|^2$ is a polynomial in $\cos\frac{\omega}{2}$.

12. Use the sequence $\{-0.102859456942, 0.477859456942, 1.205718913884, 0.544281086116, -0.102859945694, -0.022140543058\}$ as the two scale sequence $\{g0[n]\}$ in the program *iterate.m* and view the results. The resultant scaling function is a member of the Coifman wavelet system or *coiflets* [8]. The main feature of this system is that in this case the scaling functions also have vanishing moment properties. For mth order coiflets

$$\int_{-\infty}^{\infty} t^p\psi(t)\,dt = 0, \quad p = 0, \ldots, m-1;$$

$$\int_{-\infty}^{\infty} t^p\phi(t)\,dt = 0, \quad p = 1, \ldots, m-1;$$

$$\int_{-\infty}^{\infty} \phi(t)\,dt = 1.$$

13. Construct biorthogonal wavelets beginning with the two scale sequance $\{g0[n]\}$ for linear spline.

6.8 COMPUTER PROGRAMS

6.8.1 Daubechies Wavelet

```
%
% PROGRAM wavelet.m
%
% Generates Daubechies scaling functions and wavelets

g0 = [0.68301; 1.18301; 0.31699; -0.18301];
k = [0; 1; 2; 3];
g1 = flipud(g0).*(-1).^k;

ng1 = length(g1);

% Compute scaling funtion first

NIter = 10; % interation time
phi_new = 1; % initialization

for i = 1:NIter
  unit = 2^(i-1);
  phi = conv(g0,phi_new);
  n = length(phi);
  phi_new(1:2:2*n) = phi;
  length(phi_new);
  if(i == (NIter-1))
    phi2 = phi;
  end
end
%
dt = 1.0 / (2 * unit);
t = [1:length(phi)] * dt;
subplot(2,1,1), plot(t,phi)
title('Scaling Function')

% Compute wavelet using 2-scale relation

for i = 1:ng1
  a = (i-1) * unit + 1;
  b = a + length(phi2) - 1;
  psi2s(i,a:b) = phi2 * g1(i);
  psi2s(1,n) = 0;
end

psi = sum(psi2s);
```

```
dt = 1.0 / (2 * unit);
t = [0:length(phi)-1] * dt - (ng1 - 2) / 2;
subplot(2,1,2), plot(t,psi)
title('Wavelet')
```

6.8.2 Iteration Method

```
%
% PROGRAM iterate.m
%
% Iterative procedure to get scaling function
% generates Figure 6.10
%

g0 = [0.68301 1.18301 0.31699 -0.18301];

NIter = 10; % number of interation
phi_new = 1; % initialization

for i = 1:NIter
  unit = 2^(i-1);
  phi = conv(g0,phi_new);
  n = length(phi);
  phi_new(1:2:2*n) = phi;
  subplot(5,2,i), plot(phi); hold on;
  heading = sprintf('Iteration = %.4g',i)
  title(heading);
end
%
```

REFERENCES

1. C. K. Chui and J. Z. Wang, "On compactly supported spline wavelets and a duality principle," *Trans. Am. Math. Soc.*, **330**, pp. 903–915, 1992.
2. C. K. Chui and J. Z. Wang, "High-order orthonormal scaling functions and wavelets give poor time–frequency localization," *Fourier Anal. Appl.*, **2**, pp. 415–426, 1996.
3. Y. Meyer, "Principe d'incertitude, bases Hilbertiennes et algèbres d'opérateurs," *Semin. Bourbaki*, **662**, 1985–1986.
4. Y. Meyer, *Wavelets: Algorithms and Applications.* Philadelphia: SIAM, 1993.
5. G. Battle, "A block spline construction of ondelettes: I. Lemaré functions," *Commun. Math. Phys.*, **110**, pp. 601–615, 1987.
6. P. G. Lemarié, "Une nouvelle base d'ondelettes de $L^2(\mathbb{R}^n)$," *J. Math. Pures Appl.*, **67**, pp. 227–236, 1988.
7. I. Daubechies, "Orthonormal bases of compactly supported wavelets," *Commun. Pure Appl. Math.*, **41**. pp. 909–996, 1988.

8. I. Daubechies, *Ten Lectures on Wavelets.* CBMS-NSF Ser. App. Math. 61. Philadelphia: SIAM, 1992.
9. C. K. Chui, *Wavelets: A Mathematical Tool for Signal Analysis.* Philadelphia: SIAM, 1997.
10. C. K. Chui, *An Introduction to Wavelets.* San Diego, Calif.: Academic Press, 1992.
11. A. Cohen, I. Daubechies, and J. C. Feauveau, "Biorthogonal bases of compactly supported wavelets," *Commun. Pure Appl. Math.*, **45**, pp. 485–500, 1992.

CHAPTER SEVEN

Discrete Wavelet Transform and Filter Bank Algorithms

The discussion of multiresolution analysis in Chapter 5 prepares readers for an understanding of wavelet construction and algorithms for fast computation of the continuous wavelet transform (CWT). The two-scale and decomposition relations are essential for development of the fast algorithms. The need for these algorithms is obvious since a straightforward evaluation of the integral in (4.32) puts a heavy computation load on problem solving. The CWT places redundant information on the time–frequency plane. To overcome these deficiencies, the CWT is discretized and algorithms equivalent to the two-channel filter bank have been developed for signal representation and processing. The perfect reconstruction (PR) constraint is placed on these algorithm developments. In this chapter we develop these algorithms in detail. Since the semiorthogonal spline functions and compactly supported spline wavelets require their duals in the dual spaces, signal representation and the PR condition for this case are developed along with the algorithms for change of bases. Before we develop the algebra for these algorithms, we discuss the basic concepts of sampling-rate changes through decimation and interpolation.

7.1 DECIMATION AND INTERPOLATION

In signal processing we often encounter signals whose spectrum may vary with time. A linear chirp signal is a good example. To avoid aliasing, this chirp signal must be sampled at least twice at its highest frequency. For a chirp signal with a wide bandwidth, this Nyquist rate may be too high for the low-frequency portion of the chirp. Consequently, there is a lot of redundant information to be carried around if one uses a fixed rate for the entire chirp. There is the area of multirate signal processing, which deals with signal representation using more than one sampling rate. The mechanisms for changing the sample rate are decimation and interpolation. We discuss their basic characteristics in the time and spectral domains.

7.1.1 Decimation

An M-point decimation retains only every Mth sample of a given signal. In the time domain, an M-point decimation of an input sequence $\{x(n)\}$ is given by

$$y(n) = x(nM) \qquad \text{for } n \in \mathbb{Z}. \tag{7.1}$$

Figure 7.1 depicts the system diagram of an M-point decimator. The output of the decimator may be written in terms of a product of $x(n)$ and a sequence of unit impulses separated by M samples $\sum_{k\in\mathbb{Z}} \delta(n - kM)$. Let

$$u(n) = \sum_{k\in\mathbb{Z}} x(n)\delta(n - kM) \qquad \text{for } k \in \mathbb{Z}, \tag{7.2}$$

which selects only the kMth samples of $x(n)$. The Fourier series representation of the M-point period impulse sequence (7.2) is

$$\sum_{k\in\mathbb{Z}} \delta(n - kM) = \frac{1}{M}\sum_{k=0}^{M-1} e^{-j2\pi kn/M}. \tag{7.3}$$

Based on the geometric sum

$$\sum_{k=0}^{M-1} e^{-j2\pi kn/M} = \begin{cases} M & \text{for } k = \ell M, \qquad \ell \in \mathbb{Z} \\ 0, & \text{otherwise,} \end{cases}$$

the identity in (7.3) is proved. Writing $y(n) = u(nM)$, the z-transform of $y(n)$ has the form

$$\begin{aligned} Y(z) &= \frac{1}{M}\sum_{k=0}^{M-1} X(z^{1/M} e^{-j2\pi k/M}) \\ &= \frac{1}{M}\sum_{k=0}^{M-1} X(z^{1/M} W_M^k), \end{aligned} \tag{7.4}$$

where the M-point exponential basis function $W_M^k := e^{-j2\pi k/M}$ has been used. In the spectral domain, we obtain the discrete Fourier transform (DFT) of $y(n)$ simply

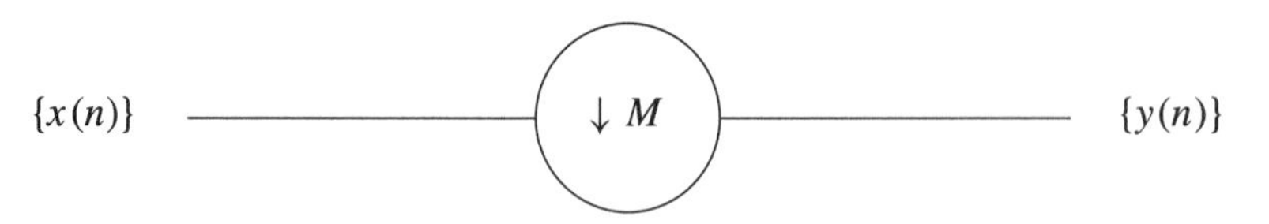

FIGURE 7.1 An M-point decimator.

by setting $z = e^{j\omega}$ to yield

$$\widehat{y}(e^{j\omega}) = \frac{1}{M}\sum_{k=0}^{M-1}\widehat{x}(e^{j[(\omega-2\pi k)/M]}). \tag{7.5}$$

The spectrum of the decimator output contains M copies of the input spectrum. The amplitude of the copy is reduced by a factor of $1/M$. In addition, the bandwidth of the copy is expanded by M times. As a result, if the spectral bandwidth of the input signal is greater than π/M (i.e., $|\omega| > \pi/M$), an M-point decimator will introduce aliasing in its output signal. We will see later that aliasing does indeed occur in a wavelet decomposition tree or a two-channel filter bank decomposition algorithm. However, the aliasing is canceled by carefully designing the reconstruction algorithm to remove the aliasing and recover the original signal.

For $M = 2$, we decimate a sequence by taking every other data point. From (7.4), we obtain

$$\begin{aligned} Y(z) &= \frac{1}{2}\sum_{k=0}^{1} X(z^{1/2}W_2^k) \\ &= \frac{1}{2}[X(z^{1/2}) + X(-z^{1/2})] \end{aligned} \tag{7.6}$$

and

$$\widehat{y}(e^{j\omega}) = \frac{1}{2}[\widehat{x}(e^{j(\omega/2)}) + \widehat{x}(-e^{j(\omega/2)})]. \tag{7.7}$$

The spectrum of $\widehat{y}(e^{j\omega})$ is shown in Figure 7.2.

For the sake of simplicity in using matrix form, we consider only the case where $M = 2$. We use $\downarrow 2$ in the subscript to represent decimation by 2. We write

$$[y] = [x]_{\downarrow 2} \tag{7.8}$$

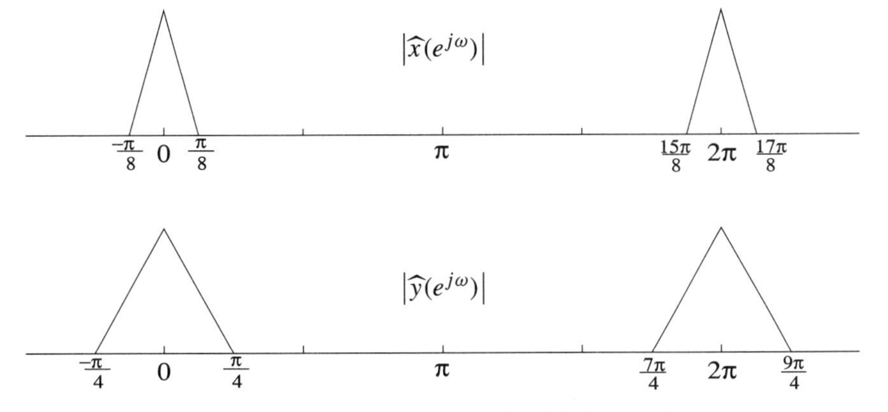

FIGURE 7.2 Spectral characteristic of decimation by 2.

as

$$
\begin{bmatrix} \cdot \\ \cdot \\ y(-2) \\ y(-1) \\ y(0) \\ y(1) \\ y(2) \\ y(3) \\ \cdot \\ \cdot \\ \cdot \end{bmatrix} = \begin{bmatrix} \cdot \\ \cdot \\ x(-4) \\ x(-2) \\ x(0) \\ x(2) \\ x(4) \\ x(6) \\ x(8) \\ \cdot \\ \cdot \end{bmatrix}. \tag{7.9}
$$

In terms of a matrix operator, we write (7.9) as

$$
\begin{bmatrix} \cdot \\ \cdot \\ y(-2) \\ y(-1) \\ y(0) \\ y(1) \\ y(2) \\ y(3) \\ \cdot \\ \cdot \\ \cdot \end{bmatrix} = \begin{bmatrix} & & & & & & & & & & \\ & & & & & & & & & & \\ 1 & 0 & 0 & \cdot & \cdot & & & & & & \\ 0 & 0 & 1 & 0 & 0 & \cdot & \cdot & & & & \\ & \cdot & 0 & 0 & 1 & 0 & 0 & & & & \\ & \cdot & \cdot & \cdot & 0 & 0 & 1 & 0 & 0 & \cdot & \\ & & & \cdot & \cdot & \cdot & 0 & 0 & 1 & 0 & 0 \\ & & & & & \cdot & \cdot & \cdot & 0 & 0 & 1 \\ & & & & & & & & & & \\ & & & & & & & & & & \\ & & & & & & & & & & \end{bmatrix} \begin{bmatrix} \cdot \\ \cdot \\ x(-2) \\ x(-1) \\ x(0) \\ x(1) \\ x(2) \\ x(3) \\ x(4) \\ \cdot \\ \cdot \end{bmatrix} \tag{7.10}
$$

or

$$
[y] = \left[\mathbf{DEC}_{\downarrow 2}\right][x].
$$

The shift-variant property of the decimator is evident when we shift the input column either up or down by a given number of position. In addition, the decimation matrix is an orthogonal matrix since

$$
\left[\mathbf{DEC}_{\downarrow 2}\right]^{-1} = \left[\mathbf{DEC}_{\downarrow 2}\right]^{t}.
$$

Consequently, decimation is an orthogonal transformation.

7.1.2 Interpolation

Interpolation of data means inserting additional data points into the sequence to increase the sampling rate. Let $y(n)$ be the input to the interpolator. If we wish to increase the number of sample by M folds, we insert $M-1$ zeros in between any two adjacent samples so that the interpolator output gives

$$
x'(n) = \begin{cases} y\left(\dfrac{n}{M}\right) & \text{for } n = kM, \quad k \in \mathbb{Z} \\ 0, & \text{otherwise.} \end{cases} \tag{7.11}
$$

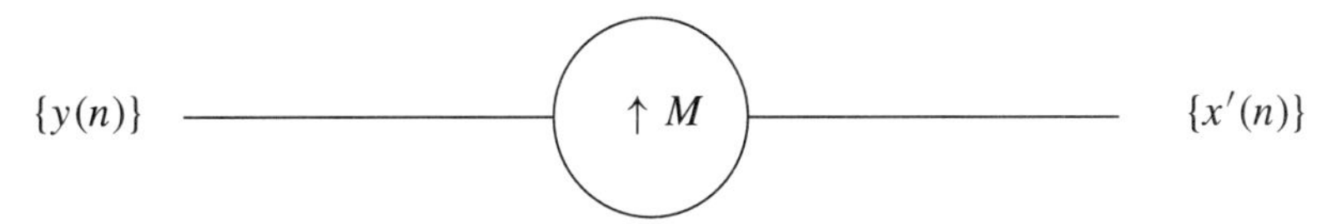

FIGURE 7.3 An M-point interpolator.

The system diagram of an M-point interpolator is shown in Figure 7.3. We can also write the expression of interpolation in the standard form of a convolution sum,

$$x'(n) = \sum_k y(k)\delta(n - kM). \tag{7.12}$$

The spectrum of the interpolator output is given by

$$\begin{aligned}\widehat{x'}(e^{j\omega}) &= \sum_n \sum_k y(k)\delta(n - kM)e^{-jn\omega} \\ &= \sum_k y(k)e^{-jkM\omega} \\ &= \widehat{y}(e^{-jM\omega}).\end{aligned} \tag{7.13}$$

The z-transform of the interpolator output is

$$X'(z) = Y(z^M). \tag{7.14}$$

Interpolation raises the sampling rate by filling zeros in between samples. The output sequence has M times more points than the input sequence, and the output spectrum is shrunk by a factor of M on the ω-axis. Unlike the decimator, there is no danger of aliasing for interpolator since the output spectrum has a narrower bandwidth than the input spectrum. The spectrum of a twofold interpolator is given in Figure 7.4.

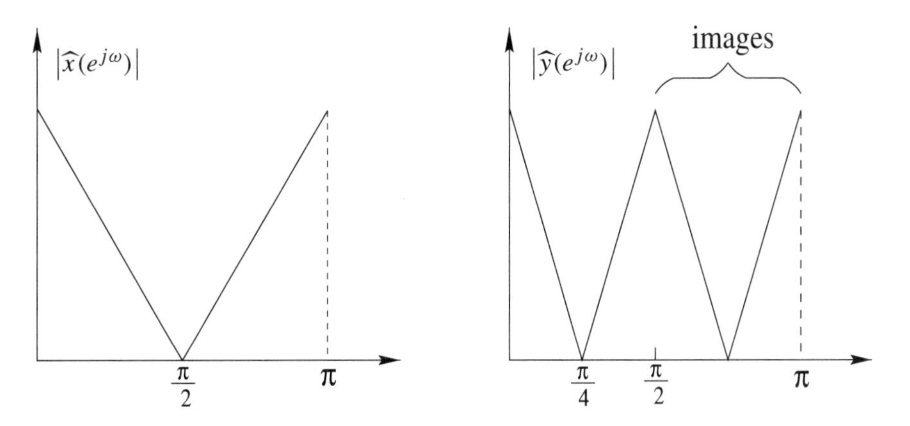

FIGURE 7.4 Spectral characteristic of interpolation by 2.

Using $M = 2$ as an example, we write

$$
\begin{aligned}
x'(n) &= y(n)_{\uparrow 2} \\
&= \begin{cases} y(n/2) & \text{for } n \text{ even} \\ 0, & \text{otherwise.} \end{cases}
\end{aligned}
\tag{7.15}
$$

In matrix form we have

$$
\begin{bmatrix} \cdot \\ \cdot \\ x'(-2) \\ x'(-1) \\ x'(0) \\ x'(1) \\ x'(2) \\ x'(3) \\ x'(4) \\ \cdot \\ \cdot \end{bmatrix}
=
\begin{bmatrix} \cdot \\ \cdot \\ y(-1) \\ 0 \\ y(0) \\ 0 \\ y(1) \\ 0 \\ y(2) \\ \cdot \\ \cdot \end{bmatrix}.
\tag{7.16}
$$

As before, we represent the interpolator by a linear matrix operator. It turns out that the interpolation matrix is the transpose of the decimation matrix

$$
\begin{bmatrix} \cdot \\ \cdot \\ x'(-2) \\ x'(-1) \\ x'(0) \\ x'(1) \\ x'(2) \\ x'(3) \\ \cdot \\ \cdot \\ \cdot \end{bmatrix}
=
\begin{bmatrix}
\cdot & \cdot & 1 & 0 & & & & \\
\cdot & \cdot & 0 & 0 & & & & \\
\cdot & \cdot & 0 & 1 & 0 & & & \\
\cdot & \cdot & 0 & 0 & 0 & 0 & \cdot & \\
\cdot & \cdot & 0 & 0 & 1 & 0 & & \\
\cdot & \cdot & 0 & 0 & 0 & 0 & 0 & \\
\cdot & \cdot & 0 & 0 & 0 & 1 & 0 & \\
\cdot & \cdot & 0 & 0 & 0 & 0 & 0 & \\
 & & & & & & 1 & 0 \\
 & & & & & & 0 & 0 \\
 & & & & & & 0 & 1
\end{bmatrix}
\begin{bmatrix} \cdot \\ \cdot \\ y(-2) \\ y(-1) \\ y(0) \\ y(1) \\ y(2) \\ y(3) \\ y(4) \\ \cdot \\ \cdot \end{bmatrix}
\tag{7.17}
$$

or we can write

$$
[y] = \left[\mathbf{INT}_{\uparrow 2} \right] [x].
\tag{7.18}
$$

The operations of convolution followed by decimation and interpolation followed by convolution are two of the most important building blocks of algorithms. They will be used to build tree algorithms for wavelets and wavelet packets as well as in two- and three-dimensional signal processing. We show only their time-domain identities in the following sections.

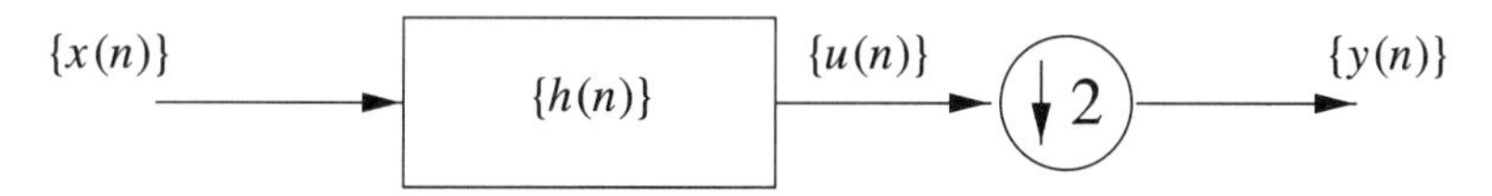

FIGURE 7.5 Convolution followed by decimation.

7.1.3 Convolution Followed by Decimation

Mathematically, we express this operation by

$$y(n) = \{h(n) * x(n)\}_{\downarrow 2}. \tag{7.19}$$

The processing block diagram is given in Figure 7.5. If we label the intermediate output as $u(n)$, it is the convolution of $x(n)$ and $h(n)$ given by

$$u(n) = \sum_k x(k)h(n-k).$$

The two-point decimation gives

$$y(n) = u(2n) = \sum_k x(k)h(2n-k). \tag{7.20}$$

7.1.4 Interpolation Followed by Convolution

The time-domain expression of this operation is given by

$$y(n) = \left\{g(n) * [x(n)]_{\uparrow 2}\right\}. \tag{7.21}$$

Using $v\,(n)$ as the intermediate output, we have

$$y(n) = \sum_k v(k)g(n-k).$$

Since $v(k) = x(k/2)$ for even k, we have

$$\begin{aligned} y(n) &= \sum_{k:\text{even}} x\left(\frac{k}{2}\right) g(n-k) \\ &= \sum_{\ell} x(\ell)g(n-2\ell). \end{aligned} \tag{7.22}$$

This process is shown in Figure 7.6.

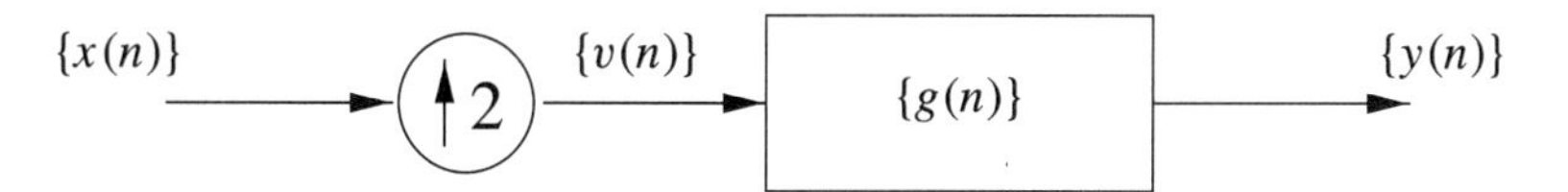

FIGURE 7.6 Interpolation followed by convolution.

7.2 SIGNAL REPRESENTATION IN THE APPROXIMATION SUBSPACE

We have shown in Chapter 5 that the approximation subspaces $\mathbf{A}_n$ are nested so that the subspace $\mathbf{A}_\infty = L^2$, $\mathbf{A}_{-\infty} = \{\mathbf{0}\}$, and $\mathbf{A}_n \subset \mathbf{A}_{n+1}$ for any $n \in \mathbb{Z}$. For an arbitrary finite-energy signal $x(t)$, there is no guarantee that this signal is in any of these approximation subspaces. That is, we may not be able to find a coefficient $a_{k,s}$ such that

$$x(t) = \sum_{k\in\mathbb{Z}} a_{k,s}\phi(2^s t - k) \qquad \text{for some } s. \tag{7.23}$$

To make use of the two-scale relations for processing, a signal must be in one of these nested approximation subspaces. One way of meeting this requirement is by projecting the signal into one of the $\mathbf{A}_s$ for some s. This is particularly important if one only knows the sampled values of the signal at $x(t = k/2^s, k \in \mathbb{Z})$ for some large value of s.

Assuming that the signal $x(t)$ is not in the approximation $\mathbf{A}_s$,we wish to find $x_s(t) \in \mathbf{A}_s$ such that

$$x(t) \cong x_s(t) = \sum_k a_{k,s}\phi_{k,s}(t) = \sum_k a_{k,s}\phi(2^s t - k), \tag{7.24}$$

where $a_{k,s}$ are the scaling function coefficients to be computed from the signal samples. We will show how one can determine $a_{k,s}$ from the sample data $x(t = k/2^s)$ using the orthogonal projection of $x(t)$ onto the $\mathbf{A}_s$ space.

Since $\mathbf{A}_s$ is a subspace of L^2 and $x(t) \in L^2$, we consider $x_s(t)$ as the orthogonal projection of $x(t)$ onto the $\mathbf{A}_s$ subspace. Then $x(t) - x_s(t)$ is orthogonal to $\mathbf{A}_s$ and therefore orthogonal to the basis function $\phi_{\ell,s}$:

$$\langle (x(t) - x_s(t)), \phi_{\ell,s}\rangle = 0 \qquad \forall\, \ell \in \mathbb{Z}. \tag{7.25}$$

Consequently, the coefficients are determined from the equation

$$\langle x_s(t), \phi_{\ell,s}\rangle = \langle x(t), \phi_{\ell,s}\rangle = \left\langle \sum_k a_{k,s}\phi_{k,s}(t), \phi_{\ell,s}(t)\right\rangle. \tag{7.26}$$

We expand the last equality, yielding

$$\begin{aligned} 2^{s/2}\int_{-\infty}^{\infty} x(t)\overline{\phi(2^s t - \ell)}\,dt &= 2^s \sum_k a_{k,s}\left[\int_{-\infty}^{\infty} \phi(2^s t - k)\overline{\phi(2^s t - \ell)}\,dt\right] \\ &= \sum_m a_{m,s}\left[\int_{-\infty}^{\infty} \phi(t)\overline{\phi(t - m)}\,dt\right], \end{aligned} \tag{7.27}$$

where we have made a change of index $m = \ell - k$. The matrix form of (7.27) is

$$\begin{bmatrix} \cdot & & & & \\ & \cdot & \cdot & & \\ & \alpha_1 & \alpha_0 & \alpha_1 & \\ & & \cdot & \alpha_0 & \alpha_1 \\ & & & \cdot & \alpha_0 \\ & & & & \cdot \end{bmatrix} \begin{bmatrix} \cdot \\ \cdot \\ a_{m,s} \\ \cdot \\ \cdot \\ \cdot \end{bmatrix} = \begin{bmatrix} \cdot \\ \cdot \\ \langle x(t), \phi_{m,s} \rangle \\ \cdot \\ \cdot \\ \cdot \end{bmatrix}, \tag{7.28}$$

where

$$\alpha_m = \int_{-\infty}^{\infty} \phi(t)\overline{\phi(t-m)}\, dt = \overline{\alpha_{-m}}$$

is the autocorrelation of the scaling function $\phi(t)$. If the scaling function is supported compactly, the autocorrelation matrix is banded with a finite-size diagonal band. If the scaling function and its translates form an orthonormal basis, then

$$\alpha_m = \delta_{m,0}.$$

By assuming an orthonormal basis, the autocorrelation matrix is the identity matrix and the coefficients are obtained by computing the inner product,

$$a_{m,s} = \langle x(t), \phi_{m,s} \rangle. \tag{7.29}$$

If we are given only the sample values of the signal $x(t)$ at $x(t = k/2^s)$, we can approximate the integral by a sum. That is,

$$\begin{aligned} a_{m,s} &= 2^{s/2} \int_{-\infty}^{\infty} x(t)\overline{\phi(2^s t - m)}\, dt \\ &\cong 2^{-s/2} \sum_k x\left(\frac{k}{2^s}\right) \overline{\phi(k-m)}. \end{aligned} \tag{7.30}$$

This equation demonstrates the difference between the scaling function coefficients and the sample values of the signal. The former are expansion coefficients of an analog signal, while the latter are samples of a discrete-time signal. For representation of a given discrete signal in terms of a spline series of orders 2 and 4, we have given formulas in Section 5.6.

7.3 WAVELET DECOMPOSITION ALGORITHM

Let us rewrite the expression of the CWT of a signal $x(t)$:

$$W_\psi x(b,a) = \frac{1}{\sqrt{a}} \int_{-\infty}^{\infty} x(t)\overline{\psi\left(\frac{t-b}{a}\right)}\, dt. \tag{7.31}$$

Let us denote the scale $a = 1/2^s$ and the translation $b = k/2^s$, where s and k belong to the integer set $\mathbb{Z}$. The CWT of $x(t)$ is a number at $(k/2^s, 1/2^s)$ on the time-scale plane. It represents the correlation between $x(t)$ and $\overline{\psi}(t)$ at that time-scale point. We call this the *discrete wavelet transform* (DWT), which generates a sparse set of values on the time-scale plane. We use

$$w_{k,s} = W_{\psi}x\left(\frac{k}{2^s}, \frac{1}{2^s}\right) = \int_{-\infty}^{\infty} x(t)\overline{\psi\left(\frac{t - k/2^s}{1/2^s}\right)}\,dt \tag{7.32}$$

to represent the wavelet coefficient at $(b = k/2^s, a = 1/2^s)$. A discrete time-scale map representing the signal $x(t)$ may look like Figure 7.7.

It is known that the CWT generates redundant information about the signal on the time-scale plane. By choosing $(b = k/2^s, a = 1/2^s)$, it is much more efficient using the DWT to process a signal. It has been shown that the DWT keeps enough information of the signal such that it reconstructs the signal perfectly from the wavelet coefficients. In fact, the number of coefficients needed for perfect reconstruction is the same as the number of data samples. Known as *critical sampling*, this minimizes redundant information.

The decomposition (analysis) algorithm is used most often in wavelet signal processing. It is used in signal compression as well as in signal identification, although in the latter case, reconstruction of the original signal is not always required. The

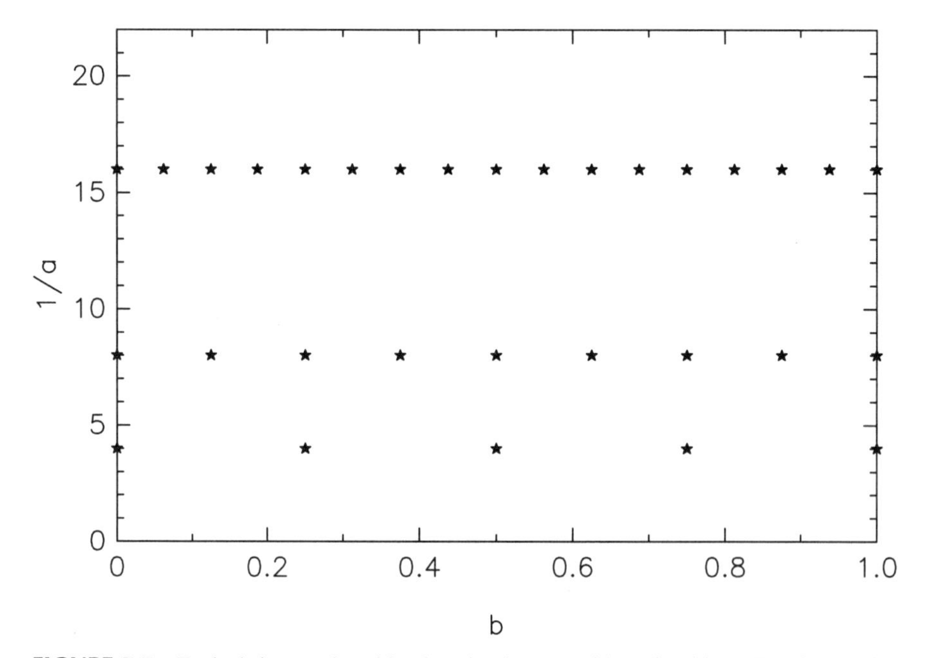

FIGURE 7.7 Typical time-scale grid using the decomposition algorithm. (Reprinted with permission from [1], copyright © 1995 by Springer-Verlag.)

algorithm separates a signal into components at various scales corresponding to successive octave frequencies. Each component can be processed individually by a different algorithm. In echo cancellation, for example, each component is processed with an adaptive filter of a different filter length to improve convergence. The important issue of this algorithm is to retain all pertinent information so that the user may recover the original signal (if needed). The algorithm is based on the decomposition relation in MRA discussed in Chapter 5. We rewrite several of these relations here for easy reference.

Let

$$x_{s+1}(t) \in \mathbf{A}_{s+1}, \Longrightarrow x_{s+1}(t) = \sum_k a_{k,s+1}\phi_{k,s+1}(t)$$

$$x_s(t) \in \mathbf{A}_s, \Longrightarrow x_s(t) = \sum_k a_{k,s}\phi_{k,s}(t)$$

$$y_s(t) \in \mathbf{W}_s, \Longrightarrow y_s(t) = \sum_k w_{k,s}\psi_{k,s}(t).$$

Since the MRA requires that

$$\mathbf{A}_{s+1} = \mathbf{A}_s + \mathbf{W}_s, \tag{7.33}$$

we have

$$\begin{aligned} x_{s+1}(t) &= x_s(t) + y_s(t) \\ \sum_k a_{k,s+1}\phi_{k,s+1}(t) &= \sum_k a_{k,s}\phi_{k,s}(t) + \sum_k w_{k,s}\psi_{k,s}(t). \end{aligned} \tag{7.34}$$

We substitute the decomposition relation

$$\phi(2^{s+1}t - \ell) = \sum_k \left\{ h_0\,[2k - \ell]\,\phi(2^s t - k) + h_1\,[2k - \ell]\,\psi(2^s t - k) \right\} \tag{7.35}$$

into (7.34) to yield an equation in which all bases are at resolution s. After interchanging the order of summations and comparing the coefficients of $\phi_{k,s}(t)$ and $\psi_{k,s}(t)$ on both sides of the equation, we obtain

$$a_{k,s} = \sum_\ell h_0\,[2k - \ell]\,a_{\ell,s+1}$$

$$w_{k,s} = \sum_\ell h_1\,[2k - \ell]\,a_{\ell,s+1}$$

where the right side of the equations corresponds to decimation by 2 after convolution (see Section 7.1.3). These formulas relate the coefficients of the scaling functions and wavelets at any scale to coefficients at the next higher scale. By repeating this algorithm, one obtains signal components at various frequency octaves. This algorithm is depicted in Figure 7.8, where we have used the vector notation

$$\mathbf{a}_s := \{a_{k,s}\}, \quad \mathbf{w}_s := \{w_{k,s}\}, \quad \mathbf{h}_0 := \{h_0[k]\}, \quad \text{and} \quad \mathbf{h}_1 := \{h_1[k]\} \tag{7.36}$$

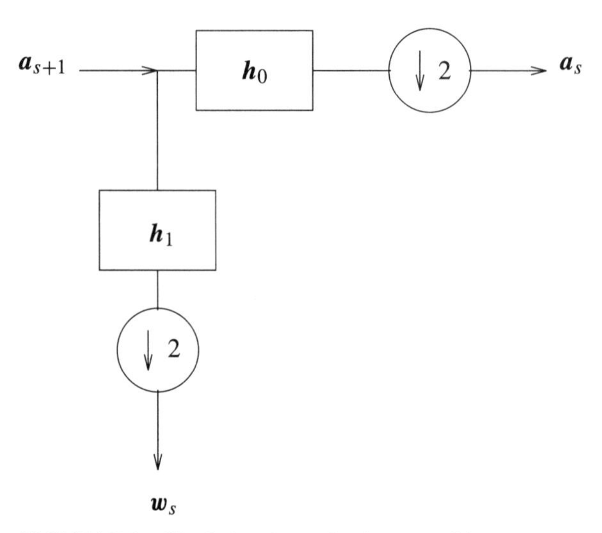

FIGURE 7.8 Single-level wavelet decomposition process.

with $k \in \mathbb{Z}$. This decomposition block can be applied repeatedly to the scaling function coefficients at lower resolution to build a wavelet decomposition tree as shown in Figure 7.9.

The reader should note that the wavelet decomposition tree is not symmetric since only the scaling function coefficients are further "decomposed" to obtain signal components at lower resolutions. A symmetric tree may be constructed by decomposing the wavelet coefficients as well. This is the wavelet packet tree discussed in Chapter 8.

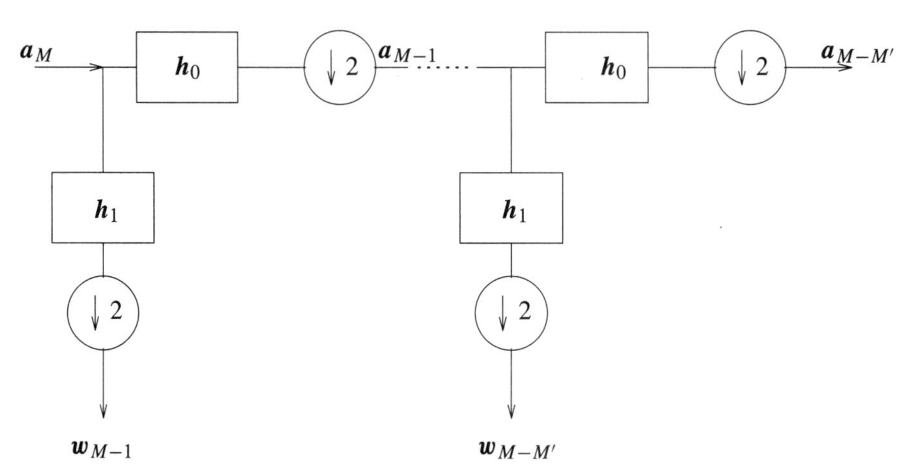

FIGURE 7.9 Wavelet decomposition tree.

7.4 RECONSTRUCTION ALGORITHM

It is important for any transform to have a unique inverse such that the original data can be recovered perfectly. For random signals, some transforms have their unique inverses in theory but cannot be implemented in reality. There exists a unique inverse discrete wavelet transform (or the synthesis transform) such that the original function can be recovered perfectly from its components at different scales. The reconstruction algorithm is based on the two-scale relations of the scaling function and the wavelet. We consider a sum of these components at the sth resolution:

$$x_s(t) + y_s(t) = \sum_k a_{k,s}\phi_{k,s}(t) + \sum_k w_{k,s}\psi_{k,s}(t) = x_{s+1}(t). \tag{7.37}$$

By a substitution of the two-scale relations into (7.37), one obtains

$$\begin{aligned}\sum_k a_{k,s} \sum_\ell g_0[\ell]\phi(2^{s+1}t - 2k - \ell) + \sum_k w_{k,s} \sum_\ell g_1[\ell]\phi(2^{s+1}t - 2k - \ell) \\ = \sum_\ell a_{\ell,s+1}\phi(2^{s+1}t - \ell).\end{aligned} \tag{7.38}$$

Comparing the coefficients of $\phi(2^{s+1}t - \ell)$ on both sides of (7.38) yields

$$a_{\ell,s+1} = \sum_k \left\{g_0[\ell - 2k]a_{k,s} + g_1[\ell - 2k]w_{k,s}\right\}, \tag{7.39}$$

where the right side of the equations corresponds to interpolation followed by convolution, as discussed in Section 7.1.4. The reconstruction algorithm of (7.39) is shown graphically in Figure 7.10.

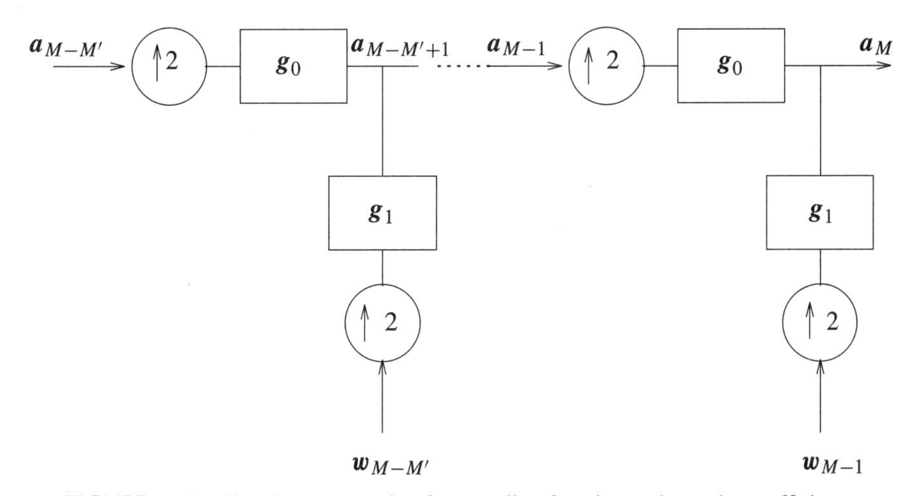

FIGURE 7.10 Signal reconstruction from scaling function and wavelet coefficients.

We emphasize here that although the mechanics of computation are carried out in digital signal processing fashion, the decomposition and reconstruction algorithms are actually processing analog signals. The fundamental idea is to represent an analog signal by its components at different scales for efficient processing.

7.5 CHANGE OF BASES

The algorithms discussed in Section 7.4 apply to all types of scaling functions and wavelets, including orthonormal, semiorthogonal, and biorthogonal systems. We have seen in Chapter 6 that the processing sequences $\{g_0[k], g_1[k]\}$, and $\{h_0[k], h_1[k]\}$ are finite and equilength sequences for compactly supported orthonormal wavelets. In the case of semiorthogonal wavelets such as compactly supported B-spline wavelets, the processing sequences $\{h_0[k], h_1[k]\}$ are infinitely long. Truncation of the sequences is necessary for efficient processing. To avoid using the infinite sequences, it is better to map the input function into the dual spline space and process the dual spline coefficients with $g_0[k]$ and $g_1[k]$ that have finite lengths. This and the next two sections are devoted to modification of the algorithm via a change of bases.

We have shown in Chapter 6 that the mth-order spline $\phi_m = N_m$ and the corresponding compactly supported spline wavelets ψ_m are semiorthogonal bases. To compute the expansion coefficients of a spline series or a spline wavelet series, it is necessary to make use of the dual spline $\widetilde{\phi}_m$ or the dual spline wavelet $\widetilde{\psi}_m$. In semiorthogonal spaces, all these bases span the same spline space $\mathbf{S}_m$. For certain real-time applications in wavelet signal processing, it is more desirable to use finite-length decomposition sequences for efficiency and accuracy. Therefore, it is necessary to represent the input signal by dual splines of the same order before the decomposition process.

Let us recall the formulation of the multiresolution analysis, in which we have the approximation subspace as an orthogonal sum of the wavelet subspaces,

$$\begin{aligned} \mathbf{A}_M &= \oplus_{s=M-M'}^{M-1} \mathbf{W}_s \oplus \mathbf{A}_{M-M'} \\ &= \mathbf{W}_{M-1} \oplus \mathbf{W}_{M-2} \oplus \cdots \oplus \mathbf{W}_{M-M'} \oplus \mathbf{A}_{M-M'} \end{aligned} \tag{7.40}$$

for any positive integer M'. Consequently, any good approximant $x_M \in \mathbf{A}_M$ of a given function $x \in L^2$ (for sufficiently large M) has a unique (orthogonal) decomposition

$$x_M = \sum_{n=1}^{M'} y_{M-n} + x_{M-M'}, \tag{7.41}$$

where $x_s \in \mathbf{A}_s$ and $y_s \in \mathbf{W}_s$. Since ϕ_m and $\widetilde{\phi}_m$ generate the same MRA while ψ_m and $\widetilde{\psi}_m$ generate the same wavelet subspace (a property not possessed by biorthogonal scaling functions and wavelets that are not semiorthogonal), we write

$$\begin{aligned} x_s(t) &= \sum_k a_{k,s}\phi(2^s t - k) = \sum_k \widetilde{a}_{k,s}\widetilde{\phi}(2^s t - k) \\ y_s(t) &= \sum_k w_{k,s}\psi(2^s t - k) = \sum_k \widetilde{w}_{k,s}\widetilde{\psi}(2^s t - k) \end{aligned} \tag{7.42}$$

for each $s \in \mathbb{Z}$. To simplify the implementation, we have not included the normalization factor $2^{s/2}$.

If we apply the decomposition formula (7.36) to the scaling function coefficients, we have

$$\begin{aligned} a_{k,s} &= \sum_\ell h_0\,[2k - \ell]\, a_{\ell,s+1} \\ w_{s,k} &= \sum_\ell h_1\,[2k - \ell]\, a_{\ell,s+1}. \end{aligned} \tag{7.43}$$

Since sequences $\{h_0[k]\}$ and $\{h_1[k]\}$ are infinitely long for semiorthogonal setting, it will be more efficient to use sequences $\{g_0[k]\}$ and $\{g_1[k]\}$ instead. This change of sequences is valid from the duality principle, which states that $\{g_0[k], g_1[k]\}$ and $\{h_0\,[k]\,, h_1\,[k]\}$ can be interchanged, in the sense that

$$\begin{aligned} \tfrac{1}{2}g_0[k] &\leftrightarrow h_0\,[-k] \\ \tfrac{1}{2}g_1[k] &\leftrightarrow h_1\,[-k] \end{aligned} \tag{7.44}$$

when ϕ_m and ψ_m are replaced by $\widetilde{\phi}_m$ and $\widetilde{\psi}_m$. With the application of the duality principle, we have

$$\begin{aligned} \widetilde{a}_{k,s} &= \sum_\ell g_0[\ell - 2k]\widetilde{a}_{\ell,s+1} \\ \widetilde{w}_{k,s} &= \sum_\ell g_1[\ell - 2k]\widetilde{a}_{\ell,s+1}. \end{aligned} \tag{7.45}$$

However, to take advantage of the duality principle, we need to transform the coefficients $\{a_{k,s}\}$ to $\{\widetilde{a}_{k,s}\}$. We recall that both ϕ and $\widetilde{\phi}$ generate the same $\mathbf{A}_s$ space, so that ϕ can be represented by a series of $\widetilde{\phi}$:

$$\phi(t) = \sum_k r_k \widetilde{\phi}(t - k) \tag{7.46}$$

for some sequence $\{r_k\}$. We observe that this change-of-bases sequence is a finite sequence if the scaling function has compact support. Indeed, by the definition of the dual, we have

$$r_k = \int_{-\infty}^{\infty} \phi(t)\phi(t - k)\, dt. \tag{7.47}$$

Therefore, at the original scale of approximation, with $s = M$, application of (7.46) yields

$$\widetilde{a}_{k,M} = \sum_\ell r_{k-\ell}\, a_{\ell,M}, \tag{7.48}$$

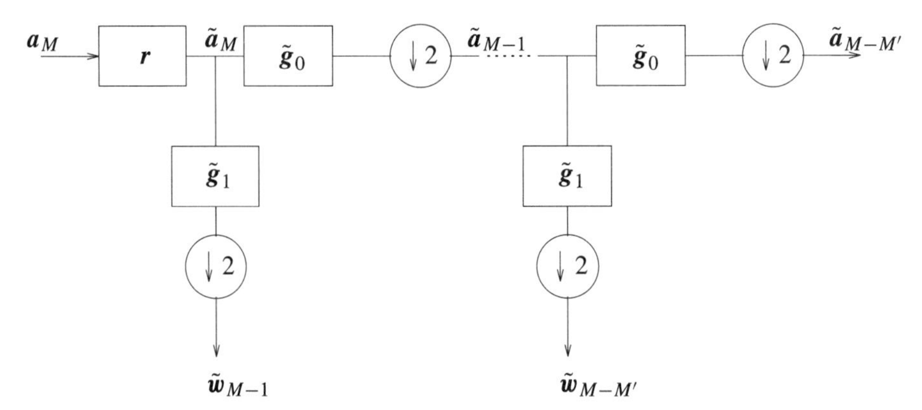

FIGURE 7.11 Standard wavelet decomposition process implemented with change of bases. (Reprinted with permission from [1], copyright © 1995 by Springer-Verlag.)

which is an FIR operation. Observe that if we take splines as scaling functions [i.e., $\phi(t) = N_m(t)$], then $r_k = N_{2m}(m-k); k = 0, \pm 1, \ldots, \pm m - 1$ [1]. As we have seen in previous discussions, the sequences $\{g_0[k]\}$ and $\{g_1[k]\}$ in the decomposition algorithm are finite sequences.

We can summarize our computation scheme as in Figure 7.11. The computation of $\widetilde{w}_{k,s}, s = M-1, \ldots, M-M'$, using $\mathbf{a}_M$ as the input sequence requires $2M'$ FIR filters. The importance of the coefficients $\widetilde{w}_{k,s}$ is that they constitute the CWT of x_M relative to the analyzing wavelet ψ_m at certain dyadic points, namely

$$\widetilde{w}_{k,s} = 2^{s/2} \left(W_\psi x_M \right) \left(\frac{k}{2^s}, \frac{1}{2^s} \right), \qquad M - M' \le s < M, \quad k \in \mathbb{Z}. \tag{7.49}$$

7.6 SIGNAL RECONSTRUCTION IN SEMIORTHOGONAL SUBSPACES

The algorithm described in Section 7.4 concerns the recovery of the original data. In that case, the original data are the set of scaling function coefficients $\{a_{\ell,M}\}$ at the highest resolution. Since the original input signal is an analog function $x(t) \approx x_M(t) = \sum_\ell a_{\ell,M} \phi(2^M t - \ell)$, it is necessary to recover the signal by performing the summation. Recall that the decomposition algorithm discussed in Section 7.5 produces the spline and wavelet coefficients in the dual spaces, namely $(\{\widetilde{a}_{k,s}\}, \{\widetilde{w}_{k,s}\})$. To use finite-length two-scale sequences for the reconstruction, we must express the coefficients in dual spaces in terms of $(\{a_{k,s}\}, \{w_{k,s}\})$ in the spline and wavelet spaces. In addition, if users need to see the signal component at any intermediate steps in the decomposition, they would have to use the dual spline and dual wavelet series. In both cases one can simplify the problem by a change of basis that maps the dual sequences back to the original space [2]. Since the sequences do not depend on the scale, the second subscript of the coefficients can be arbitrary. Such sequences are applicable to mapping between any two different scales.

7.6.1 Change of Basis for Spline Functions

Our objective is to write

$$s(t) = \sum_k \widetilde{a}_k \widetilde{N}_m(t-k) = \sum_k a_k N_m(t-k). \tag{7.50}$$

By taking the Fourier transform of (7.50), we get

$$\widetilde{A}(e^{j\omega})\hat{\widetilde{N}}_m(\omega) = A(e^{j\omega})\hat{N}_m(\omega), \tag{7.51}$$

where, as usual, the hat over a function implies its Fourier transform, and $A(e^{j\omega})$ and $\widetilde{A}(e^{j\omega})$ are the symbols of $\{a_k\}$ and $\{\widetilde{a}_k\}$, respectively, defined as

$$\widetilde{A}(e^{j\omega}) := \sum_k \widetilde{a}_k e^{jk\omega}, \qquad A(e^{j\omega}) := \sum_k a_k e^{jk\omega}. \tag{7.52}$$

The dual scaling function $\widetilde{N}_m$ is given by

$$\hat{\widetilde{N}}_m(\omega) = \frac{\hat{N}_m(\omega)}{E_{N_m}(z^2)}, \quad z = e^{j\omega/2}, \tag{7.53}$$

where $E_{N_m}(z^2) = |\hat{N}_m(\omega + 2\pi k)|^2 \neq 0$ for almost all ω since $\{N_m(\cdot - k)\}$ is a stable or Riesz basis of $\mathbf{A}_0$. As discussed in Chapter 6, $E_{N_m}(\omega)$ is the Euler–Frobenius–Laurent series and is given by

$$\begin{aligned} E_{N_m}(z^2) &= |\hat{N}_m(\omega + 2\pi k)|^2 \\ &= \sum_{k=-m+1}^{m-1} N_{2m}(m+k)z^{2k}. \end{aligned} \tag{7.54}$$

It is clear that by multiplying (7.54) by z^{2m-2}, we can get a polynomial of degree $2m-1$ in z^2. The last equality in (7.54) is a consequence of the relation

$$\sum_{k=-\infty}^{\infty} |\hat{f}(\omega + 2\pi k)|^2 = \sum_{k=-\infty}^{\infty} \left[\int_{-\infty}^{\infty} f(t+k)\overline{f(t)}\, dt \right] e^{jk\omega}. \tag{7.55}$$

PROOF FOR (7.55): Using Parseval's identity, we have

$$\begin{aligned} F(\ell) &:= \int_{-\infty}^{\infty} f(t+\ell)\overline{f(t)}\, dt \\ &= \frac{1}{2\pi} \int_{-\infty}^{\infty} \left|\widehat{f}(\omega)\right|^2 e^{-j\omega\ell}\, d\omega \\ &= \frac{1}{2\pi} \sum_{k=-\infty}^{\infty} \int_{2k\pi}^{2\pi(k+1)} \left|\widehat{f}(\omega)\right|^2 e^{-j\omega\ell}\, d\omega \end{aligned}$$

$$= \frac{1}{2\pi} \int_0^{2\pi} \sum_{k=-\infty}^{\infty} \left| \widehat{f}(\omega + 2k\pi) \right|^2 e^{-j\omega\ell} \, d\omega. \tag{7.56}$$

It is clear the $F(\ell)$ is the ℓth Fourier coefficient of a 2π-periodic function $\sum_{k=-\infty}^{\infty} \left| \widehat{f}(\omega + 2k\pi) \right|^2$. With this relation, (7.55) follows directly. It is easy to show that

$$\int_{-\infty}^{\infty} N_m(t+k)\overline{N_m(t)}\, dt = N_{2m}(m+k) \tag{7.57}$$

with $\mathrm{supp} N_{2m}(t) = [0, 2m]$.

Combining (7.51), (7.53), and (7.54) and taking the inverse Fourier transform, we get

$$a_k = (\{\widetilde{a}_n\} * p[n])\,(k), \tag{7.58}$$

where

$$\frac{1}{E_{N_m}(z)} = \sum_k p[k] z^k, \qquad |z| = 1. \tag{7.59}$$

It can be shown that

$$p[k] = u_m \sum_{i=1}^{p_m} \Lambda_i \lambda_i^{k+p_m}, \qquad k \geq 0, \tag{7.60}$$

where

$$\Lambda_i = \frac{1}{\lambda_i \prod_{j=1, j\neq i}^{2p_m} (\lambda_i - \lambda_j)} \tag{7.61}$$

and $\lambda_i : i = 1, \ldots, 2p_m$ are the roots of (7.54) with $|\lambda_i| < 1$ and $\lambda_i \lambda_{2p_m+1-i} = 1$ for $i = 1, \ldots, p_m$. Here $u_m = (2m-1)!$ and $p_m = m - 1$. Observe from (7.54) and (7.59) that

$$\sum_k p[k] = \frac{1}{E_{N_m}(1)} = \frac{1}{\sum_{k=-m+1}^{m-1} N_{2m}(m+k)} = 1, \tag{7.62}$$

where the last equality is a consequence of the partition of unity property of cardinal B-splines, described in Chapter 5.

Roots λ_i for linear and cubic splines are given below. The coefficients $\{p[k]\}$ are given in Tables 7.1 and 7.2. The coefficients $p[k]$ have better decay than $\{h_0[k]\}$ (see Figure 7.12).

TABLE 7.1 Coefficients $\{p[k]\}$ for the Linear Spline Case ($p[k] = p[-k]$)

k	$p[k]$	k	$p[k]$
0	1.7320510	8	$0.46023608 \times 10^{-4}$
1	−0.46410170	9	$-0.12331990 \times 10^{-4}$
2	0.12435570	10	$0.33043470 \times 10^{-5}$
3	$-0.33321008 \times 10^{-1}$	11	$-0.88539724 \times 10^{-6}$
4	$0.89283381 \times 10^{-2}$	12	$0.23724151 \times 10^{-6}$
5	$-0.23923414 \times 10^{-2}$	13	$-0.63568670 \times 10^{-7}$
6	$0.64102601 \times 10^{-3}$	14	$0.17033177 \times 10^{-7}$
7	$-0.17176243 \times 10^{-3}$	15	$-0.45640265 \times 10^{-8}$

Linear spline ($m = 2$)*:*

$$\lambda_1 = -2 + \sqrt{3} = \frac{1}{\lambda_2} \tag{7.63}$$

$$p[k] = (-1)^{|k|}\sqrt{3}\left(2 - \sqrt{3}\right)^{|k|}. \tag{7.64}$$

Cubic spline ($m = 4$)*:*

$$\begin{aligned} \lambda_1 &= -9.1486946 \times 10^{-3} = \frac{1}{\lambda_6} \\ \lambda_2 &= -0.1225546 = \frac{1}{\lambda_5} \\ \lambda_3 &= -0.5352805 = \frac{1}{\lambda_4}. \end{aligned} \tag{7.65}$$

TABLE 7.2 Coefficients $\{p[k]\}$ for Cubic Spline Case ($p[k] = p[-k]$)

k	$p[k]$	k	$p[k]$
0	0.49647341	15	$-0.51056378 \times 10^{-3}$
1	−0.30910430	16	$0.27329483 \times 10^{-3}$
2	0.17079600	17	$-0.14628941 \times 10^{-3}$
3	−0.92078239	18	$0.78305879 \times 10^{-4}$
4	0.49367899	19	$-0.41915609 \times 10^{-4}$
5	−0.26435509	20	$0.22436609 \times 10^{-4}$
6	0.14151619	21	$-0.12009880 \times 10^{-4}$
7	$-0.75752318 \times 10^{-1}$	22	$0.64286551 \times 10^{-5}$
8	$0.40548921 \times 10^{-1}$	23	$-0.34411337 \times 10^{-5}$
9	$-0.21705071 \times 10^{-1}$	24	$0.18419720 \times 10^{-5}$
10	$0.11618304 \times 10^{-1}$	25	$-0.98597172 \times 10^{-6}$
11	$-0.62190532 \times 10^{-2}$	26	$0.52777142 \times 10^{-6}$
12	$0.33289378 \times 10^{-2}$	27	$-0.28250579 \times 10^{-6}$
13	$-0.17819155 \times 10^{-2}$	28	$0.15121984 \times 10^{-6}$
14	$0.95382473 \times 10^{-3}$	29	$-0.80945043 \times 10^{-7}$

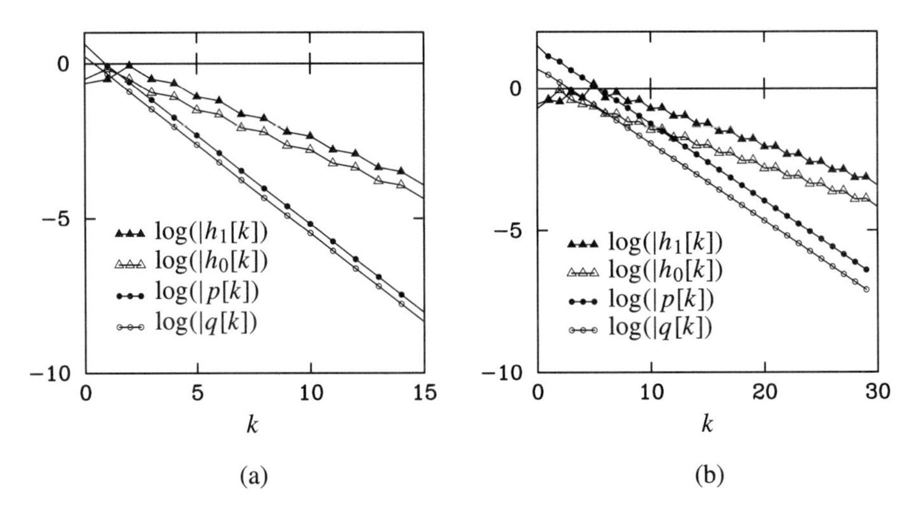

FIGURE 7.12 Plots of $h_0[k]$, $h_1[k]$, $p[k]$, and $q[k]$ versus k for (*a*) linear and (*b*) cubic spline cases.

7.6.2 Change of Basis for Spline Wavelets

Here our objective is to write

$$r(t) = \sum_k \widetilde{w}_k \widetilde{\psi}_m(t-k) = \sum_k w_k \psi_m(t-k). \tag{7.66}$$

Replacing N_m by ψ_m in (7.53), we get the relationship between ψ_m and $\widetilde{\psi}_m$. Proceeding in the same way as before, we get

$$w_k = (\{\widetilde{w}_n\} * q[n])(k), \tag{7.67}$$

where

$$\frac{1}{\sum_k |\hat{\psi}_m(\omega + 2\pi k)|^2} = \sum_k q[k] e^{-j\omega k}. \tag{7.68}$$

Furthermore, we have

$$\sum_k |\hat{\psi}_m(\omega + 2\pi k)|^2 = E_{N_m}(z^2) E_{N_m}(z) E_{N_m}(-z), \qquad |z| = 1. \tag{7.69}$$

PROOF FOR (7.69): With the help of a two-scale relation, we can write

$$\sum_k |\hat{\psi}_m(\omega + 2\pi k)|^2 = \sum_k \left| G_1 \left[\exp\left(j \frac{\omega + 2\pi k}{2} \right) \right] \hat{N}_m \left(\frac{\omega + 2\pi k}{2} \right) \right|^2 \tag{7.70}$$

with

$$G_1 e^{j\omega/2} = \frac{1}{2}\sum_k g_1[k]e^{jk\omega/2}. \tag{7.71}$$

Now separating the right-hand side of (7.70) into parts with even k and odd k and making use of the relation (7.54), we can write

$$\sum_k |\hat{\psi}_m(\omega + 2\pi k)|^2 = |G_1(z)|^2 E_{N_m}(z) + |G_1(-z)|^2 E_{N_m}(-z). \tag{7.72}$$

From the relation $|G_1(z)| = |G_0(-z)E_{N_m}(-z)|$, with $G_0(z)$ defined in a similar way as in (7.71) with $g_1[k]$ replaced by $g_0[k]$, we can write

$$\sum_k |\hat{\psi}_m(\omega + 2\pi k)|^2 = \left\{|G_0(-z)|^2\,\overline{E_{N_m}(-z)} + |G_0(z)|^2\,\overline{E_{N_m}(z)}\right\} \times E_{N_m}(z)E_{N_m}(-z). \tag{7.73}$$

Following the steps used to arrive at (7.72), it can be shown that

$$|G_0(-z)|^2\,\overline{E_{N_m}(-z)} + |G_0(z)|^2\,\overline{E_{N_m}(z)} = E_{N_m}(z^2). \tag{7.74}$$

which, together with (7.72), gives the desired relation (7.69).

The expression for $q[k]$ has the same form as that of $p[k]$ with $u_m = -[(2m-1)!]^3$, $p_m = 2m-2$, and λ_i being the roots of (7.69). Observe from (7.68) and (7.69) that

$$\sum_k q[k] = \frac{1}{E_{N_m}(-1)} \tag{7.75}$$

since $E_{N_m}(1) = 1$. Roots λ_i and $\sum_k q[k]$ for linear and cubic splines are given below. The coefficients are given in Tables 7.3 and 7.4. The coefficients $q[k]$ have better decay than $\{h_1[k]\}$ (see Figure 7.12).

TABLE 7.3 Coefficients $\{q[k]\}$ for the Linear Spline Case ($q[k] = q[-k]$)

k	$q[k]$	k	$q[k]$
0	4.3301268	8	$0.92047740 \times 10^{-4}$
1	−0.86602539	9	$-0.24663908 \times 10^{-4}$
2	0.25317550	10	$0.66086895 \times 10^{-5}$
3	$-0.66321477 \times 10^{-1}$	11	$-0.17707921 \times 10^{-5}$
4	$0.17879680 \times 10^{-1}$	12	$0.47448233 \times 10^{-6}$
5	$-0.47830273 \times 10^{-2}$	13	$-0.12713716 \times 10^{-6}$
6	$0.12821698 \times 10^{-2}$	14	$0.34066300 \times 10^{-7}$
7	$-0.34351606 \times 10^{-3}$	15	$-0.91280379 \times 10^{-8}$

TABLE 7.4 Coefficients $\{q[k]\}$ for the Cubic Spline Case ($q[k] = q[-k]$)

k	q_k	k	q_k
0	33.823959	18	$0.39035085 \times 10^{-3}$
1	−13.938340	19	$-0.20894629 \times 10^{-3}$
2	9.0746698	20	$0.11184511 \times 10^{-3}$
3	−4.4465132	21	$-0.59868424 \times 10^{-4}$
4	2.5041881	22	$0.32046413 \times 10^{-4}$
5	−1.3056690	23	$-0.17153812 \times 10^{-4}$
6	0.70895731	24	$0.91821012 \times 10^{-5}$
7	−0.37662071	25	$-0.49149990 \times 10^{-5}$
8	0.20242150	26	$0.26309024 \times 10^{-5}$
9	−0.10811640	27	$-0.14082705 \times 10^{-5}$
10	$0.57940185 \times 10^{-1}$	28	$0.75381962 \times 10^{-6}$
11	$-0.30994879 \times 10^{-1}$	29	$-0.40350486 \times 10^{-6}$
12	$0.16596500 \times 10^{-1}$	30	$0.21598825 \times 10^{-6}$
13	$-0.88821910 \times 10^{-2}$	31	$-0.11561428 \times 10^{-6}$
14	$0.47549186 \times 10^{-2}$	32	$0.61886055 \times 10^{-7}$
15	$-0.25450843 \times 10^{-2}$	33	$-0.33126394 \times 10^{-7}$
16	$0.13623710 \times 10^{-2}$	34	$0.17731910 \times 10^{-7}$
17	$-0.72923984 \times 10^{-3}$	35	$-0.94915444 \times 10^{-8}$

Linear spline ($m = 2$):

$$\begin{aligned} \lambda_1 &= 7.1796767 \times 10^{-2} = \frac{1}{\lambda_4} \\ \lambda_2 &= -0.2679492 = \frac{1}{\lambda_3} \end{aligned} \tag{7.76}$$

$$\sum_k q[k] = 3.0. \tag{7.77}$$

Cubic spline ($m = 4$):

$$\begin{aligned} \lambda_1 &= 8.3698615 \times 10^{-5} = \frac{1}{\lambda_{12}} \\ \lambda_2 &= -9.1486955 \times 10^{-3} = \frac{1}{\lambda_{11}} \\ \lambda_3 &= 1.5019634 \times 10^{-2} = \frac{1}{\lambda_{10}} \\ \lambda_4 &= -0.1225546 = \frac{1}{\lambda_9} \\ \lambda_5 &= 0.2865251 = \frac{1}{\lambda_8} \end{aligned} \tag{7.78}$$

$$\lambda_6 = -0.5352804 = \frac{1}{\lambda_7}$$

$$\sum_k q[k] = 18.5294121.$$

7.7 EXAMPLES

In Figure 7.13 we have shown decomposition of a music signal with some additive noise. Here the music data are considered to be at integer points. Intermediate approximate functions s_j and detail functions r_j have been plotted after mapping the

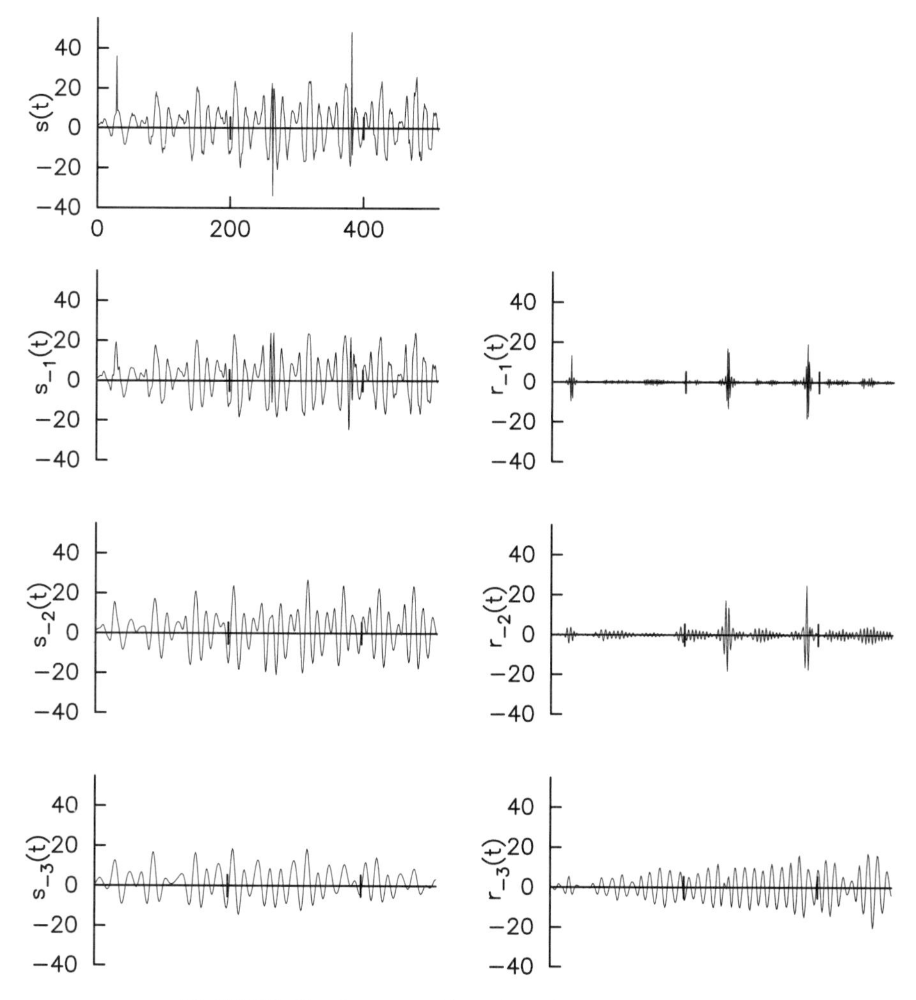

FIGURE 7.13 Decomposition of music signal with noise.

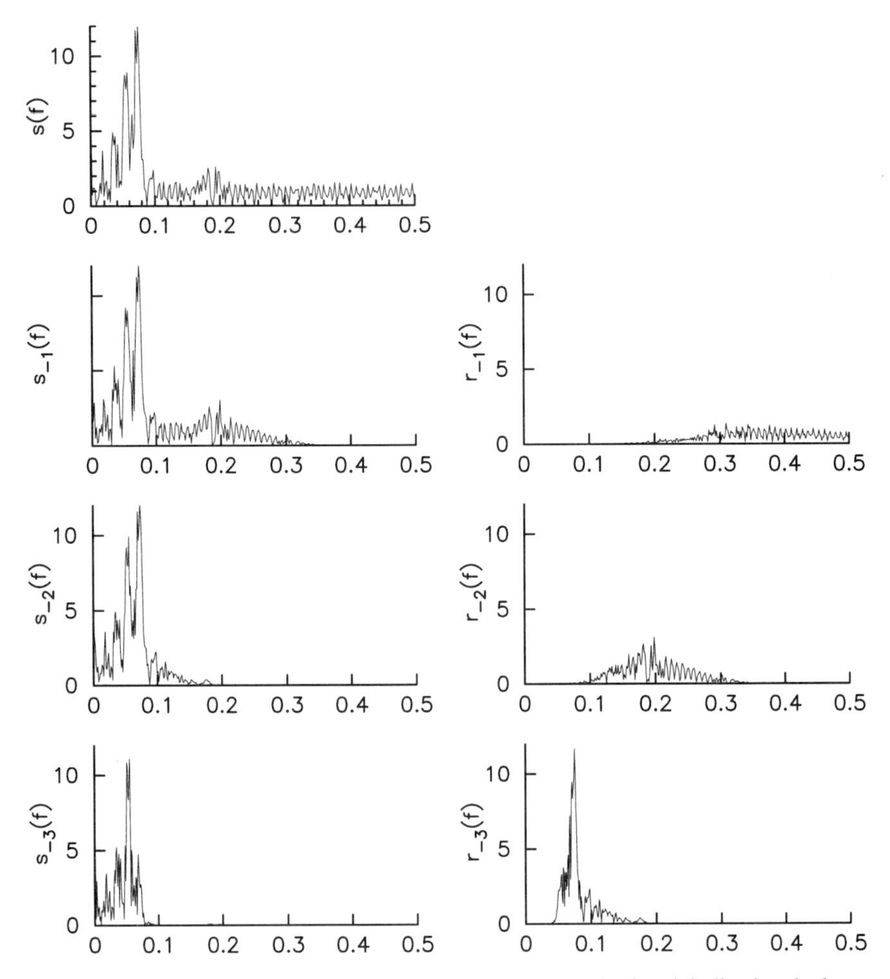

FIGURE 7.14 Magnitude spectrum of the decomposed music signal, indicating the lowpass and bandpass filter characteristics of scaling functions and wavelets, respectively.

dual spline and wavelet coefficients into the original space with the help of coefficients $p[k]$ and $q[k]$ derived in this chapter. To illustrate the lowpass and bandpass characteristics of splines and wavelets, respectively, in Figure 7.14 we show the magnitude spectra of the decomposed signals at various scales. The reconstruction process is shown in Figure 7.15 using the same sequences $(\{g_0[k]\}, \{g_1[k]\})$ as were used for the decomposition. The original signal $s(t)$ is also plotted next to the reconstructed signal $s_0(t)$ for the purpose of comparison.

To further expound the process of separating a complicated function into several simple ones with the help of wavelet techniques, we consider a function composed of three sinusoids with different frequencies. These frequencies are chosen such that

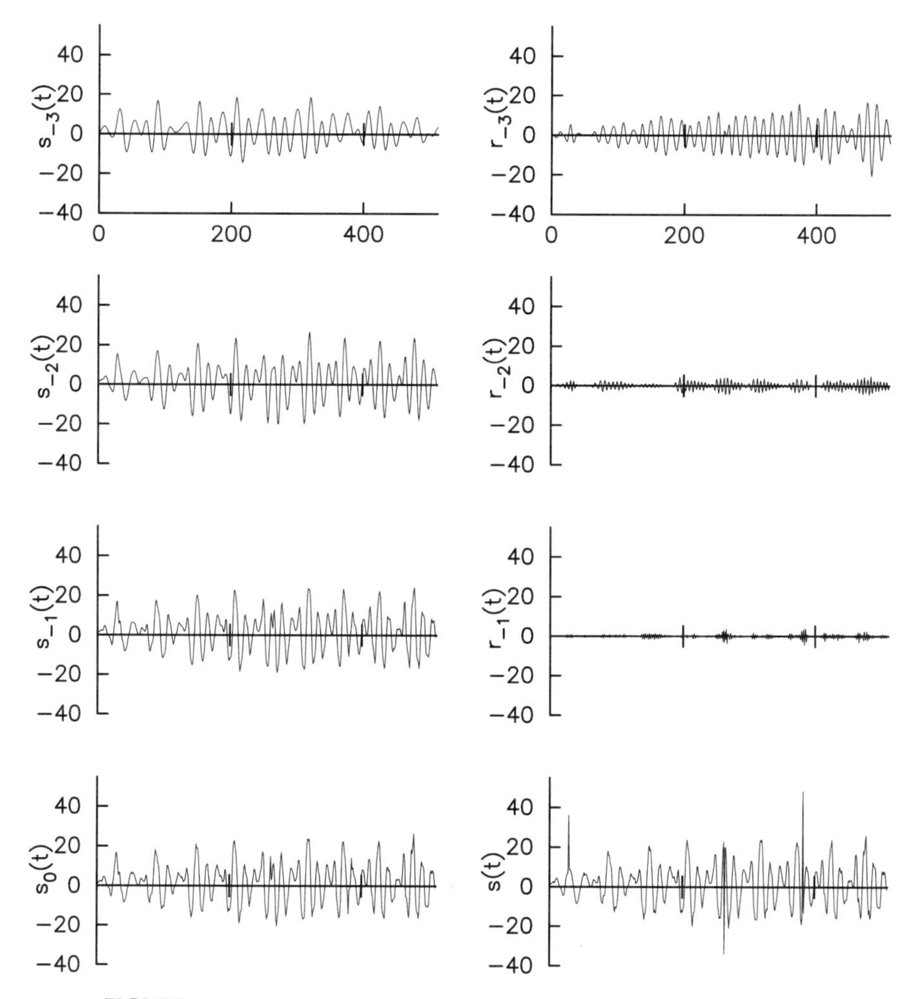

FIGURE 7.15 Reconstruction of the music signal after removing the noise.

they correspond to octave scales. As can be seen from Figures 7.16 and 7.17, standard wavelet decomposition separates the frequency components fairly well.

7.8 TWO-CHANNEL PERFECT RECONSTRUCTION FILTER BANK

Many applications in digital signal processing require multiple bandpass filters to separate a signal into components whose spectra occupy different segments of the frequency axis. Examples of these applications include a filter bank for Doppler frequencies in radar signal processing and tonal equalizer in music signal processing. Figure 7.18 demonstrates the concept of multiband filtering. In this mode of multi-

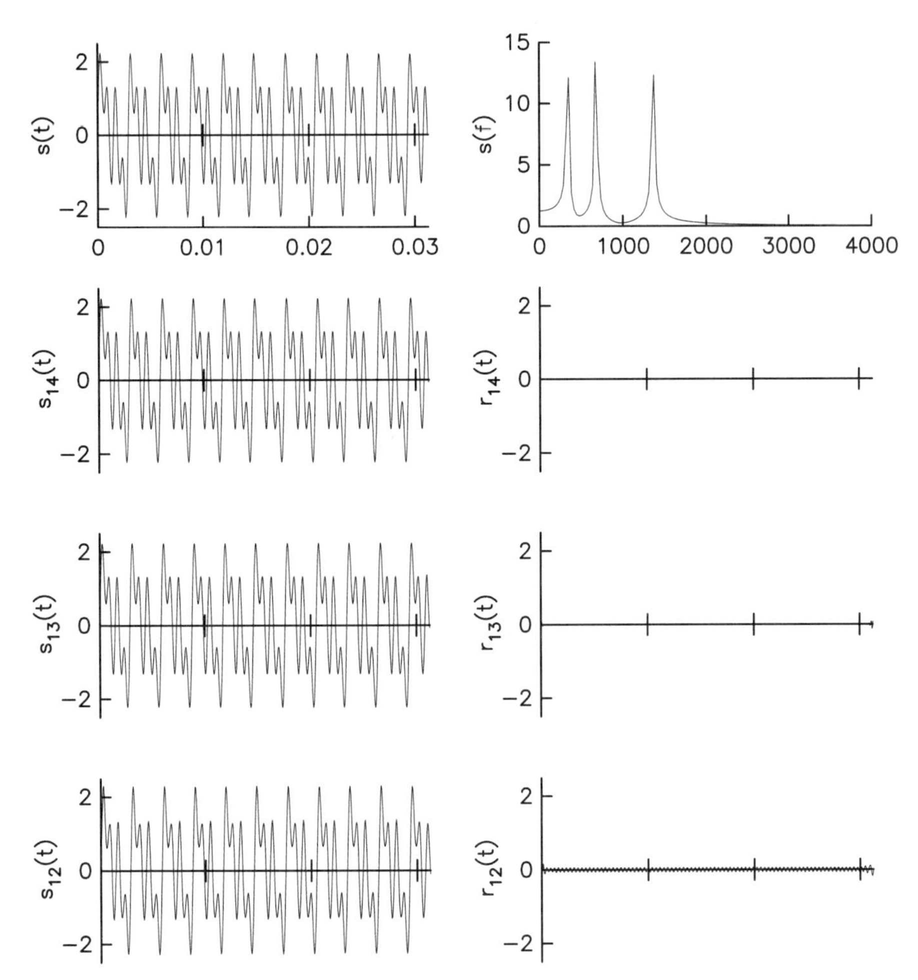

FIGURE 7.16 Decomposition of a signal composed of three sinusoids with different frequencies corresponding to octave scales.

band filtering, the spectral bands corresponding to components of the signal may be processed with a different algorithm to achieve a desirable effect on the signal. For Doppler processing and tonal equalizers, there is no need to reconstruct the original signal from the components processed. However, there is another form of filtering that requires the original signal to be recovered from its component: the subband filter bank. The major application of subband filtering is in signal compression, in which the subband components are coded for archiving or transmission purposes. The original signal can be recovered from the coded components with various degrees of fidelity.

We use a basic two-channel perfect reconstruction (PR) filter bank to illustrate the main features of this algorithm. Filter bank tree structures can be constructed using

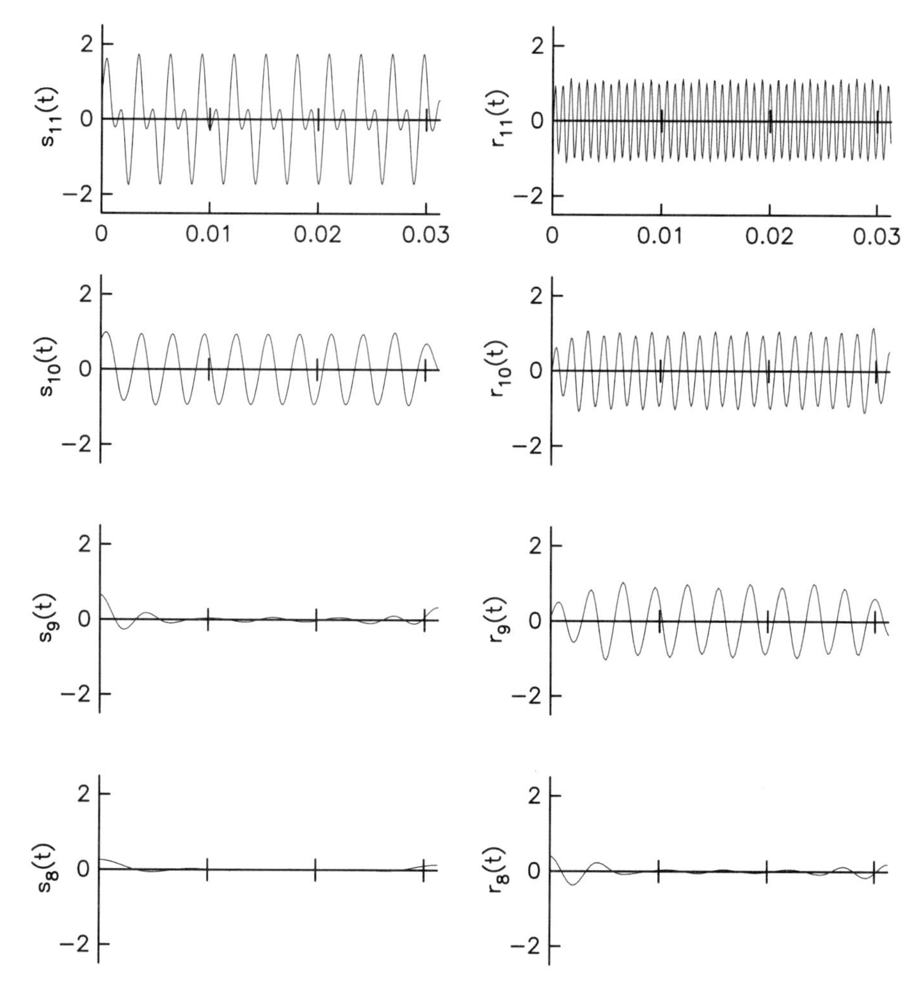

FIGURE 7.17 Decomposition of a signal with three frequency components (continued from Figure 7.16).

this basic two-channel filter bank. A two-channel filter bank consists of an analysis section and a synthesis section, each consisting of two filters. The analysis section includes a highpass and a lowpass filter that are complementary to each other so that information in the input signal is processed by either one of the two filters. The block diagram for a two-channel PR filter bank is shown in Figure 7.19.

The perfect reconstruction condition is an important condition in filter bank theory. It establishes the unique relationship between the lowpass and highpass filters of the analysis section. Removal of the aliasing caused by decimation defines the relations between analysis and synthesis filters. We elaborate on these conditions in much greater detail below.

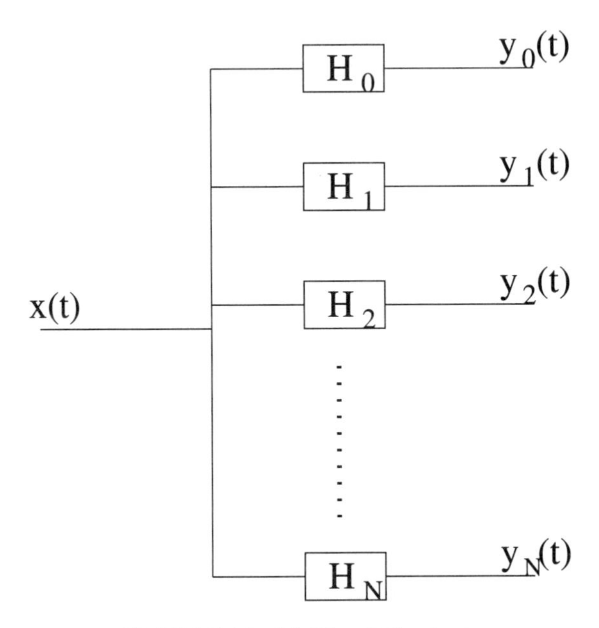

FIGURE 7.18 Multiband filter bank.

The filters in a two-channel PR filter bank are specially designed so that the component signals may be reconstructed perfectly with no loss of information. The output of the filter bank is simply a delayed version of the input signal. For a two-channel filter bank, the filtering operation is exactly the same as the wavelet algorithm. Because of the PR condition and the need to remove the aliasing components in the output, one needs to design only one of the four filters. For further detail on filter banks and how they relate to wavelet theory, readers are referred to [3–7].

7.8.1 Spectral-Domain Analysis of a Two-Channel PR Filter Bank

Let a discrete signal $X(z)$ be the input to a two-channel PR filter bank as shown in Figure 7.19 in terms of z-transforms with intermediate output signals. The analysis

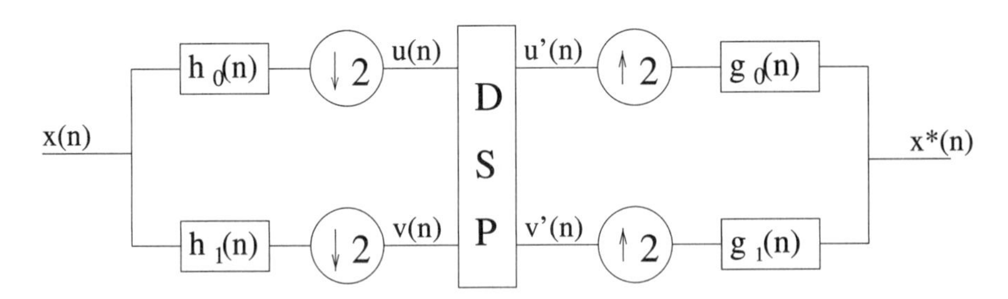

FIGURE 7.19 Two-channel perfect reconstruction filter bank.

section of the filter bank consists of a lowpass filter $H_0(z)$ and a highpass filter $H_1(z)$. The convolved output of the lowpass filter $H_0(z)$ followed by a two-point decimation $(\downarrow 2)$ is

$$U(z) = \frac{1}{2}[X(z^{1/2})H_0(z^{1/2}) + X(-z^{1/2})H_0(-z^{1/2})], \tag{7.79}$$

while the highpass filter $H_1(z)$ with decimation yields

$$V(z) = \frac{1}{2}[X(z^{1/2})H_1(z^{1/2}) + X(-z^{1/2})H_1(-z^{1/2})]. \tag{7.80}$$

For analysis purposes, we assume that the outputs of the analysis bank are not processed, so that the outputs of the processor labeled $U'(z)$ and $V'(z)$ are

$$U'(z) = U(z)$$
$$V'(z) = V(z).$$

After the interpolator $(\uparrow 2)$ and the synthesis filter bank $G_0(z)$ and $G_1(z)$, the outputs of the filters are

$$U''(z) = \frac{1}{2}[X(z)H_0(z)G_0(z) + X(-z)H_0(-z)G_0(z)] \tag{7.81}$$

and

$$V''(z) = \frac{1}{2}[X(z)H_1(z)G_1(z) + X(-z)H_1(-z)G_1(z)]. \tag{7.82}$$

These outputs are combined synchronously so that the processed output $X^*(z)$ is

$$\begin{aligned} X^*(z) &= U''(z) + V''(z) \\ &= \frac{1}{2}X(z)[H_0(z)G_0(z) + H_1(z)G_1(z)] \\ &\quad + \frac{1}{2}X(-z)[H_0(-z)G_0(z) + H_1(-z)G_1(z)]. \end{aligned} \tag{7.83}$$

The second term of the expression contains the alias version of the input signal [one that contains $X(-z)$]. For perfect reconstruction, we may choose the filters $G_0(z)$ and $G_1(z)$ to eliminate the aliasing component. We obtain the aliasing-free condition for the filter bank

$$\begin{aligned} G_0(z) &= \pm H_1(-z) \\ G_1(z) &= \mp H_0(-z). \end{aligned} \tag{7.84}$$

Once the analysis filters have been designed, the synthesis filters are determined automatically. Choosing the upper signs in (7.84), the output of the filter bank becomes

$$X^*(z) = \frac{1}{2} X(z)[H_0(z)H_1(-z) - H_1(z)H_0(-z)]. \tag{7.85}$$

The perfect reconstruction condition requires that $X^*(z)$ can only be a delayed version of the input $X(z)$ [i.e., $X^*(z) = X(z)z^{-m}$ for some integer m]. We obtain the following relations:

$$\begin{aligned} H_0(z)G_0(z) + H_1(z)G_1(z) &= H_0(z)H_1(-z) - H_1(z)H_0(-z) && (7.86)\\ &= H_0(z)G_0(z) - H_0(-z)G_0(-z) && (7.87)\\ &= 2z^{-m}. && (7.88)\end{aligned}$$

We define the transfer function of the filter bank:

$$\begin{aligned} T(z) &= \frac{X^*(z)}{X(z)} = \frac{1}{2}\left[H_0(z)G_0(z) + H_1(z)G_1(z)\right]\\ &= z^{-m}. \end{aligned}$$

To simplify the analysis, let us also define composite filters $C_0(z)$ and $C_1(z)$ as product filters for the two filtering paths

$$\begin{aligned} C_0(z) &= H_0(z)G_0(z) = -H_0(z)H_1(-z)\\ C_1(z) &= H_1(z)G_1(z) = H_1(z)H_0(-z)\\ &= -H_0(-z)G_0(-z)\\ &= -C_0(-z), \end{aligned} \tag{7.89}$$

where we have made use of the aliasing-free condition. In terms of the composite filters, the PR condition becomes

$$C_0(z) - C_0(-z) = 2z^{-m} \tag{7.90}$$

and

$$T(z) = \frac{1}{2}\left[C_0(z) - C_0(-z)\right]. \tag{7.91}$$

If we design a composite filter $C_0(z)$ that satisfies the condition in (7.90), the analysis filters $H_0(z)$ and $G_0(z)$ can be obtained through spectral factorization. We will have numerical examples to demonstrate this procedure in later sections.

We note that the transfer function $T(z)$ is an odd function since

$$\begin{aligned} T(-z) &= \frac{1}{2}\left[C_0(-z) - C_0(z)\right]\\ &= -T(z). \end{aligned} \tag{7.92}$$

The integer m in (7.90) must be odd, which implies that $C_0(z)$ must contain only even-indexed coefficients except $c_m = 1$, where m is odd. Finding $H_0(z)$ and $H_1(z)$

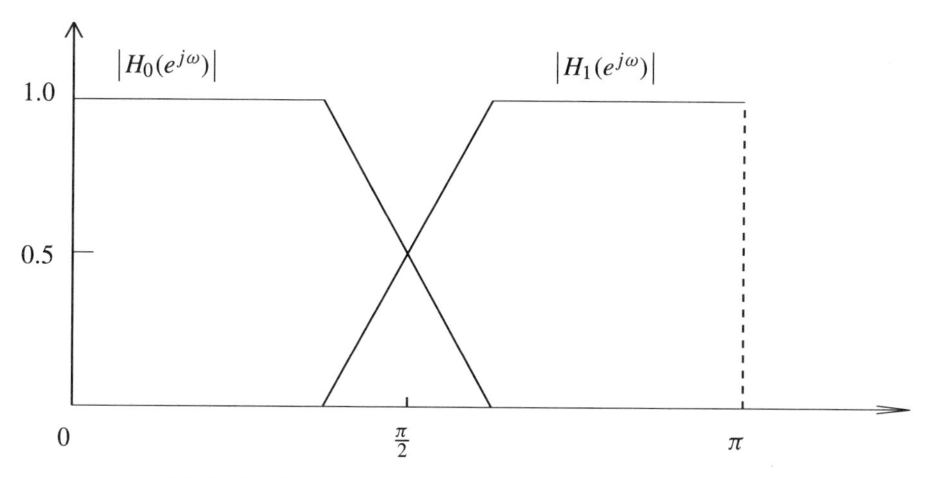

FIGURE 7.20 Spectral characteristic of quadrature mirror filter.

[or $H_0(z)$ and $G_0(z)$] to meet the PR requirement is the subject of filter bank design. Two basic approaches emerged in the early development of PR filter bank theory: (1) the quadrature mirror filter (QMF) approach, and (2) the half-band filter (HBF) approach. In this section we discuss the fundamental ideas in these two approaches.

Quadrature Mirror Filter Approach Let us choose $H_1(z) = H_0(-z)$. We have, in the spectral domain,

$$\begin{aligned} H_1(e^{j\omega}) &= H_0(-e^{j\omega}) \\ &= H_0(e^{j(\omega+\pi)}). \end{aligned} \tag{7.93}$$

The spectrum of the highpass filter $H_1(e^{j\omega})$ is the mirror image of that of the lowpass filter with the spectral crossover point at $\omega = \pi/2$, as shown in Figure 7.20. The transfer function becomes

$$T(z) = \frac{1}{2}[H_0^2(z) - H_1^2(z)] = \frac{1}{2}[H_0^2(z) - H_0^2(-z)] = z^{-m}. \tag{7.94}$$

Suppose that $H_0(z)$ is a linear phase† FIR filter of order N, so that

†A function $f \in L^2(\mathbb{R})$ has *linear phase* if

$$\widehat{f}(\omega) = \pm|\widehat{f}(\omega)|e^{-ja\omega},$$

where a is some real constant. The function f has *generalized linear phase* if

$$\widehat{f}(\omega) = \widehat{g}(\omega)e^{-ja\omega+b},$$

where $\widehat{g}(\omega)$ is a real-valued function and constants a and b are also real-valued. To avoid distortion in signal reconstruction, a filter must have a linear or generalized linear phase.

$$H_0(e^{j\omega}) = e^{-j(\omega/2)(N-1)} \left| H_0(e^{j\omega}) \right|$$

$$H_1(e^{j\omega}) = e^{-j[(\omega+\pi)/2](N-1)} \left| H_0(e^{j(\omega+\pi)}) \right|.$$

The spectral response of the transfer function becomes

$$T(e^{j\omega}) = \frac{1}{2} e^{-j(\omega/2)(N-1)} \left[\left| H_0(e^{j\omega}) \right|^2 - (-1)^{N-1} \left| H_1(e^{j\omega}) \right|^2 \right]. \tag{7.95}$$

If $(N-1)$ is even, $Te^{j\omega} = 0$ at the crossover point $\omega = \pi/2$! The transfer function produces severe amplitude distortion at this point, and that violates the PR requirement. Therefore, N must be even. If we wish to eliminate all amplitude distortion for even N, we must have the spectral amplitude of $H_0(z)$ and $H_1(z)$, satisfying

$$\left| H_0(e^{j\omega}) \right|^2 + \left| H_1(e^{j\omega}) \right|^2 = 2. \tag{7.96}$$

Observe that the condition (7.96) differs from the normalized form by a factor of 2 on the right-hand side. This happens because previously we used a normalizing factor in the definition of a z-transform with two-scale and decomposition sequences.

The trivial solution to (7.96) is the sine and cosine function for $H_0(e^{j\omega})$ and $H_1(e^{j\omega})$, which contradict our initial assumption for an FIR filter. Any nontrivial linear-phase FIR filter H_0 causes amplitude distortion. If the right-hand side of (7.96) is normalized to unity, the type of filter that satisfies this normalization is the *power complementary filter*, an IIR filter that can be used in IIR-PR filter banks.

Returning to (7.94), if we restrict the filters to be FIR, $H_0(z)$ can have at most two coefficients, so that $H_0^2(z)$ has only one term with an odd power of z^{-1}. It is easy to see that this solution leads to Haar filters. We discuss these filters further when we discuss orthogonal filter banks.

Half-Band Filter Approach Observe from (7.89) that if we allow only causal FIR filters for the analysis filter bank, the composite filter C_0 is also causal FIR with only one odd-indexed coefficient. To overcome this restriction, we can design anticausal or noncausal filters and then add a delay to make them causal. We first simplify the analysis by adding an "advance" to the composite filter and by making use of the properties of a half-band filter defined below. The composite filter C_0 is advanced by m taps, so that

$$S(z) = z^m C_0(z), \tag{7.97}$$

where $S(z)$ is a noncausal filter symmetric with respect to the origin. The PR condition becomes

$$S(z) + S(-z) = 2 \tag{7.98}$$

since $S(-z) = (-z)^m C_0(-z) = -z^m C_0(-z)$ for odd m. All even-indexed coefficients in $S(z)$ are zero except $s(0) = 1$. $S(z)$ is a half-band filter satisfying the following conditions:

1. $s(n) = 0$ for all even n except $n = 0$.
2. $s(0) =$ constant.
3. $s(n) = s(-n)$.
4. $S(e^{j\omega}) + S(-e^{-j\omega}) =$ constant.

This half-band filter is capable of being spectrally factorized into a product of two filters. We use examples to discuss the HBF.

To find the solution to (7.98), let $H_1(z) = -z^{-m} H_0(-z^{-1})$; the transfer function becomes

$$\begin{aligned} T(z) &= \frac{1}{2}[H_0(z)H(-z) - H_1(z)H_0(-z)] \\ &= \frac{1}{2}z^{-m}\left[-H_0(z)H_0(z^{-1})(-1)^{-m} + H_0(-z)H_0(-z^{-1})\right]. \end{aligned} \tag{7.99}$$

In view of (7.90), m must be odd. We have the expression

$$T(z) = \frac{1}{2}z^{-m}\left[H_0(z)H_0(z^{-1}) + H_0(-z)H_0(-z^{-1})\right]. \tag{7.100}$$

The filter bank has been designed once the half-band filter has been designed. The resultant filters are listed as follows:

$$\begin{aligned} S(z) &= H_0(z)H_0(z^{-1}) \\ C_0(z) &= H_0(z)H_0(z^{-1})z^{-m} \\ C_1(z) &= -H_0(z)H_0(z^{-1})z^{-m} \\ H_1(z) &= -z^{-m}H_0(-z^{-1}) \\ G_0(z) &= H_1(-z) \\ G_1(z) &= -H_0(-z) \\ T(z) &= \tfrac{1}{2}\left[S(z) + S(-z)\right]. \end{aligned} \tag{7.101}$$

The lowpass filter $H_0(z)$ comes from the spectral factorization of $S(z)$.

Example 1: We use the derivation of the Daubechies [5] scaling function coefficients as an example. Let us recall the conditions on the half-band filter,

$$S(z) + S(-z) = 2.$$

The simplest form of $S(z)$ other than the Haar filter is

$$S(z) = (1+z)^2(1+z^{-1})^2 R(z). \tag{7.102}$$

All even coefficients of $S(z)$ must be zero except at 0, where $s(0) = 1$. Let

$$R(z) = az + b + az^{-1}$$

be a noncausal symmetric filter so that $S(z)$ remains symmetric. By carrying out the algebra in (7.102) and using conditions on $S(z)$, we have

$$\begin{aligned} S(0) &= 1 & &\Longrightarrow & 8a + 6b &= 1 \\ S(2) &= S(-2) = 0 & &\Longrightarrow & 4a + b &= 0, \end{aligned}$$

giving $a = -\frac{1}{16}$ and $b = \frac{1}{4}$. The symmetric filter $R(z)$ becomes

$$\begin{aligned} R(z) &= -\frac{1}{16}z + \frac{1}{4} - \frac{1}{16}z^{-1} \\ &= \left(\frac{1}{4\sqrt{2}}\right)^2 \left[1 + \sqrt{3} + (1 - \sqrt{3})z^{-1}\right]\left[1 + \sqrt{3} + (1 - \sqrt{3})z\right]. \end{aligned}$$

This expression is substituted into (7.102) so that we can factor $S(z)$ into a product of two filters, $H_0(z)$ and $H_0(z^{-1})$. The result of this spectral factorization gives a causal filter:

$$\begin{aligned} H_0(z) &= \frac{1}{4\sqrt{2}}(1 + z^{-1})^2 \left[1 + \sqrt{3} + (1 - \sqrt{3})z^{-1}\right] \\ &= \frac{1}{4\sqrt{2}}\left[\left(1 + \sqrt{3}\right) + \left(3 + \sqrt{3}\right)z^{-1} + \left(3 - \sqrt{3}\right)z^{-2} + \left(1 - \sqrt{3}\right)z^{-3}\right] \\ &= 0.4929 + 0.8365z^{-1} + 0.2241z^{-2} - 0.1294z^{-3}. \end{aligned}$$

Note that these coefficients need to be multiplied by $\sqrt{2}$ to get the values given in Chapter 6.

Biorthogonal Filter Bank A linear-phase FIR filter bank is desirable because it minimizes phase distortion in signal processing. On the other hand, an orthogonal FIR filter bank is also desirable because of its simplicity. One has to design only one filter, namely $H_0(z)$, and all other filters in the entire bank are specified. Biorthogonal filter banks are designed to satisfy the linear-phase requirement.

Let us recall the PR and antialiasing conditions with synthesis and analysis filters. They are

$$\begin{aligned} H_0(z)G_0(z) + H_1(z)G_1(z) &= 2z^{-m} \\ G_0(z)H_0(-z) + G_1(z)H_1(-z) &= 0. \end{aligned}$$

We can solve for the synthesis filter's $G_0(z)$ and $G_1(z)$ in terms of the analysis filter's $H_0(z)$ and $H_1(z)$. The result is

$$\begin{bmatrix} G_0(z) \\ G_1(z) \end{bmatrix} = \frac{2z^{-m}}{\det[\mathrm{Tr}]} \begin{bmatrix} H_1(-z) \\ -H_0(-z) \end{bmatrix}, \tag{7.103}$$

where the transfer matrix is

$$[\mathrm{Tr}] = \begin{bmatrix} H_0(z) & H_1(z) \\ H_0(-z) & H_1(-z) \end{bmatrix}.$$

If we allow symmetric filters

$$H_0(z) = H_0(z^{-1}) \Leftrightarrow h_0(n) = h_0(-n)$$

and are not concerned with causality at the moment, we may safely ignore the delay z^{-m}. This is equivalent to designing all filters to be symmetric or antisymmetric about the origin. We also recall the definitions of the composite filters

$$\begin{aligned} C_0(z) &= H_0(z)G_0(z) \\ C_1(z) &= H_1(z)G_1(z). \end{aligned}$$

Using the result of (7.103), we write

$$\begin{aligned} C_0(z) = H_0(z)G_0(z) &= \frac{2H_0(z)H_1(-z)}{\det[\mathrm{Tr}]} \\ C_1(z) = H_1(z)G_1(z) &= \frac{-2H_1(z)H_0(-z)}{\det[\mathrm{Tr}]}. \end{aligned} \tag{7.104}$$

If we replace $-z$ for z in the second equation and note that

$$\det[\mathrm{Tr}(-z)] = -\det[\mathrm{Tr}(z)],$$

we have

$$C_1(z) = C_0(-z). \tag{7.105}$$

The final result is

$$C_0(z) + C_0(-z) = 2. \tag{7.106}$$

We now have a half-band filter for $C_0(z)$, from which we can use spectral factorization to obtain $H_0(z)$ and $G_0(z)$. There are many choices for spectral factorization and the resulting filters are also correspondingly different. They may have different filter lengths for the synthesis and analysis banks. The resulting filters have linear phase. The user can make a judicious choice to design the analysis bank or the synthesis bank to meet the requirements of the problem on hand. We use the example in [3] to show different ways of spectral factorization to obtain $H_0(z)$ and $G_0(z)$.

Let the product filter

$$\begin{aligned} C_0(z) = H_0(z)G_0(z) &= (1+z^{-1})^4 Q(z) \\ &= \frac{1}{16}(-1 + 9z^{-2} + 16z^{-3} + 9z^{-4} - z^{-6}). \end{aligned} \tag{7.107}$$

Since the binomial $(1 + z^{-1})^n$ is symmetrical, $Q(z)$ must be symmetrical to make $C_0(z)$ symmetrical. An advance of z^3 makes $S(z)$ a half-band filter. The choices of spectral factorization include

$$\begin{array}{lll}
1. & H_0(z) = (1+z^{-1})^0 & G_0(z) = (1+z^{-1})^4 Q(z) \\
2. & H_0(z) = (1+z^{-1})^1 & G_0(z) = (1+z^{-1})^3 Q(z) \\
3. & H_0(z) = (1+z^{-1})^2 & G_0(z) = (1+z^{-1})^2 Q(z) \\
 & \quad \text{or } (1+z^{-1})(2-\sqrt{3}-z^{-1}) & \quad \text{or } (1+z^{-1})^3(2+\sqrt{3}-z^{-1}) \\
4. & H_0(z) = (1+z^{-1})^3 & G_0(z) = (1+z^{-1})Q(z) \\
5. & H_0(z) = (1+z^{-1})^2(2-\sqrt{3}-z^{-1}) & G_0(z) = (1+z^{-1})^2(2+\sqrt{3}-z^{-1}).
\end{array} \tag{7.108}$$

The last choice corresponds to Daubechies' orthogonal filters, which do not have linear phase. The 3/5 filter in the upper line of factorization choice 3 gives a linear-phase filter, whereas the lower one does not.

7.8.2 Time-Domain Analysis

The development of filter bank theory is based primarily on spectral analysis. We discuss the time-domain equivalent of the theory for enhancement of the understanding as well as for the digital implementation of the algorithm. Thus it suffices to illustrate the meaning of the terms, filter requirements, and filter systems in terms of time-domain variables.

Causality An FIR filter is causal if the impulse response

$$h(n) = 0 \qquad \forall\, n < 0.$$

The z-transform of $h(n)$ is a right-sided polynomial of z^{-1}:

$$H(z) = h(0) + h(1)z^{-1} + h(2)z^{-2} + \cdots + h(m)z^{-m}.$$

If $H(z)$ is a causal filter, $H(z^{-1})$ is anticausal since

$$H(z^{-1}) = h(0) + h(1)z + h(2)z^2 + \cdots + h(m)z^m,$$

which is a left-sided polynomial of z. As a result, $H(-z^{-1})$ is also anticausal since the polynomial is the same as that of $H(z^{-1})$ except that the signs of odd coefficients have been changed:

$$H(-z^{-1}) = h(0) - h(1)z + h(2)z^2 - \cdots - h(m)z^m.$$

The last term has a negative sign if we assume that m is odd. To realize the anticausal FIR filter, we must delay the filter by the length of the filter to make it causal. Hence

$$-z^{-m}H(-z^{-1}) = h(m) - h(m-1)z^{-1} + \cdots + h(1)z^{-m+1} - h(0)z^{-m}$$

is a causal filter. If we choose

$$\begin{aligned} H_0(z) &= H(z) \\ H_1(z) &= -z^{-m} H_0(-z^{-1}) \\ G_0(z) &= H_1(-z) \\ G_1(z) &= -H_0(-z), \end{aligned}$$

we have a filter bank consisting of causal filters.

PR Requirements Perfect reconstruction demands that

$$S(z) + S(-z) = 2.$$

In terms of the lowpass filter $H_0(z)$, the equation becomes

$$H_0(z)H_0(z^{-1}) + H_0(-z)H_0(-z^{-1}) = 2. \tag{7.109}$$

Let us consider the PR condition in (7.109). In the time domain we have

$$\begin{aligned} S(z) + S(-z) &= \sum_n h_0(n)z^{-n} \sum_m h_0(m)z^m \\ &\quad + \sum_n h_0(n)z^{-n} \sum_m (-1)^{-(n+m)} h_0(m)z^m \\ &= \sum_{n,m} h_0(n)h_0(m)z^{-n}z^m + \sum_{n,m} (-1)^{-(n+m)} h_0(n)h_0(m)z^{-n}z^m \\ &= 2. \end{aligned} \tag{7.110}$$

Satisfaction of (7.110) requires that $(m+n)$ be even and we have

$$\sum_{n,m} h_0(n)h_0(m)z^{-n}z^m = 1. \tag{7.111}$$

The left side of (7.111) is the z-transform of the autocorrelation function of the sequence $h_0(n)$. To show this relation, we denote

$$\kappa(n) = \sum_k h_0(k)h_0(k+n) = \kappa(-n) \tag{7.112}$$

as the autocorrelation function. Its z-transform is written as

$$\begin{aligned} K(z) &= \sum_n \sum_k h_0(k)h_0(k+n)z^{-n} \\ &= \sum_k h_0(k) \sum_n h_0(k+n)z^{-n} \\ &= \sum_k h_0(k) \sum_m h_0(m)z^{-(m-k)} \end{aligned} \tag{7.113}$$

which implies that

$$\kappa(n) = h_0(n) * h_0(-n). \tag{7.114}$$

Comparing (7.113) and (7.110) and making the substitution

$$\begin{aligned} K(z) &\longrightarrow S(z) \\ \kappa(n) &\longrightarrow s(n), \end{aligned}$$

we have

$$S(z) = \sum_n \sum_k h_0(k) h_0(k+n) z^{-n}.$$

From (7.98) and the fact that $s(2n) = 0$ for all integer n, we have the orthonormality condition required for PR:

$$\sum_k h_0(k) h_0(k+2n) = \delta_{n,0}. \tag{7.115}$$

This implies the orthogonality of the filter on all its even translates. We apply the same analysis to the highpass filter $h_1(n)$ and get the same condition for $h_1(n)$:

$$\sum_k h_1(k) h_1(k+2n) = \delta_{n,0} \tag{7.116}$$

$$\sum_k h_0(k) h_1(k+2n) = 0. \tag{7.117}$$

In terms of wavelet and approximation function basis, the orthonormality conditions given above are expressed as inner products:

$$\begin{aligned} \langle h_0(k), h_0(k+2n) \rangle &= \delta_{n,0} \\ \langle h_1(k), h_1(k+2n) \rangle &= \delta_{n,0} \\ \langle h_0(k), h_1(k+2n) \rangle &= 0, \end{aligned} \tag{7.118}$$

where the approximation basis $h_0(k)$ and the wavelet basis $h_1(k)$ are orthonormal to their even translates. They are also orthogonal to each other. If we construct an infinite matrix $[H_0]$ using the FIR sequence $h_0(n)$ such that

$$[H_0] = \begin{bmatrix} h_0[0] & h_0[1] & h_0[2] & h_0[3] & 0 & 0 & 0 \\ 0 & 0 & h_0[0] & h_0[1] & h_0[2] & h_0[3] & 0 \\ & 0 & 0 & 0 & h_0[0] & h_0[1] & h_0[2] \\ & & & 0 & 0 & 0 & h_0[0] \\ & & & & & 0 & 0 \\ & & & & & & 0 \\ & & & & & & 0 \end{bmatrix}, \tag{7.119}$$

it is obvious that

$$[H_0]\,[H_0]^t = I \tag{7.120}$$

using the orthonormality conditions in (7.118). Therefore, $[H_0]$ is an orthogonal matrix. We define $[H_1]$ in a similar way using the FIR sequence of $h_1(n)$ and show that

$$[H_1]\,[H_1]^t = I. \tag{7.121}$$

In addition, the reader can also show that

$$[H_1]\,[H_0]^t = [H_0]\,[H_1]^t = [0]\,. \tag{7.122}$$

Equations in (7.118) constitute the orthogonal conditions imposed on the FIR filters. This type of filter bank is called an *orthogonal filter bank.* The processing sequences for the Haar scaling function and Haar wavelets are the simplest linear-phase orthogonal filter bank. Indeed, if we denote

$$h_0^H(n) = \left\{\frac{1}{\sqrt{2}}, \frac{1}{\sqrt{2}}\right\}$$

and

$$h_1^H(n) = \left\{\frac{1}{\sqrt{2}}, -\frac{1}{\sqrt{2}}\right\},$$

these two sequences satisfy the orthogonal conditions in (7.118). We recall that linear-phase FIR filters must be either symmetric or antisymmetric, a condition not usually satisfied by orthogonal filters. This set of Haar filters is the only orthogonal set that has linear phase.

Two-Channel Biorthogonal Filter Bank in the Time Domain We have shown that the biorthogonal condition on the analysis and synthesis filters is

$$\begin{aligned} C_0(z) + C_0(-z) &= H_0(z)G_0(z) + H_0(-z)G_0(-z) \\ &= 2. \end{aligned}$$

Writing this equation in the time domain and using the convolution formula yields the time-domain biorthogonal condition

$$\sum_k h_0(k)g_0(\ell - k) + (-1)^\ell \sum_k h_0(k)g_0(\ell - k) = 2\delta_{\ell,0}. \tag{7.123}$$

For nontrivial solution, the equality holds only if ℓ is even. This results in a biorthogonal relation between the analysis and synthesis filters:

$$\begin{aligned} \sum_k h_0(k)g_0(2n - k) &= \langle h_0(k), g_0(2n - k)\rangle \\ &= \delta_{n,0}. \end{aligned} \tag{7.124}$$

The biorthogonal condition can also be expressed in terms of $H_1(z)$ to yield

$$\sum_k h_1(k)g_1(2n-k) = \langle h_1(k), g_1(2n-k)\rangle$$
$$= \delta_{n,0}. \tag{7.125}$$

The additional biorthogonal relations are

$$\begin{aligned}\langle h_1(k), g_0(2n-k)\rangle &= 0,\\ \langle h_0(k), g_1(2n-k)\rangle &= 0.\end{aligned} \tag{7.126}$$

If we consider the filters as discrete bases, we have

$$\widetilde{g}_m(k) = g_m(-k). \tag{7.127}$$

The biorthogonal relations become

$$\begin{aligned}\langle h_0(k), \widetilde{g}_0(k-2n)\rangle &= \delta_{n,0}\\ \langle h_1(k), \widetilde{g}_1(k-2n)\rangle &= \delta_{n,0}\\ \langle h_1(k), \widetilde{g}_0(k-2n)\rangle &= 0\\ \langle h_0(k), \widetilde{g}_1(k-2n)\rangle &= 0.\end{aligned} \tag{7.128}$$

7.9 POLYPHASE REPRESENTATION FOR FILTER BANKS

Polyphase representation of a signal is an alternative approach to discrete signal representation other than in the spectral and time domains. It is an efficient representation for computation. Consider the process of convolution and decimation by 2; we compute all the resulting coefficients and then cast out half of them. In the polyphase approach we decimate the input signal and then convolve with only half of the filter coefficients. This approach increases the computational efficiency by reducing the redundancy.

7.9.1 Signal Representation in the Polyphase Domain

Let the z-transform of a discrete causal signal separated into segments of M points be written as

$$\begin{aligned}X(z) = x(0) &+ x(1)z^{-1} + x(2)z^{-2} + x(3)z^{-3} + \cdots + x(M-1)z^{-M+1}\\ &+ x(M)z^{-M} + x(M+1)z^{-(M+1)} + x(M+2)z^{-(M+2)} + \cdots\\ &+ x(2M)z^{-2M} + x(2M+1)z^{-(2M+1)} + x(2M+2)z^{-(2M+2)} + \cdots\\ &+ x(3M)z^{-3M} + x(3M+1)z^{-(3M+1)} + \cdots\end{aligned} \tag{7.129}$$

$$= \sum_{\ell=0}^{M-1} z^{-\ell} X_\ell(z^M), \tag{7.130}$$

where $X_\ell(z^M)$ is the z-transform of $x(n)$ decimated by $M(\downarrow M)$. The index ℓ indicates the number of sample shifts. For the case of $M = 2$, we have

$$X(z) = X_0(z^2) + z^{-1}X_1(z^2). \tag{7.131}$$

7.9.2 Filter Bank in the Polyphase Domain

For a filter $H(z)$ in a two-channel setting, the polyphase representation is exactly the same as in (7.131):

$$H(z) = H_e(z^2) + z^{-1}H_o(z^2), \tag{7.132}$$

where $H_e(z^2)$ consists of the even samples of $h(n)$ and $H_o(z^2)$ has all the odd samples. The odd and even parts of the filter are used to process the odd and even coefficients of the signal separately. To formulate the two-channel filter bank in the polyphase domain, we need the help of two identities:

$$\begin{aligned} &1. \quad (\downarrow M)\, G(z) = G(z^M)\, (\downarrow M)\,. \\ &2. \quad (\uparrow M)\, G(z^M) = G(z)\, (\uparrow M)\,. \end{aligned} \tag{7.133}$$

A filter $G(z^2)$ followed by a two-point decimator is equivalent to a two-point decimator followed by $G(z)$. The second identity is useful for the synthesis filter bank.

Let us consider first the time-domain formulation of the lowpass branch of the analysis filter. Assuming a causal input sequence and causal filter, the output $y(n) = [x(n) * f(n)]_{\downarrow 2}$ is expressed in matrix form as

$$\begin{bmatrix} \cdot \\ y(0) \\ y(1) \\ y(2) \\ y(3) \\ y(4) \\ y(5) \\ y(6) \end{bmatrix} = \begin{bmatrix} \cdot & \cdot & \cdot & \cdot & \cdot & \cdot \\ f(0) & 0 & 0 & 0 & 0 & 0 \\ f(1) & f(0) & 0 & 0 & 0 & 0 \\ f(2) & f(1) & f(0) & 0 & 0 & 0 \\ f(3) & f(2) & f(1) & f(0) & 0 & 0 \\ f(4) & f(3) & f(2) & f(1) & f(0) & 0 \\ f(5) & f(4) & f(3) & f(2) & f(1) & \\ \cdot & \cdot & \cdot & \cdot & \cdot & \cdot \end{bmatrix} \begin{bmatrix} \cdot \\ x(0) \\ x(1) \\ x(2) \\ x(3) \\ x(4) \\ x(5) \\ x(6) \end{bmatrix}. \tag{7.134}$$

The output coefficients are represented separately by the odd and even parts as

$$[y(n)] = [y_e(n)] + (\text{delay})\, [y_o(n)]\,,$$

where

$$[y(n)]_{\downarrow 2} = [y_e(n)] = \begin{bmatrix} y(0) \\ y(2) \\ y(4) \\ y(6) \end{bmatrix}. \tag{7.135}$$

The even part of $y(n)$ is made up of the products of $f_e(n)$ with $x_e(n)$ and $f_o(n)$ with $x_o(n)$ plus a delay. The signal $x(n)$ is divided into the even and odd parts and they are processed by the even and odd parts of the filter, respectively. In the same way, the highpass branch of the analysis section can be seen exactly as we demonstrate above. In the polyphase domain, the intermediate output from the analysis filter is given by

$$\begin{aligned}\begin{bmatrix} U_0(z) \\ U_1(z) \end{bmatrix} &= \begin{bmatrix} H_{00}(z) & H_{01}(z) \\ H_{10}(z) & H_{11}(z) \end{bmatrix} \begin{bmatrix} X_0(z) \\ z^{-1}X_1(z) \end{bmatrix} \\ &= [H] \begin{bmatrix} X_0(z) \\ z^{-1}X_1(z) \end{bmatrix}, \end{aligned} \tag{7.136}$$

where $[H]$ is the analysis filter in the polyphase domain. In the same manner, we obtain the reconstructed sequence $X'(z)$ from the synthesis filter bank as

$$\begin{aligned} X'(z) &= \begin{bmatrix} z^{-1} & 1 \end{bmatrix} \begin{bmatrix} G_{00}(z^2) & G_{01}(z^2) \\ G_{10}(z^2) & G_{11}(z^2) \end{bmatrix} \begin{bmatrix} U_0(z^2) \\ U_1(z^2) \end{bmatrix} \\ &= \begin{bmatrix} z^{-1} & 1 \end{bmatrix} [G] \begin{bmatrix} U_0(z^2) \\ U_1(z^2) \end{bmatrix}. \end{aligned} \tag{7.137}$$

The PR condition for the polyphase processing matrices is $[H][G] = I$.

7.10 COMMENTS ON DWT AND PR FILTER BANKS

We have shown the parallel between the algorithms of the DWT and the two-channel filter bank. In terms of numerical computation, the algorithms of both disciplines are exactly the same. We would like to point out several fundamental differences between the two disciplines.

1. *Processing domain.* Let us represent an analog signal $f(t) \in L^2$ by an orthonormal wavelet series

$$f(t) = \sum_k \sum_s w_{k,s} \psi_{k,s}(t). \tag{7.138}$$

The coefficients $w_{k,s}$ are computed via the inner product

$$w_{k,s} = \langle f(t), \psi_{k,s}(t) \rangle. \tag{7.139}$$

In much the same way as the Fourier series coefficients, the wavelet series coefficients are time (or analog)-domain entities. From this point of view we see that the DWT is a fast algorithm to compute the CWT at a sparse set of points on the time-scale plane, much like the FFT is a fast algorithm to compute the discrete Fourier transform. The DWT is a time-domain transform for analog signal processing. On the other hand, the filter bank algorithms are designed from spectral domain considerations (i.e., highpass and lowpass design) for processing of signal "samples" (instead of coefficients).

2. *Processing goal.* We have shown that the wavelet series coefficients are essentially the components (from projection) of the signal in the "direction" of the wavelet ψ at the scale $a = 2^{-s}$ and at the time point $b = k2^{-s}$. This concept of component is similar to that of the Fourier component. The magnitude of the wavelet series coefficient represents the strength of the correlation between the signal and the wavelet at that particular scale and point in time. The processing goal of the filter bank is to separate the high- and low-frequency components of the signal so that they may be processed or refined by different DSP algorithms. Although the DWT algorithms inherently have the same function, the focus of DWT is on finding the similarity between the signal and the wavelet at a given scale.
3. *Design origin.* A wavelet is designed primarily via the two-scale relation to satisfy the MRA requirements. Once the two-scale sequences are found, the DWT processing sequences have been set. A wavelet can be constructed and its time and scale window widths can be computed. In general, a filter bank is designed in the spectral domain via spectral factorization to obtain the processing filters. These sequences may or may not serve as the two-scale sequences for the approximation function and the wavelet. The time-scale or time–frequency characteristics of these filters may not be measurable.
4. *Application areas.* Most signal and image processing applications can be carried out with either DWT or filter bank algorithms. In some application areas, such as in non-Fourier magnetic resonance imaging, where the processing pulse required is in the analog domain, a wavelet is more suitable for the job because the data set is obtained directly via projection.
5. *Flexibility.* Since filter banks may be designed in the spectral domain via spectral factorization, a given half-band filter may result in several sets of filters, each having its own merit vis-à-vis the signal given. In this regard, the filter bank is much more adaptable than wavelets to the processing need.

Wavelet or filter bank? Users must decide for themselves based on the problem at hand and the efficiency and accuracy of using either one or the other.

7.11 EXERCISES

1. For a positive integer $M \geq 2$, set $w_M^k = \exp(j\frac{2\pi k}{M})$ for $k = 1, \ldots, M$. Show that

$$\frac{1}{M}\sum_{k=1}^{M} w_M^{k\ell} = \begin{cases} 0 & \text{if } M \nmid \ell \\ 1 & \text{if } M / \ell \end{cases}. \tag{7.140}$$

Using this relation, prove that

$$Y\left(e^{-j\omega}\right) = \frac{1}{M}\sum_{k=1}^{M} X\left(w_M^k \exp\left(-j\frac{\omega}{M}\right)\right), \tag{7.141}$$

where $X(z) = \sum_k x[k]z^k$ and $Y(z) = \sum_k y[k]z^k$ are the z-transform of sequences $\{x[k]\}$ and $\{y[k]\}$.

2. If the sequence $\{y[k]\}$ is generated from $\{x[k]\}$ by upsampling by M, that is,

$$y[k] = \begin{cases} x[\frac{k}{M}] & \text{if } k \in M\mathbb{Z} \\ 0 & \text{otherwise,} \end{cases} \tag{7.142}$$

show that

$$\left(Y\left(e^{-j\omega}\right) = X\left(e^{-jM\omega}\right)\right) \tag{7.143}$$

for the respective z-transforms.

3. In the QMF solution to the PR condition, it is found that the only solution that can satisfy the condition is the use of Haar filters. Why can no other FIR filters satisfy the PR condition?

4. Use the antialiasing condition and the PR condition, find the filter sequences $h_0(n), h_1(n), g_1(n)$ if $g_0(n)$ is the D_2 sequence given the example discussed in this chapter.

5. Show the validity of the identities given in Section 7.9.2.

7.12 COMPUTER PROGRAMS

7.12.1 Algorithms

```
%
% PROGRAM algorithm.m
%
% Decomposes and reconstructs a function using Daubechies'
% wavelet (m = 2). The initial coefficients are taken as
% the function values themselves.
%

% Signal

v1 = 100;            % frequency
v2 = 200;
v3 = 400;
r = 1000;            %sampling rate
```

```
k = 1:100;
t = (k-1) / r;
s = sin(2*pi*v1*t) + sin(2*pi*v2*t) + sin(2*pi*v3*t);

% Decomposition and reconstruction filters

g0 = [0.68301; 1.18301; 0.31699; -0.18301];
k = [0; 1; 2; 3];
g1 = flipud(g0).*(-1).^k;
h0 = flipud(g0) / 2;
h1 = flipud(g1) / 2;

% Decomposition process

% First level decomposition

x = conv(s,h0);
a0 = x(1:2:length(x)); %downsampling
x = conv(s,h1);
w0 = x(1:2:length(x)); %downsmapling

% Second level decomposition

x = conv(a0,h0);
a1 = x(1:2:length(x));
x = conv(a0,h1);
w1 = x(1:2:length(x));

% Plot

subplot(3,2,1), plot(s)
ylabel('Signal')
subplot(3,2,3), plot(a0)
ylabel('a_0')
subplot(3,2,4), plot(w0)
ylabel('w_0')
subplot(3,2,5), plot(a1)
ylabel('a_{-1}')
subplot(3,2,6), plot(w1)
ylabel('w_{-1}')
set(gcf,'paperposition',[0.5 0.5 7.5 10])

% Reconstuction process

% Second level reconstruction

x = zeros(2*length(a1),1);
x(1:2:2*length(a1)) = a1(1:length(a1));
y = zeros(2*length(w1),1);
y(1:2:2*length(w1)) = w1(1:length(w1));
```

```
x = conv(x,g0) + conv(y,g1);
a0_rec = x(4:length(x)-4);

% First level reconstruction

y = zeros(2*length(w0), 1);
y(1:2:2*length(w0)) = w0(1:length(w0));
x = zeros(2*length(a0_rec), 1);
x(1:2:2*length(a0_rec)) = a0_rec;

x = conv(x,g0);
y = conv(y,g1);
y = x(1:length(y))+y;
s_rec = y(4:length(y)-4);

% Plot

figure(2)
subplot(3,2,1), plot(a1)
ylabel('a_{-1}')
subplot(3,2,2), plot(w1)
ylabel('w_{-1}')
subplot(3,2,3), plot(a0_rec)
ylabel('Reconstructed a_0')
subplot(3,2,4), plot(w0)
ylabel('w_0')
subplot(3,2,5), plot(s_rec)
ylabel('Reconstructed Signal')
set(gcf,'paperposition',[0.5 0.5 7.5 10])
```

REFERENCES

1. C. K. Chui, J. C. Goswami, and A. K. Chan, "Fast integral wavelet transform on a dense set of time-scale domain," *Numer. Math.*, **70**, pp. 283–302, 1995.
2. J. C. Goswami, A. K. Chan, and C. K. Chui, "On a spline-based fast integral wavelet transform algorithm," in *Ultra-Wideband Short Pulse Electromagnetics 2*, L. Carin and L. B. Felsen (Eds.). New York: Plenum Press, 1995, pp. 455–463.
3. G. Strang and T. Nguyen, *Wavelets and Filter Banks.* Wellesley, Mass.: Wellesley-Cambridge Press, 1996.
4. P. P. Vaidyanathan, *Multirate Systems and Filter Banks.* Upper Saddle River, N.J.: Prentice Hall, 1993.
5. M. Vetterli and J. Kovacevic, *Wavelets and Subband Coding.* Upper Saddle River, N.J.: Prentice Hall, 1995.
6. A. N. Akansu and R. A. Haddad, *Multiresolution Signal Decomposition.* San Diego, Calif.: Academic Press, 1992.
7. A. N. Akansu and M. J. Smith (Eds.), *Subband and Wavelet Transform.* Boston: Kluwer Academic, 1995.

CHAPTER EIGHT

Fast Integral Transform and Applications

In Chapter 7 we discussed standard wavelet decomposition and reconstruction algorithms. By applying an optimal-order local spline interpolation scheme as described in Section 5.6, we obtain the coefficient sequence $\boldsymbol{a}^M$ of the desired B-spline series representation. Then, depending on the choice of linear or cubic spline interpolation, we apply the change-of-bases sequences (Section 7.5) to obtain the coefficient sequence $\tilde{\boldsymbol{a}}^M$ of the dual series representation for the purpose of FIR wavelet decomposition.

A typical time-scale grid obtained by following the implementation scheme described in Chapter 7 is shown in Figure 7.7. In other words, the integral wavelet transform (IWT) values of the given signal at the time-scale positions shown in Figure 7.7 can be obtained (in real time) by following this scheme. However, in many signal analysis applications, such as wideband correlation processing [1] used in some radar and sonar applications, this information on the IWT of f on such a sparse set of dyadic points (as shown in Figure 7.7) is insufficient for the desired time–frequency analysis of the signal. It becomes necessary to compute the IWT at nondyadic points as well. By maintaining the same time resolution at all the binary scales, the aliasing and the time variance difficulties associated with the standard wavelet decomposition algorithm can be circumvented. Furthermore, as will be shown in this chapter, computation only at binary scales may not be appropriate to separate all the frequency contents of a function.

An algorithm for computing the IWT with finer time resolution has been introduced and studied by Rioul and Duhamel [2] and Shensa [3]. In addition, there have been some advances in fast computation of the IWT with finer frequency resolution, such as the multivoice per octave (MVPO) scheme, first introduced in [4] (see also [5, pp. 71–72]) and later improved with the help of FFT by Rioul and Duhamel [2]. However, the computational complexity of the MVPO scheme, with or without FFT, increases with the number of values of the scale parameter a. For example, in the

FFT-based computational scheme, both the signal and the analyzing wavelet have to be sampled at the same rate, with the sampling rate determined by the highest frequency content (or the smallest scale parameter) of the signal, and this sampling rate cannot bc changed at the subsequent larger scale values for any fixed signal discretization. Furthermore, even at the highest frequency level, where the width of the wavelet is narrowest in the time domain, the number of sampled data required for the wavelet will be significantly larger than the number of decomposition coefficients in the pyramid algorithm.

In this chapter we discuss the fast integral wavelet transform (FIWT) algorithm [6–8] to compute the integral (continuous) wavelet transform on a dense set of points in the time-scale domain.

8.1 FINER TIME RESOLUTION

In this section we are concerned with maintaining the same time resolution on each scale by filling in the "holes" along the time axis on each scale; that is, we want to compute $W_\psi x_M(n/2^M, 1/2^s),\ n \in \mathbb{Z}, s < M$. Recall that the standard algorithms discussed in Chapter 7 give the IWT values only at dyadic points $\{n/2^s, 1/2^s ; n \in \mathbb{Z}, s < M\}$. For finer time resolution, we first observe that for each fixed n, by introducing the notation

$$x_{M,n}(t) := x_M\left(t + \frac{n}{2^M}\right), \tag{8.1}$$

we have

$$\begin{aligned} W_\psi x_M\left(\frac{n}{2^M}, \frac{1}{2^s}\right) &= 2^{s/2} \int_{-\infty}^{\infty} x_M(t) \overline{\psi\left[2^s\left(t - \frac{n}{2^M}\right)\right]}\, dt \\ &= 2^{s/2} \int_{-\infty}^{\infty} x_M\left(t + \frac{n}{2^M}\right) \overline{\psi\left(2^s t\right)}\, dt \\ &= W_\psi x_{M,n}\left(0, \frac{1}{2^s}\right) \end{aligned} \tag{8.2}$$

Now, since

$$x_M(t) = \sum_k \tilde{a}_{k,M} \tilde{\phi}\left(2^M t - k\right), \tag{8.3}$$

we have

$$\begin{aligned} x_{M,n}(t) &= \sum_k \tilde{a}_{k,M} \tilde{\phi}(2^M t + n - k) \\ &= \sum_k \tilde{a}_{n+k,M} \tilde{\phi}(2^M t - k). \end{aligned} \tag{8.4}$$

Hence we observe from (8.2) that the IWT of x_M at $(n/2^M, 1/2^s)$ is the same as that of $x_{M,n}$ at $(0, 1/2^s)$. In general, for every $k \in \mathbb{Z}$, we even have

$$
\begin{aligned}
W_\psi x_{M,n}\left(\frac{k}{2^s}, \frac{1}{2^s}\right) &= 2^{s/2}\int_{-\infty}^{\infty} x_{M,n}(t)\overline{\psi\left(2^s t - k\right)}\, dt \\
&= 2^{s/2}\int_{-\infty}^{\infty} x_M\left(t + \frac{n}{2^M}\right)\overline{\psi\left(2^s t - k\right)}\, dt \\
&= 2^{s/2}\int_{-\infty}^{\infty} x_M(t)\overline{\psi\left(2^s t - k - \frac{n2^s}{2^M}\right)}\, dt \\
&= W_\psi x_M\left(\frac{k2^{M-s} + n}{2^M}, \frac{1}{2^s}\right),
\end{aligned}
\tag{8.5}
$$

where $s < M$. Hence, for any fixed s and M with $s < M$, since every integer ℓ can be expressed as $k2^{M-s} + n$, where $n = 0, \ldots, 2^{M-s} - 1$ and $k \in \mathbb{Z}$, we obtain all the IWT values.

$$
W_\psi f_M\left(\frac{\ell}{2^M}, \frac{1}{2^s}\right) =: 2^{-s/2}\tilde{w}_{\ell 2^{s-M}, s}
\tag{8.6}
$$

of x_M at $(\ell/2^M, 1/2^s)$, $\ell \in \mathbb{Z}$ and $s < M$, by applying the standard wavelet decomposition algorithm of Chapter 7 to the function $x_{M,n}$. The time-scale grid for $s = M - 1, M - 2,$ and $M - 3$, but only $\ell = 0, \ldots, 3$, is given in Figure 8.1.

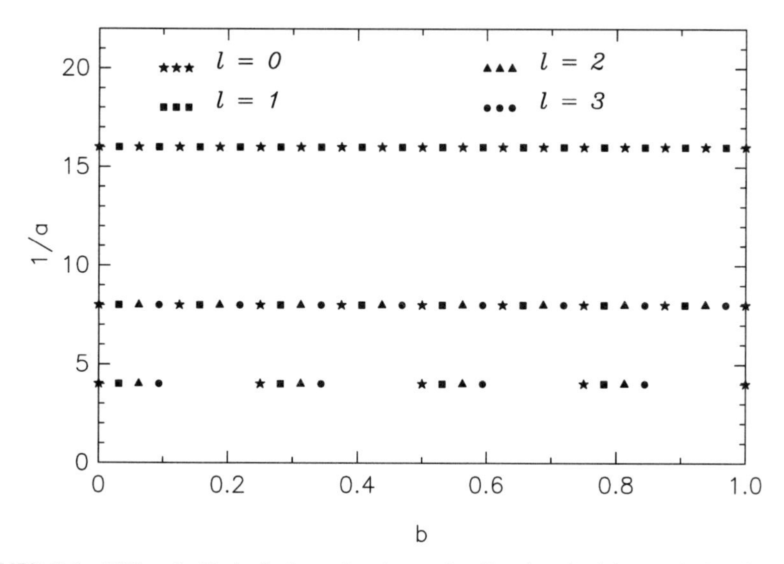

FIGURE 8.1 Filling in "holes" along the time axis. (Reprinted with permission from [6], copyright © 1995 by Springer-Verlag.)

For implementation, we need the notations

$$\tilde{\boldsymbol{w}}_s = \left\{\tilde{w}_{k2^{s-M},s}\right\}_{k\in\mathbb{Z}} \quad \text{and} \quad \tilde{\boldsymbol{a}}_s = \left\{\tilde{a}_{k2^{s-M},s}\right\}_{k\in\mathbb{Z}}, \tag{8.7}$$

and thc notation for the upsampling operations

$$\boldsymbol{\sigma}^p = \begin{cases} \text{identity operator} & \text{for } p = 0 \\ \boldsymbol{\sigma}^{p-1}\boldsymbol{\sigma} & \text{for } p \geq 1, \end{cases} \tag{8.8}$$

where

$$\boldsymbol{\sigma}\{x_n\} = \{y_n\} \text{ with } y_n = \begin{cases} x_{n/2} & \text{for even } n \\ 0 & \text{for odd } n\,. \end{cases} \tag{8.9}$$

As a consequence of (8.2) and (7.45), we have, for $s = M - 1$,

$$\begin{aligned}
(\tilde{\boldsymbol{g}}_1 * \tilde{\boldsymbol{a}}_M)_n &:= \sum_k \tilde{g}_1[k]\tilde{a}_{n-k,M} \\
&= \sum_k \tilde{g}_1[-k]\tilde{a}^M_{n+k} \\
&= 2^{(M-1)/2} W_\psi x_{M,n}\left(0, \frac{1}{2^{M-1}}\right) \\
&= \tilde{a}_{n/2,M-1}.
\end{aligned} \tag{8.10}$$

In a similar way it can be shown that

$$\tilde{a}_{n/2,M-1} = (\tilde{\boldsymbol{g}}_0 * \tilde{\boldsymbol{a}}_M)_n. \tag{8.11}$$

That is, in terms of the notations in (8.7) and (8.8), we have

$$\begin{aligned}
\tilde{\boldsymbol{a}}_{M-1} &= (\boldsymbol{\sigma}^0\tilde{\boldsymbol{g}}_0) * \tilde{\boldsymbol{a}}_M \\
\tilde{\boldsymbol{w}}_{M-1} &= (\boldsymbol{\sigma}^0\tilde{\boldsymbol{g}}_1) * \tilde{\boldsymbol{a}}_M.
\end{aligned} \tag{8.12}$$

To extend this to other lower levels, we rely on the method given in [3], yielding the algorithm

$$\left.\begin{aligned}
\tilde{\boldsymbol{a}}_{s-1} &= (\boldsymbol{\sigma}^{M-s}\tilde{\boldsymbol{g}}_0) * \tilde{\boldsymbol{a}}_s \\
\tilde{\boldsymbol{w}}_{s-1} &= (\boldsymbol{\sigma}^{M-s}\tilde{\boldsymbol{g}}_1) * \tilde{\boldsymbol{w}}_s\,,
\end{aligned}\right. \quad \text{with } s = M, M-1, \ldots, M-M'+1. \tag{8.13}$$

A schematic diagram for implementing this algorithm is shown in Figure 8.2.

8.2 FINER SCALE RESOLUTION

For the purpose of computing the IWT at certain interoctave scales, we define an *interoctave parameter*:

$$\alpha_n = \alpha_{n,N} := \frac{2^N}{n + 2^N}, \qquad N > 0 \ \text{ and } \ n = 1, \ldots, 2^N - 1, \tag{8.14}$$

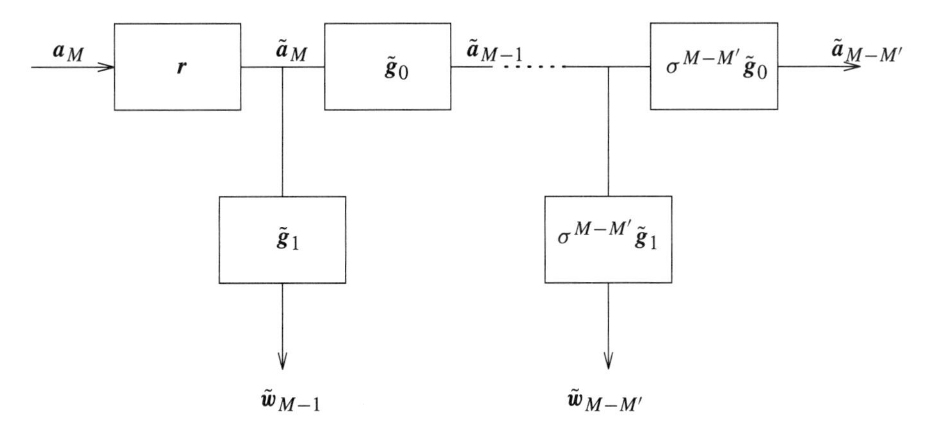

FIGURE 8.2 Wavelet decomposition process with finer time resolution. (Reprinted with permission from [6], copyright © 1995 by Springer-Verlag.)

which gives $2^N - 1$ additional levels between any two consecutive octave levels, as follows.

For each $k \in \mathbb{Z}$, $s < M$, to add $2^N - 1$ levels between the $(s-1)$st and sth octaves, we introduce the notations

$$\begin{aligned} \phi_{k,s}^n(t) &= (2^s\alpha_n)^{1/2}\phi(2^s\alpha_n t - k) \\ \psi_{k,s}^n(t) &= (2^s\alpha_n)^{1/2}\psi(2^s\alpha_n t - k). \end{aligned} \tag{8.15}$$

Observe that since $1/2 < \alpha_n < 1$, we have

$$\begin{cases} \text{supp } \phi_{k,s} \subset \text{supp } \phi_{k,s}^n \subset \text{supp } \phi_{k,s-1}^n \\ \text{supp } \psi_{k,s} \subset \text{supp } \psi_{k,s}^n \subset \text{supp } \psi_{k,s-1}^n. \end{cases} \tag{8.16}$$

As a consequence of (8.16), the RMS bandwidths of $\hat{\phi}_{0,0}^n$ and $\hat{\psi}_{0,0}^n$ are narrower than those of $\hat{\phi}$ and $\hat{\psi}$ and wider than those of $\hat{\phi}(2\cdot)$ and $\hat{\psi}(2\cdot)$, respectively.

The interoctave scales are described by the subspaces

$$V_s^n = \text{clos}_{L^2}\langle \phi_{k,s}^n : k \in \mathbb{Z}\rangle. \tag{8.17}$$

It is clear that for each n, these subspaces also constitute an MRA of L^2. In fact, the two-scale relation remains the same as that of the original scaling function ϕ, with the two-scale sequence $\{g_0[k]\}$, namely

$$\phi_{0,s}^n(t) = \sum_k g_0[k]\phi_{0,s}^n\left(2t - \frac{k}{\alpha_n}\right). \tag{8.18}$$

It is also easy to see that $\psi_{k,s}^n$ is orthogonal to V_s^n. Indeed,

$$\langle \phi_{\ell,s}^n, \psi_{k,s}^n\rangle = \langle \phi_{\ell,s}, \psi_{k,s}\rangle = 0, \quad \ell, k \in \mathbb{Z}, \tag{8.19}$$

for any $s \in \mathbb{Z}$. Hence the spaces

$$W_s^n = \text{clos}_{L^2}\langle \psi_{k,s}^n : k \in \mathbb{Z}\rangle \tag{8.20}$$

arc thc orthogonal complementary subspaces of the MRA spaces $V_{s,n}$. In addition, analogous to (8.18), the two-scale relation of $\psi_{0,s}^n$ and $\phi_0^{s,n}$ remains the same as that of ψ and ϕ, namely

$$\psi_0^{s,n}(t) = \sum_k g_1[k]\phi_{0,s}^n\left(2t - \frac{k}{\alpha_n}\right). \tag{8.21}$$

Since $(\{g_0[k]\}, \{g_1[k]\})$ remain unchanged for any interoctave scale, we can use the same implementation scheme, as shown in Figure 7.11, to compute the IWT values at $(k/2^s\alpha_n, 1/2^s\alpha_n)$. However, there are still two problems. First, we need to map x_M to $V_{M,n}$; and second, we need to compute the IWT values at $\left(k/2^M\alpha_n, 1/2^s\alpha_n\right)$ instead of the coarser grid $(k/2^s\alpha_n, 1/2^s\alpha_n)$.

Let us first consider the second problem. That is, suppose that $x_M^n \in V_M^n$ has already been determined. Then we may write

$$x_M^n = \sum_k a_{k,M}^n \phi(2^M\alpha_n t - k) = \sum_k \tilde{a}_{k,M}^n \tilde{\phi}(2^M\alpha_n t - k) \tag{8.22}$$

for some sequences $\{a_{k,M}^n\}$ and $\{\tilde{a}_{k,M}^n\} \in \ell^2$. Then the decomposition algorithm as described by Figure 7.11 yields

$$\begin{aligned}\tilde{w}_{k,s}^n &= (2^s\alpha_n)^{1/2}\langle x_M^n, \psi_n^{k,s}\rangle \\ &= 2^s\alpha_n \int_{-\infty}^{\infty} x_M^n(t)\overline{\psi(2^s\alpha_n t - k)}\, dt \\ &= (2^s\alpha_n)^{1/2} W_\psi x_M^n\left(\frac{k}{2^s\alpha_n}, \frac{1}{2^s\alpha_n}\right).\end{aligned} \tag{8.23}$$

Now by following the algorithm in (8.13), we can also maintain the same time resolution along the time axis on each interoctave scale for any fixed n. More precisely, by introducing the notations

$$\tilde{\boldsymbol{w}}_s^n = \left\{\tilde{w}_{k2^{s-M},s}^n\right\}_{k\in\mathbb{Z}} \quad \text{and} \quad \tilde{\boldsymbol{a}}_s^n = \left\{\tilde{a}_{k2^{s-M},s}^n\right\}_{k\in\mathbb{Z}}, \tag{8.24}$$

we have an algorithm for computing the IWT at the interoctave scale levels:

$$\left.\begin{aligned}\tilde{\boldsymbol{a}}_{s-1}^n &= (\sigma^{M-s}\tilde{\boldsymbol{g}}_0) * \tilde{\boldsymbol{a}}_s^n \\ \tilde{\boldsymbol{w}}_{s-1}^n &= (\sigma^{M-s}\tilde{\boldsymbol{g}}_1) * \tilde{\boldsymbol{a}}_s^n\,,\end{aligned}\right. \quad \text{with } s = M, M-1, \ldots, M - M' + 1. \tag{8.25}$$

However, it is clear from (8.23) that the time resolution for each fixed n is $1/2^M\alpha_n$, which is less than the one for the original octave scales, in which case the time resolution is $1/2^M$. As discussed in Chapter 7, the highest attainable time resolution

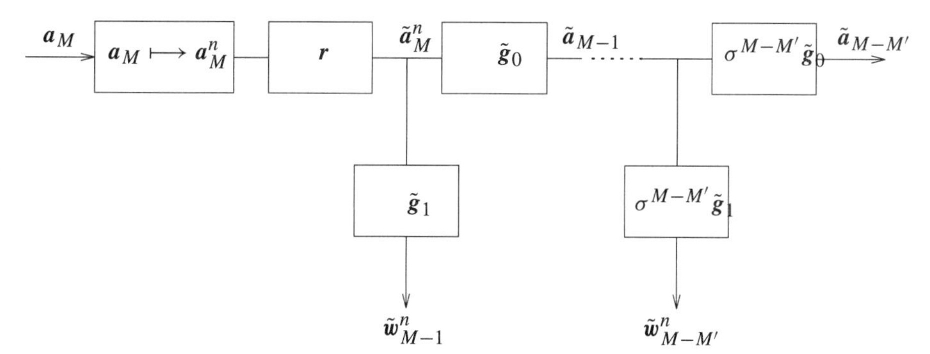

FIGURE 8.3 Wavelet decomposition process with finer time-scale resolution. (Reprinted with permission from [6], copyright © 1995 by Springer-Verlag.)

in the case of the standard (pyramid) decomposition algorithm is $1/2^{M-1}$. It should be pointed out that the position along the time axis on the interoctave scales is not the same as the original octave levels (i.e., we do not get a rectangular time-scale grid) (see Figure 8.4). The schematic diagram of (8.25) is shown in Figure 8.3. If we begin the index n of (8.14) from 0, then $n = 0$ corresponds to the original octave level. Figure 8.4 represents a typical time-scale grid for $s = M-1, M-2$, and $M-3$ with $N = 2$ and $n = 0, \ldots, 3$.

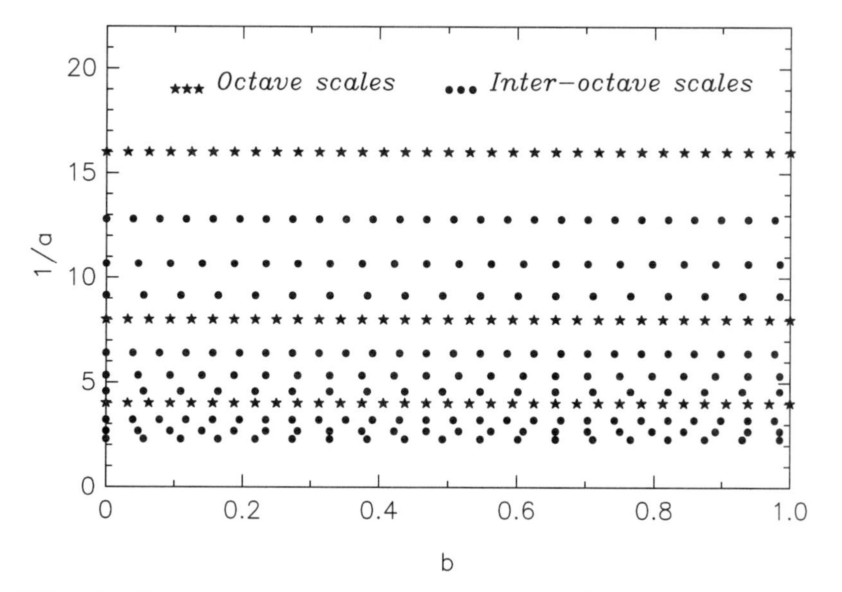

FIGURE 8.4 Time-scale grid using the scheme described in Figure 8.3. (Reprinted with permission from [6], copyright © 1995 by Springer-Verlag.)

8.3 FUNCTION MAPPING INTO THE INTEROCTAVE APPROXIMATION SUBSPACES

Now going back to the first problem of mapping x_M to x_M^n, we observe that since $V_M \neq V_M^n$, we cannot expect to have $x_M^n = x_M$ in general. However, if the MRA spaces $\{V_s\}$ contain locally all the polynomials up to order m in the sense that for each ℓ, $0 \leq \ell \leq m-1$,

$$t^\ell = \sum_k a_{\ell,k}\phi(t-k) \tag{8.26}$$

pointwise for some sequence $\{a_{\ell,k}\}_{k\in\mathbb{Z}}$, it is clear that $\{V_s^n\}$ possesses the same property. Consequently, the vanishing moment properties of the interoctave scale wavelets $\psi_{0,s}^n$ are the same as those of the original ψ. Hence, in constructing the mapping of x_M to x_M^n, we must ensure that this transformation preserves all polynomials up to order m. For the case of linear splines, such mapping can easily be obtained based on the fact that the coefficients in the linear spline representation of a function are the function values evaluated at appropriate locations.

From the points of symmetry of $N_2(2^M t)$ and $N_2(2^M \alpha_n t)$, we obtain the magnitude of the shift ξ in the centers (see Figure 8.5):

$$\xi = \frac{1}{2^M}\left(\frac{1}{\alpha_n} - 1\right) = \frac{n}{2^{M+N}}, \tag{8.27}$$

and therefore $a_{0,M}^n$ as

$$a_{0,M}^n = (1 - 2^M\xi)a_{0,M} + 2^M\xi a_{1,M}. \tag{8.28}$$

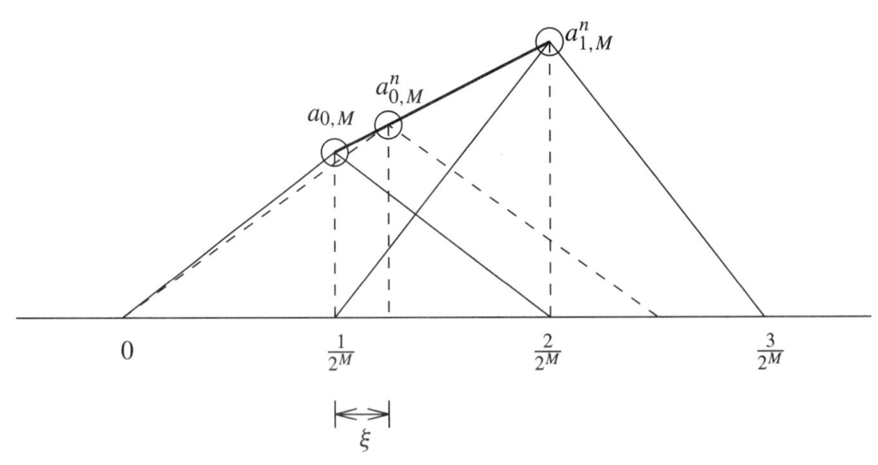

FIGURE 8.5 Mapping $\{a_{k,M}\}$ to $\{a_{k,M}^n\}$. (Reprinted with permission from [6], copyright © 1995 by Springer-Verlag.)

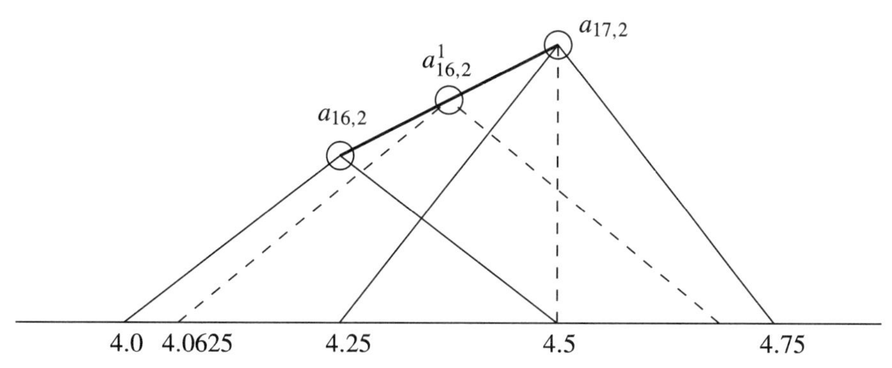

FIGURE 8.6 First term of $\{a^n_{k,M}\}$ when $\{a_{k,M}\}$ starts with $k \neq 0$. (Reprinted with permission from [6], copyright © 1995 by Springer-Verlag.)

However, if the lowest index of $\boldsymbol{a}_M$ is other than zero, $\boldsymbol{a}^n_M$ will not start with the same index as that of $\boldsymbol{a}_M$. To illustrate this situation, suppose that $x(t)$ has been discretized beginning with $t = 4.25$ with 0.25 as the step size (mapping into V_2). Then we have

$$x_2(t) = \sum_k a_{k,2} N_2(2^2 t - k) = \sum_k a^1_{k,2} N_2\left(\frac{16}{5}t - k\right), \tag{8.29}$$

with $n = 1$, $N = 2\alpha_n = \frac{4}{5}$. As is clear from Figure 8.6, the index for $\boldsymbol{a}^1_2$ does not start with the same index as $\boldsymbol{a}_2$. It should also be observed that some of the coefficients $a^n_{k,2}$ will coincide with $c_{\ell,2}$. The next index, $c^n_{k+1,2}$, will then lie between $c_{\ell+1,2}$ and $c_{\ell+2,2}$.

Taking all of the points above into account, we can obtain $\boldsymbol{a}^n_M$ from $\boldsymbol{a}_M$ by following these steps:

1. Based on the given discretized function data, determine the starting index of $\boldsymbol{a}^n_M$. Let it be $a^n_{i,M}$.
2. Let $a^n_{i,M}$ lie between $a_{s,M}$ and $a_{s+1,M}$.
3. Let $a^n_{i,M}$ be shifted from $a_{s,M}$ toward the right by ξ in time. Then starting with $r = 0$, compute

$$a^n_{i,M} = (1 - 2^M \xi) a_{s+r,M} + 2^M \xi a_{s+1+r,M}. \tag{8.30}$$

4. Increment i, s by 1 and ξ by $n/2^{M+N}$.
5. Repeat steps 3 and 4 until $1 - 2^M \xi < 0$. When $1 - 2^M \xi < 0$, increment r by 1 and reset ξ to $n/2^{M+N}$. Increment i, s by 1.
6. Repeat steps 3 to 5 until $a_{s+1+r,M}$ takes the last index of $\boldsymbol{a}_M$.

For a general case, the mapping of x_M to x^n_M can be obtained following the method described in Sections 5.6 and 7.2. For instance, to apply the linear spline interpolatory algorithm or the cubic spline interpolatory algorithm, we need to compute the function values of $x_M\left(k/2^M\alpha_n\right)$ or $x_M\left(k/2^{M-1}\alpha_n\right)$, $k \in \mathbb{Z}$. These values can easily be determined by using any spline evaluation scheme. More precisely, we have the following:

1. For $m = 2$ (linear splines), it is clear that

$$x_M^n(t) = \sum_k a_{k,M}^n N_2(2^M\alpha_n t - k) \qquad \text{with } a_{k,M}^n = x_M\left(\frac{k+1}{2^M\alpha_n}\right). \tag{8.31}$$

2. For $m = 4$ (cubic splines), we have

$$x_M^n(t) = \sum_k a_{k,M}^n N_4(2^M\alpha_n t - k) \tag{8.32}$$

with

$$a_{k,M}^n = \sum_{n=k-2}^{k+6} v_{k+2-2n} x_M\left(\frac{n}{2^{M-1}\alpha_n}\right), \tag{8.33}$$

where the weight sequence $\{v_n\}$ is given in Section 5.6. Finally, to obtain the input coefficient sequence $\{\tilde{a}_{k,M}^n\}$ from $\{a_{k,M}^n\}$ for the interoctave scale algorithm (8.19), we use the same change-of-bases sequence $\boldsymbol{r}$ as in (7.47).

8.4 EXAMPLES

In this section we present a few examples to illustrate the FIWT algorithm discussed in this chapter. The graphs shown are the centered integral wavelet transform (CIWT), defined with respect to the spline wavelet ψ_m as

$$W_{\psi_m} f(b,a) := a^{-1/2} \int_{-\infty}^{\infty} f(t)\overline{\psi_m\left(\frac{t-b}{a} + t^*\right)}\, dt, \tag{8.34}$$

where

$$t^* = \frac{2m-1}{2}. \tag{8.35}$$

Observe that the IWT as defined by (4.32) does not indicate the location of the discontinuity of a function properly, since the spline wavelets are not symmetrical with respect to the origin. The CIWT circumvents this problem by shifting the location of the IWT in the time axis toward the right by at^*.

The integral wavelet transform of a function gives local time-scale information. To get the time-frequency information we need to map the scale parameter to frequency. There is no general way of doing so. However, as a first approximation, we may consider the following mapping:

$$a \longmapsto f := \frac{c}{a}, \tag{8.36}$$

where $c > 0$ is a calibration constant. In this book the constant c has been determined based on the one-sided center (ω_+^*) and one-sided radius $(\Delta\hat{\psi}_+)$ of the wavelet $\hat{\psi}(\omega)$, which are defined in Chapter 4.

For the cubic spline wavelet we get $\omega_+^* = 5.164$ and $\Delta\hat{\psi}_+ = 0.931$. The corresponding figures for the linear spline wavelet are 5.332 and 2.360, respectively. Based on these parameters, we choose values of c as 1.1 for cubic spline and 1.5 for linear spline cases. It is important to point out that these values of c may not be suitable for all cases. Further research in this direction is required. We have chosen c by taking the lower cutoff frequency of $\hat{\psi}(\omega)$.

8.4.1 IWT of a Linear Function

To compare the results obtained by the method presented in this chapter with the results obtained by evaluating the integral of (8.34), we first take the linear function which changes slope as shown in Figure 8.7. The function is sampled with 0.25 as the step size. So for linear splines, it means that the function is mapped into V_2, whereas for cubic splines the function is mapped into V_3. We choose $N = 1$, which gives one additional scale between two consecutive octaves. It is clear from Figures 8.8 and 8.9 that the FIWT algorithm and direct integration give identical results for wavelet coefficients for octave levels, but there are errors in the results for interoctave levels as discussed before.

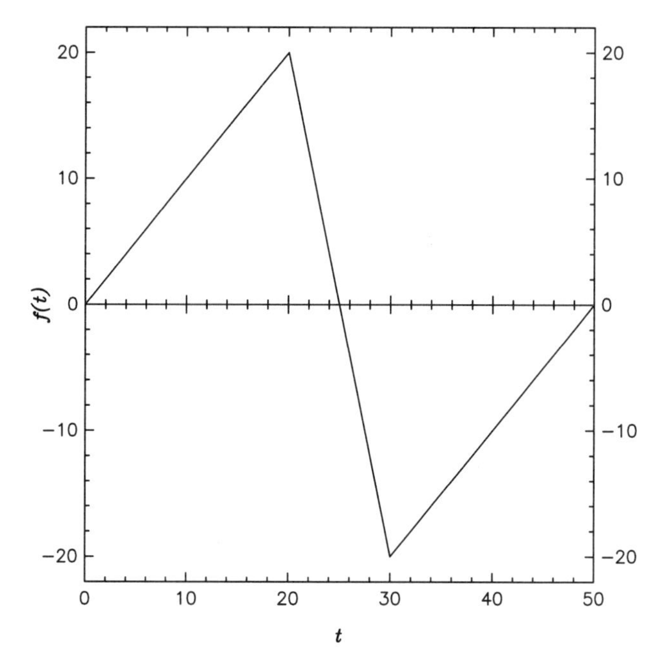

FIGURE 8.7 Linear function whose WT is shown in Figures 8.8 to 8.10. (Reprinted with permission from [8], copyright © 1995 by IEEE.)

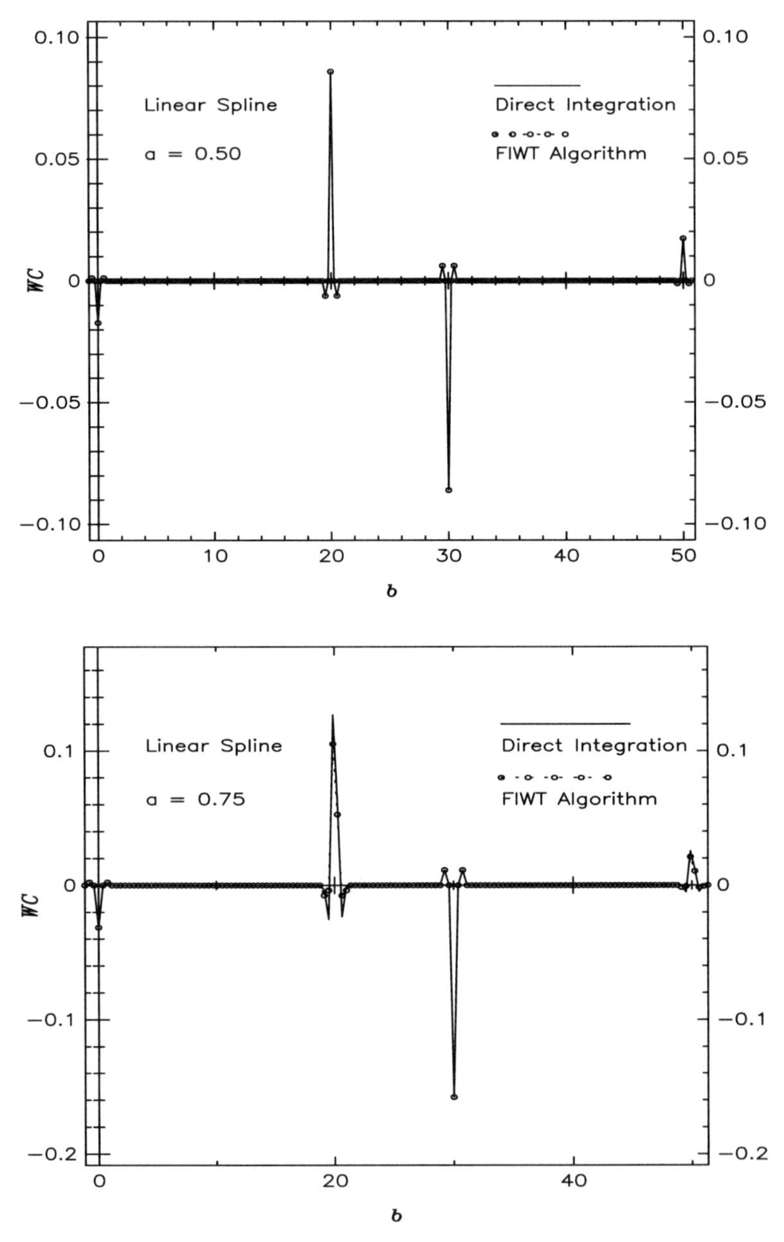

FIGURE 8.8 IWT of the function shown in Figure 8.7 using the linear spline wavelet for $a = 0.50$ and $a = 0.75$. Direct integration is performed with $f_2(t)$. (Reprinted with permission from [8], copyright © 1995 by IEEE.)

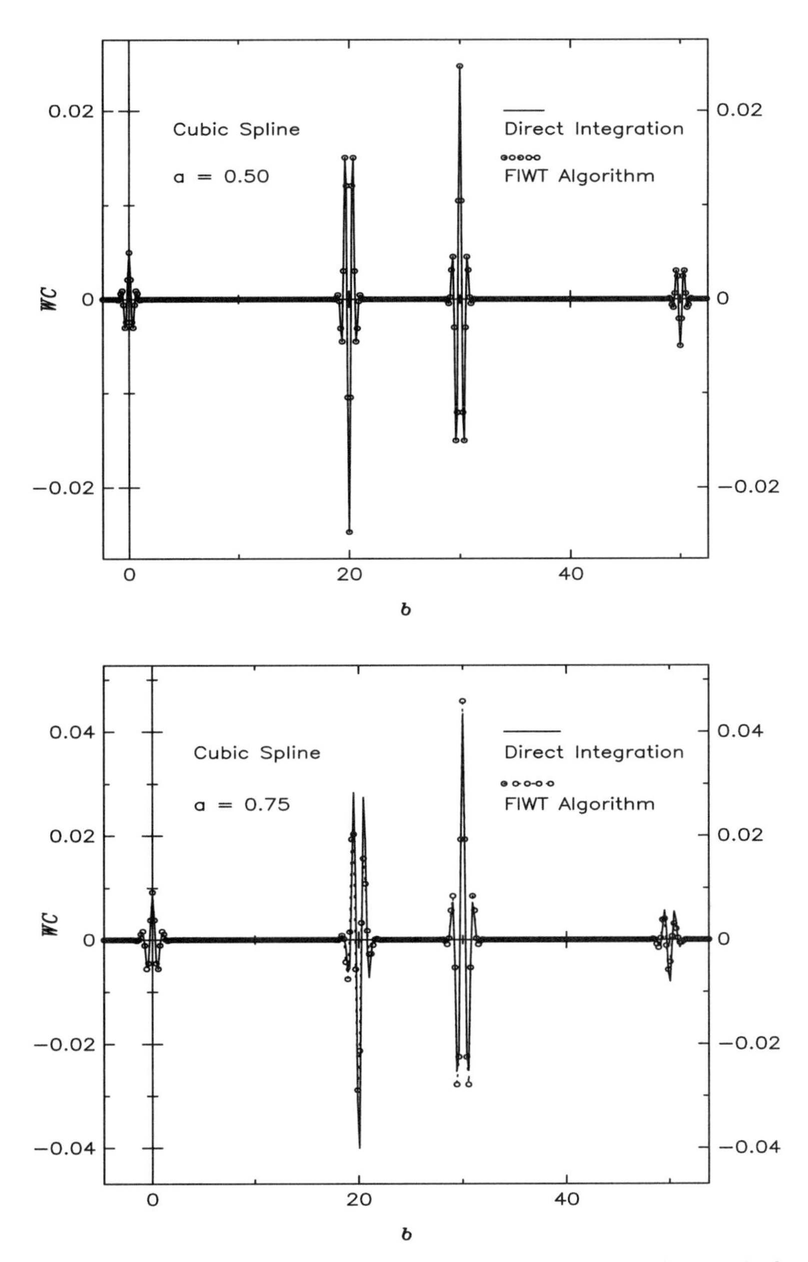

FIGURE 8.9 IWT of the function shown in Figure 8.7 using the cubic spline wavelet for $a = 0.50$ and $a = 0.75$. Direct integration is performed with $f_3(t)$. (Reprinted with permission from [8], copyright © 1995 by IEEE.)

The importance of the moment property becomes clear from Figures 8.8 and 8.9. In both the linear and cubic cases, when the wavelet is completely inside the smooth region of the function, the wavelet coefficients (WC) are close to zero since the function is linear. Wherever the function changes the slope, the wavelet coefficients have larger magnitudes. We also observe the edge effects near $t = 0$ and $t = 50$. The edge effects can be avoided by using special wavelets near the boundaries. Such boundary wavelets are discussed in Chapter 10. If we use the IWT instead of the CIWT, the entire plot will be shifted toward the left, and the shift will continue to become larger for lower levels. For Figures 8.8 and 8.9, direct evaluation of (8.34) is done with $f_2(t)$ and $f_3(t)$, respectively. In Figure 8.10, direct integration is done with $f_{3,1}(t)$, which indicates that for interoctave levels also, the FIWT algorithm gives results identical to those obtained with the corresponding approximation function.

For Figure 8.9, 440 wavelet coefficients have been computed. Direct integration takes about 300 times the CPU time of the FIWT algorithm. We wish to emphasize that the ratio of 300:1 is minimal, since with the increase in scale parameter a, the complexity of the direct integration method increases exponentially, while for the FIWT it remains almost constant. Furthermore, in the FFT-based algorithm, the complexity increases with a.

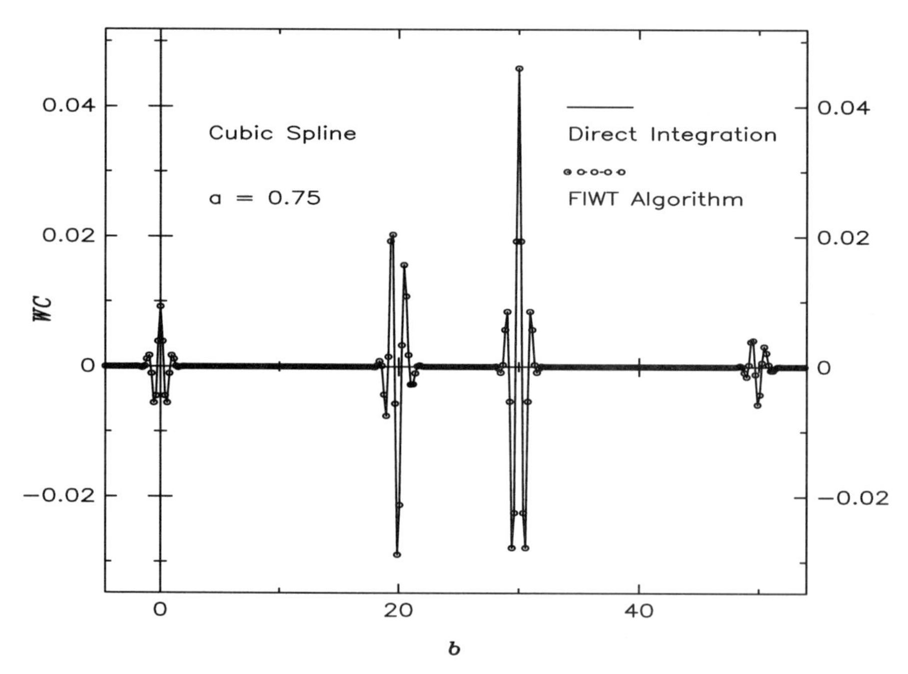

FIGURE 8.10 IWT of the function shown in Figure 8.7 using cubic spline wavelet for $a = 0.75$. Direct integration is performed with $f_{3,1}(t)$, the approximation of the function of Figure 8.7 at $s = 3$, $n = N = 1$. (Reprinted with permission from [8], copyright © 1995 by IEEE.)

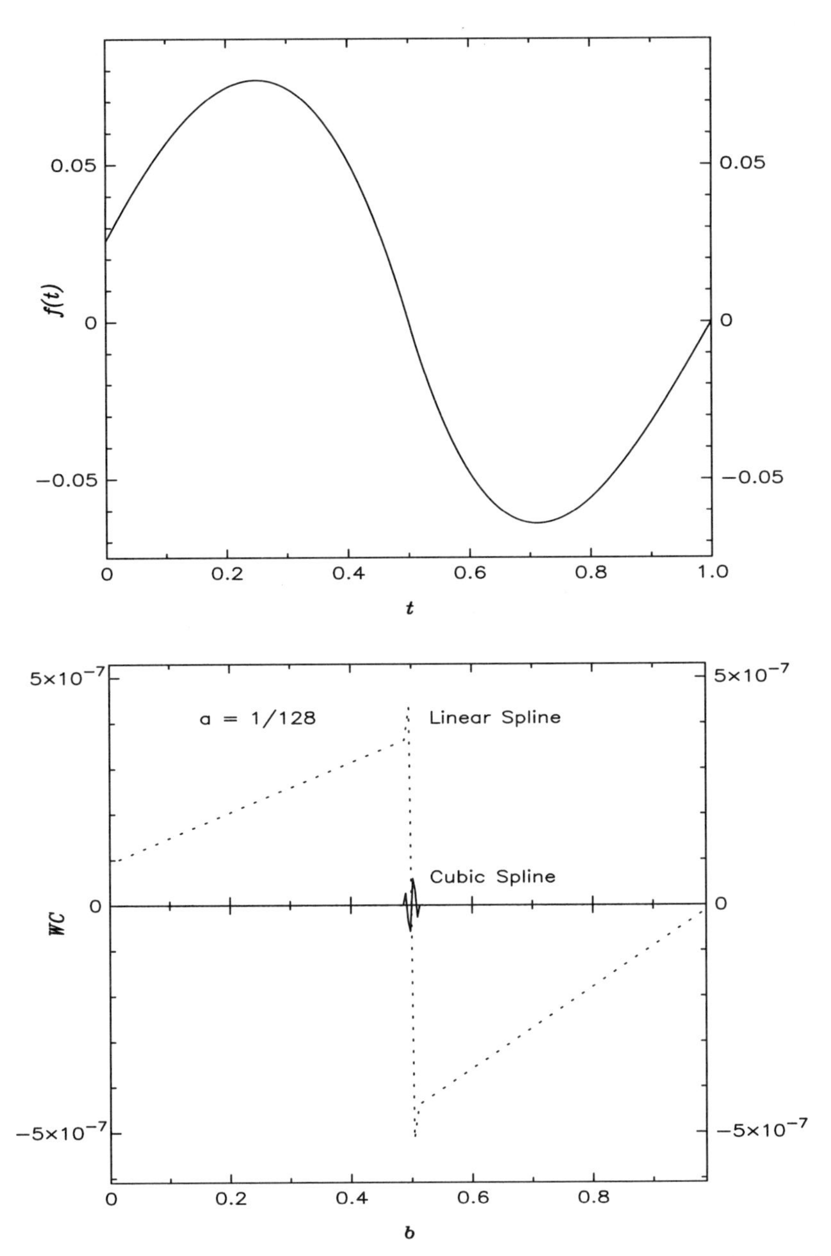

FIGURE 8.11 Function given by (8.37) and its IWT using linear and cubic spline wavelets for $a = 1/128$. (Reprinted with permission from [8], copyright © 1995 by IEEE.)

8.4.2 Crack Detection

As a further example to highlight the importance of the IWT in identifying the change in function behavior, we consider the following function. For $y := 2t - 1$

$$f(t) := \begin{cases} -\frac{3}{117} y(4y^2 + 16y + 13), & t \in [0, 1/2] \\ -\frac{1}{6} y(y-1)(y-2), & t \in (1/2, 1]\,. \end{cases} \tag{8.37}$$

Figure 8.11 shows the function and its wavelet coefficients for linear and cubic spline cases. The edge effect has not been shown. Once again, here we observe that for the cubic spline case, the wavelet coefficients are close to zero in the smooth region of the function; however, for the linear spline case, the wavelet coefficients are nonzero in this region since the function is of degree three in both intervals. This example shows that even a physically unnoticeable discontinuity can be detected using the wavelet transform.

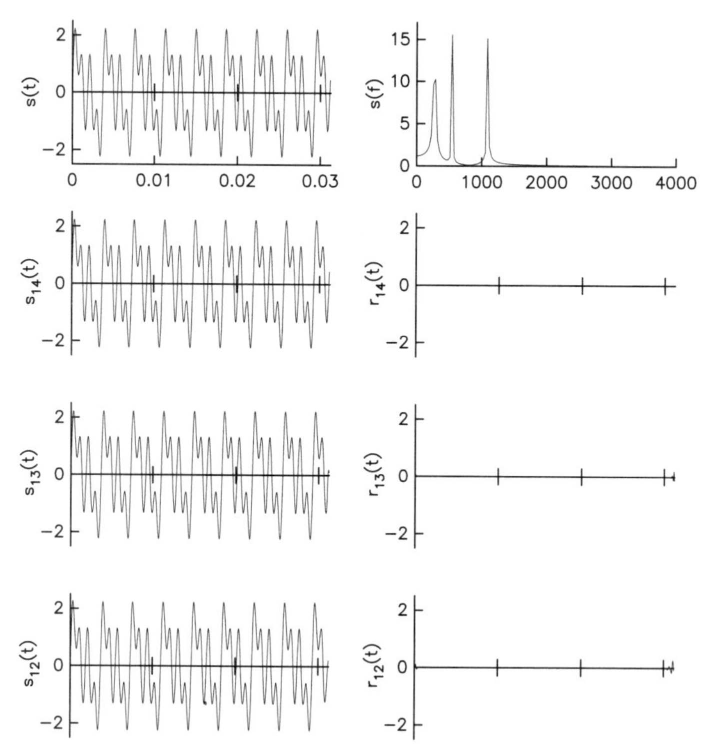

FIGURE 8.12 Decomposition of a signal composed of three sinusoids with different frequencies corresponding to nonoctave scales using the standard algorithm of Chapter 6. (Reprinted with permission from [7], copyright © 1995 by Plenum Press.)

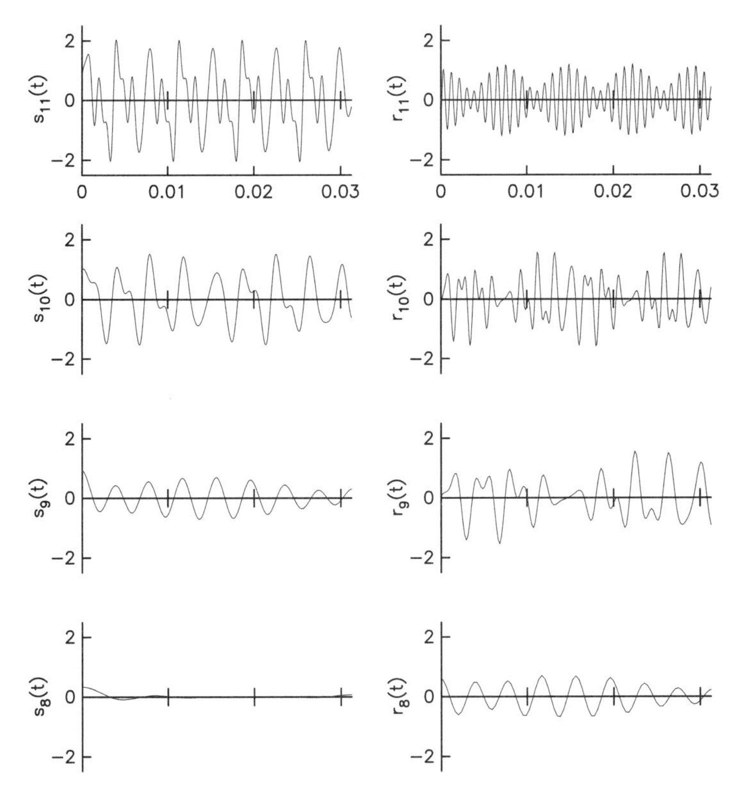

FIGURE 8.13 Decomposition of a signal with three frequency components (continued from Figure 8.12). (Reprinted with permission from [7], copyright © 1995 by Plenum Press.)

8.4.3 Decomposition of Signals with Nonoctave Frequency Components

To further emphasize the importance of the FIWT algorithm, we consider a composite function similar to that used in Chapter 7, but with slightly different frequencies that do not correspond to octave scales. Figures 8.12 and 8.13 indicate the inability of the standard decomposition algorithm of Chapter 7 to separate those frequencies that do not correspond to octave scale. Figures 8.14 and 8.15 show, on the other hand, that by properly selecting the values of n and N, we can separate any frequency band that we desire.

8.4.4 Perturbed Sinusoidal Signal

Figure 8.16 gives the time–frequency representation of a function that is composed of two sinusoids and two delta functions, represented as sharp changes in some data values. Observe that two sinusoids appear as two bands parallel to the time axis, whereas the delta functions are indicated by two vertical bands parallel to the fre-

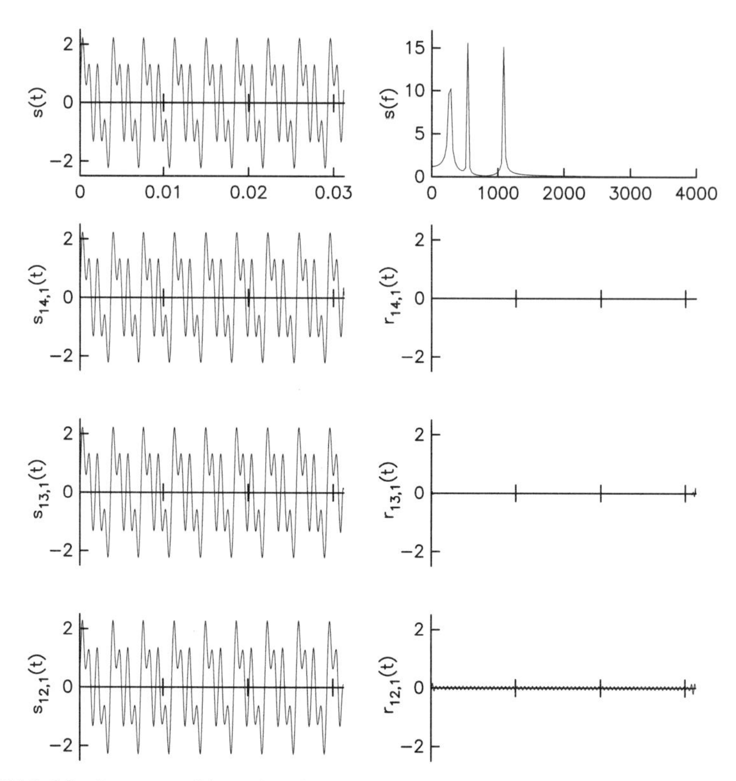

FIGURE 8.14 Decomposition of a signal composed of three sinusoids with different frequencies corresponding to nonoctave scales using the FIWT algorithm with $n = 1$ and $N = 2$. (Reprinted with permission from [7], copyright © 1995 by Plenum Press.)

quency axis. As discussed in Chapter 4, the frequency spread is due to the finite window width of the wavelets.

8.4.5 Chirp Signal

Figures 8.17 and 8.18 show the CIWT of a chirp signal with respect to linear and cubic spline wavelets, respectively. In Figure 8.19 we have shown the CIWT of a chirped signal by applying the standard wavelet decomposition algorithm. Here the interoctave scales have been filled with values at the previous octave scales. Similarly, on the time axis, "holes" are filled with values from previous locations.

8.4.6 Music Signal with Noise

In Figure 8.20 we show the CIWT of a portion of a music signal with additive noise using the cubic spline wavelet as the analyzing wavelet. Here the music data have been assumed to be at the integer points.

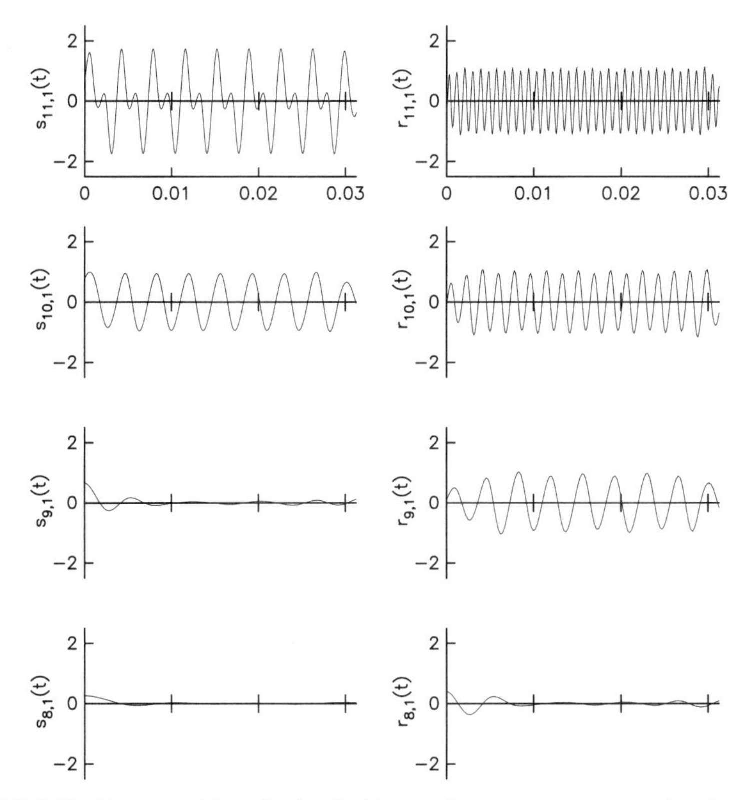

FIGURE 8.15 Decomposition of a signal with three frequency components (continued from Figure 8.14). (Reprinted with permission from [7], copyright © 1995 by Plenum Press.)

8.4.7 Dispersive Nature of the Waveguide Mode

As a final example, we find the wavelet transform of experimental data obtained for the transmission coefficient of an X-band rectangular waveguide. The waveguide is excited by a coaxial-line probe inserted through the center of the broad side of the waveguide. The scattering parameter S_{21}, of the waveguide is measured using an HP-8510 network analyzer by sweeping the input frequency from 2 to 17 GHz. The time-domain waveform is obtained by inverse Fourier-transforming the frequency-domain data. The time response (up to a constant multiplier) and the magnitude (in dB) of the frequency response are shown in Figure 8.21. It should be pointed out here that several low-amplitude impulses appeared in the negative time axis, but they have not been taken into account while performing the wavelet decomposition since they represent some unwanted signals and can be removed from the plot by proper thresholding. Furthermore, such an omission will not have any significant effect on the WC plot of Figure 8.21 because of the local nature of wavelet analysis.

The cutoff frequency and dispersive nature of the dominant TE_{10} is well observed from its time–frequency plot. Because of the guide dimension and excitation, the

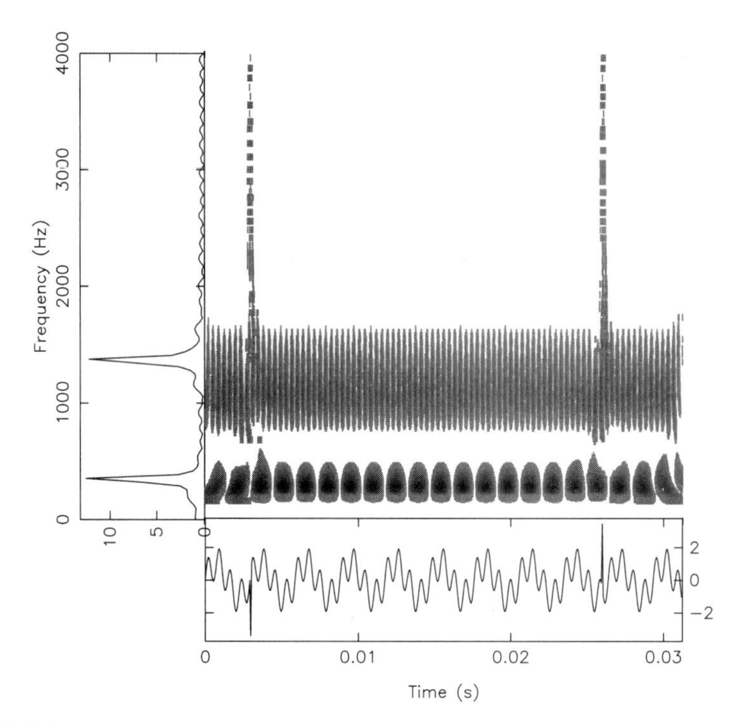

FIGURE 8.16 CIWT of a composite signal with some perturbed data (cubic spline).

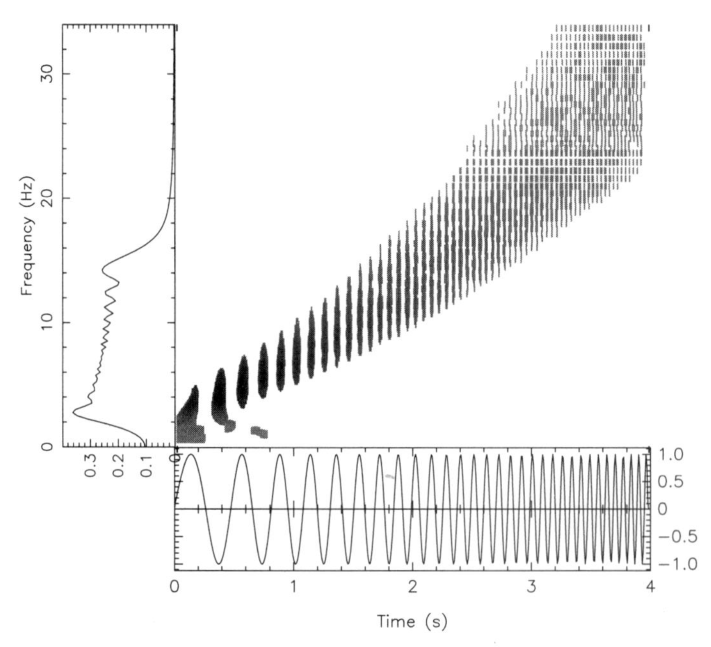

FIGURE 8.17 CIWT of a chirp signal (linear spline). (Reprinted with permission from [6], copyright © 1995 by Springer-Verlag.)

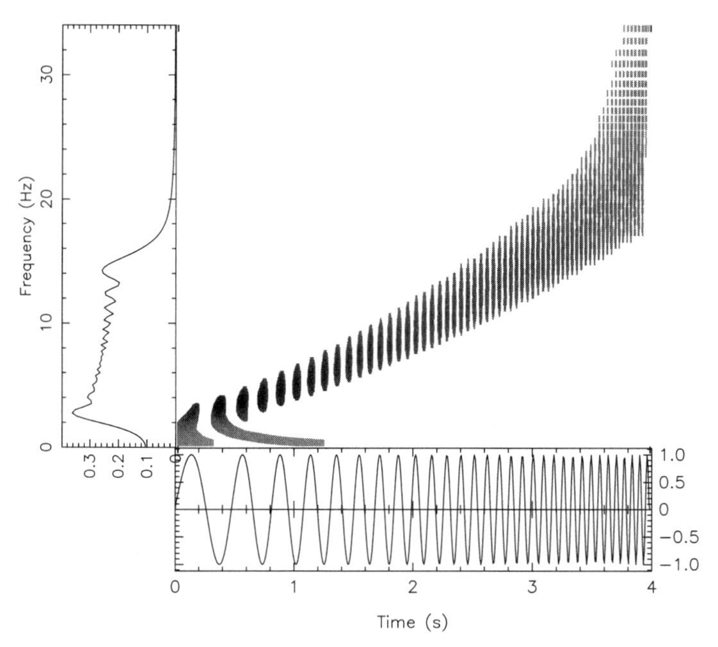

FIGURE 8.18 CIWT of a chirp signal (cubic spline). (Reprinted with permission from [6], copyright © 1995 by Springer-Verlag.)

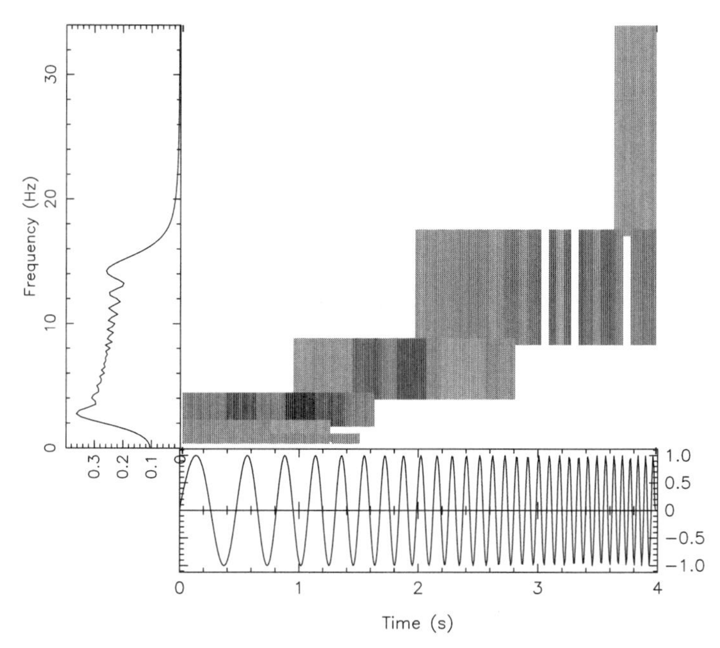

FIGURE 8.19 CIWT of a chirp signal using standard wavelet decomposition algorithm (cubic spline).

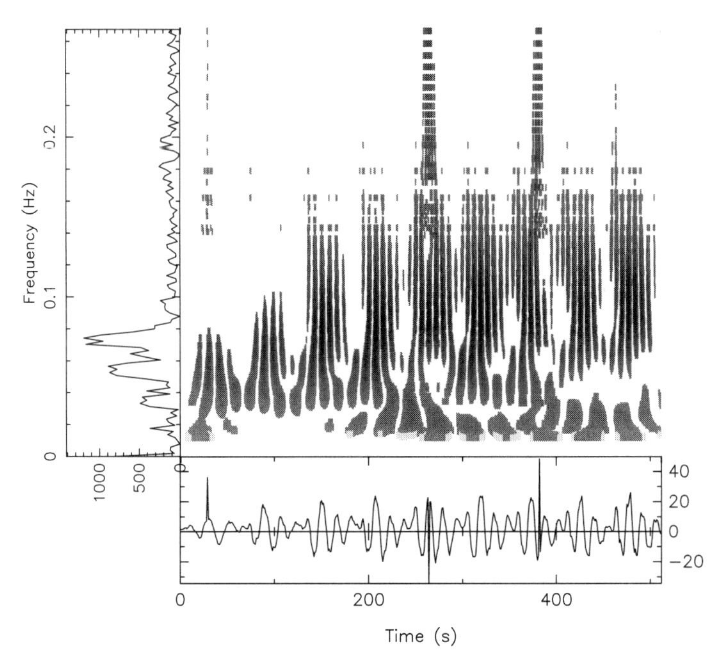

FIGURE 8.20 CIWT of a music signal with additive noise (cubic spline). (Reprinted with permission from [6], copyright © 1995 by Springer-Verlag.)

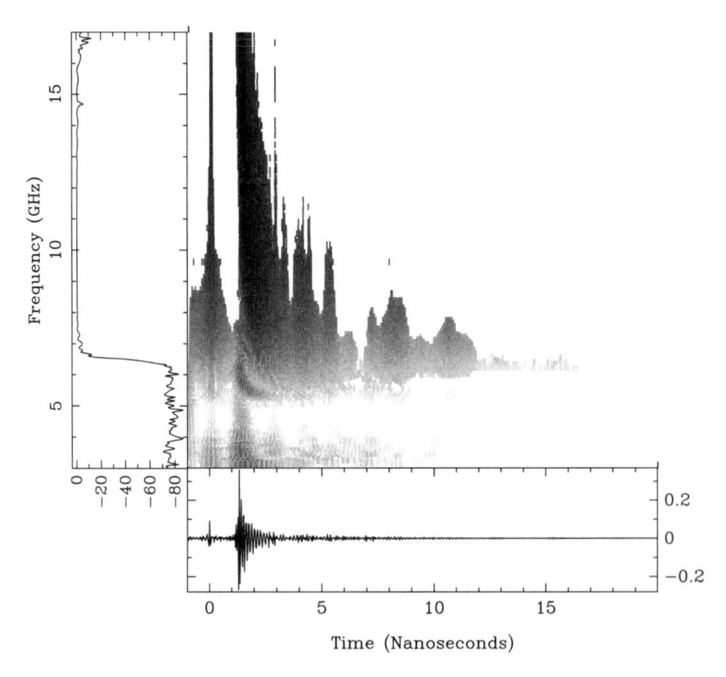

FIGURE 8.21 CIWT of the experimental data for the transmission coefficient of an X-band rectangular waveguide (cubic spline). (Reprinted with permission from [8], copyright © 1995 by IEEE.)

next-higher-order degenerate modes are TE_{11} and TM_{11} with the cutoff frequency 16.156 GHz. This does not appear on the plot. The plot indicates some transmission taking place below the lower-frequency operation. There is a short pulse at $t = 0$ which contains all the frequency components and is almost nondispersive. These can be attributed to system noise. No further attempt has been made to isolate the effects of various transitions used in the experiment. The thresholding for Figure 8.21 has been done with respect to the relative magnitude (in dB) of the local maximum of each frequency and the global maximum. Finally, the magnitude of the wavelet coefficients has been mapped to 8-bit gray-scale levels. Readers are referred to [9,10] for more applications of continuous wavelet transform to electromagnetic scattering data.

REFERENCES

1. L. G. Weiss, "Wavelets and wideband correlation processing," *IEEE Signal Process. Magazine*, pp. 13–32, January 1994.
2. O. Rioul and P. Duhamel, "Fast algorithms for discrete and continuous wavelet transforms," *IEEE Trans. Inform. Theory*, **38**, pp. 569–586, March 1992.
3. M. J. Shensa, "The discrete wavelet transform: Wedding the À trous and Mallat algorithms," *IEEE Trans. Signal Process.*, **40**, pp. 2464–2482, October 1992.
4. A. Grossmann, R. Kronland-Martinet, and J. Morlet, "Reading and understanding continuous wavelet transform," in *Wavelets, Time-Frequency Methods and Phase Space*, J. N. M. Combes, A. Grossmann, and Ph. Tchamitchian (Eds.), Berlin: Springer-Verlag, 1989, pp. 2–20.
5. I. Daubechies, *Ten Lectures on Wavelets.* CBMS-NSF Ser. Appl. Math. 61. Philadelphia: SIAM, 1992.
6. C. K. Chui, J. C. Goswami, and A. K. Chan, "Fast integral wavelet transform on a dense set of time-scale domain," *Numer. Math.*, **70**, pp. 283–302, 1995.
7. J. C. Goswami, A. K. Chan, and C. K. Chui, "On a spline-based fast integral wavelet transform algorithm," in *Ultra-Wideband Short Pulse Electromagnetics 2*, H. L. Bertoni, L. Carin, L. B. Felsen, and S. U. Pillai (Eds.). New York: Plenum Press, 1995, pp. 455–463.
8. J. C. Goswami, A. K. Chan, and C. K. Chui, "An application of fast integral wavelet transform to waveguide mode identification," *IEEE Trans. Microwave Theory Tech.*, **43**, pp. 655–663, March 1995.
9. H. Kim and H. Ling, "Wavelet analysis of radar echo from finite-size target," *IEEE Antennas Propag. Magazine*, **41**, pp. 200–207, Febuary 1993.
10. L. Carin and L. B. Felsen, "Wave-oriented data processing for frequency- and time-domain scattering by nonuniform truncated arrays," *IEEE Trans. Antennas Propag. Magazine*, pp. 29–43, June 1994.

CHAPTER NINE

Digital Signal Processing Applications

The introduction of wavelets to signal and image processing has provided a very flexible tool for engineers to use to create innovative techniques for solving various engineering problems. A survey of recent literature on wavelet signal processing shows the focus on using the wavelet algorithms for processing one-dimensional (1-D) and two-dimensional (2-D) signals. Acoustic, speech, music, and electrical transient signals are popular in 1-D wavelet signal processing. The 2-D wavelet signal processing involves mainly image compression and target identification. Problem areas include noise reduction, signature identification, target detection, signal and image compression, and interference suppression. We make no attempt to detail techniques in these areas; neither are we trying to provide readers with processing techniques at the research level. Several examples are given in this chapter to demonstrate the advantages and flexibility of using wavelets in signal and image processing.

In these examples, wavelet algorithms are working in synergy with other processing techniques to yield a satisfactory solution to the problem. Wavelet decomposition plays the vital role in separating the signal into components before other DSP techniques are applied. Algorithms include wavelet tree, wavelet-packet tree decomposition, 2-D wavelet or wavelet packet tree decomposition, pyramid or direction decomposition, and others. In signature recognition and target detection, the corresponding reconstruction algorithm is not needed since the signal components are either destroyed or rendered useless after processing. In the last two examples, the orthogonality between wavelet packets is applied to multicarrier communication systems, and the wavelet algorithms are extended to the third dimension for 3-D medical image visualization. We discuss the extension of the wavelet algorithms to wavelet packet algorithms and their 2-D versions before we discuss various application examples.

9.1 WAVELET PACKETS

Because of the two-scale relation and the choice of the scale parameter $a = 2^s$, the hierarchical wavelet decomposition produces signal components whose spectra form consecutive octave bands. Figure 9.1 depicts this concept graphically. In certain applications, the wavelet decomposition may not generate a spectral resolution fine enough to meet the problem requirements. One approach is to use the CWT to obtain the necessary finer resolution by substituting a smaller increment for the scale parameter a. This approach increases the computation load by orders of magnitude. Another approach has been discussed in Chapter 8. The use of wavelet packets also helps to avoid this problem. A wavelet packet is a generalization of a wavelet in that each octave frequency band of the wavelet spectrum is further subdivided into finer frequency bands by using the two-scale relations repeatedly. In other words, the development of wavelet packets is a refinement of wavelets in the frequency domain and is based on a mathematical theorem proven by Daubechies [1] (*splitting trick*). The theorem is stated as follows:

If $f(\cdot - k)\,|_{k\in\mathbb{Z}}$ forms an orthonormal basis and

$$F_1(x) = \sum_k g_0[k] f(x-k) \tag{9.1}$$

$$F_2(x) = \sum_k g_1[k] f(x-k), \tag{9.2}$$

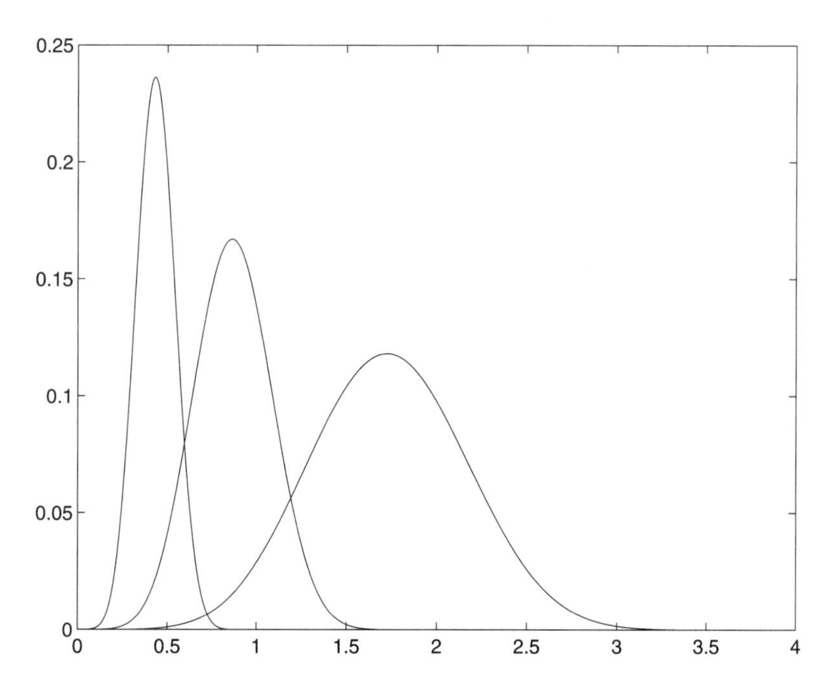

FIGURE 9.1 Constant Q spectra for wavelets at different resolutions.

then $\{F_1(\cdot - 2k), F_2(\cdot - 2k); k \in \mathbb{Z}\}$ is an orthonormal basis of $E = \text{span}\{f(\cdot - n); n \in \mathbb{Z}\}$.

This theorem is obviously true when f is the scaling function ϕ since the two-scale relations for ϕ and the wavelet ψ give

$$\mathbf{A}_j \ni \phi(2^j t) = \sum_k g_0[k]\phi(2^{j+1}t - k)$$

$$\mathbf{W}_j \ni \psi(2^j t) = \sum_k g_1[k]\phi(2^{j+1}t - k).$$

If we apply this theorem to the $\mathbf{W}_j$ spaces, we generate wavelet packet subspaces. The general recursive formulas for wavelet packet generation are

$$\mu_{2\ell}(t) = \sum_k g_0[k]\mu_\ell(2t - k) \tag{9.3}$$

$$\mu_{2\ell+1}(t) = \sum_k g_1[k]\mu_\ell(2t - k) \; k \in \mathbb{Z}, \tag{9.4}$$

where $\mu_0 = \phi$ and $\mu_1 = \psi$ are the scaling function and the wavelet, respectively. For $\ell = 1$, we have the wavelet packets μ_2 and μ_3 generated by the wavelet $\mu_1 = \psi$. This process is repeated so that many wavelet packets can be generated from the two-scale relations. The first eight wavelet packets for the Haar function and $\phi_{D;3}$ (also referred to as D_3) together with their spectra are shown in Figures 9.2 to 9.5. The translates of each of these wavelet packets form an orthogonal basis and the wavelet packets are orthogonal to one another within the same family generated by an orthonormal scaling function. We can decompose a signal into many wavelet packet components. We remark here that a signal may be represented by a selected set of wavelet packets without using every wavelet packet for a given level of resolution. An engineering practitioner may construct an algorithm to choose the packets for optimizing a certain measure (such as energy, entropy, variance, etc.). Best-basis and best-level are two popular algorithms for signal representations. Readers can find these algorithms in [2].

9.2 WAVELET PACKET ALGORITHMS

The decomposition tree for wavelet packets uses the same decomposition block of two parallel filtering channels followed by decimation by 2 ($\downarrow 2$) as in the wavelet algorithm. Any coefficient set in the tree may be processed by this block. In the wavelet decomposition tree, only the approximation coefficient sets $\{\mathbf{a}\}$ in Figure 7.9 are processed for different resolutions s, while the wavelet coefficient sets $\{\mathbf{w}\}$ are outputs of the algorithm. In wavelet packet decomposition, the wavelet coefficient sets $\{\mathbf{w}\}$ are also processed by the same building block to produce wavelet packet coefficient sets $\{\boldsymbol{\pi}\}$. We see from Figure 9.6 that for each set of N coefficients, we obtain two coefficient sets of $N/2$ length after processing by the decomposition block. The number

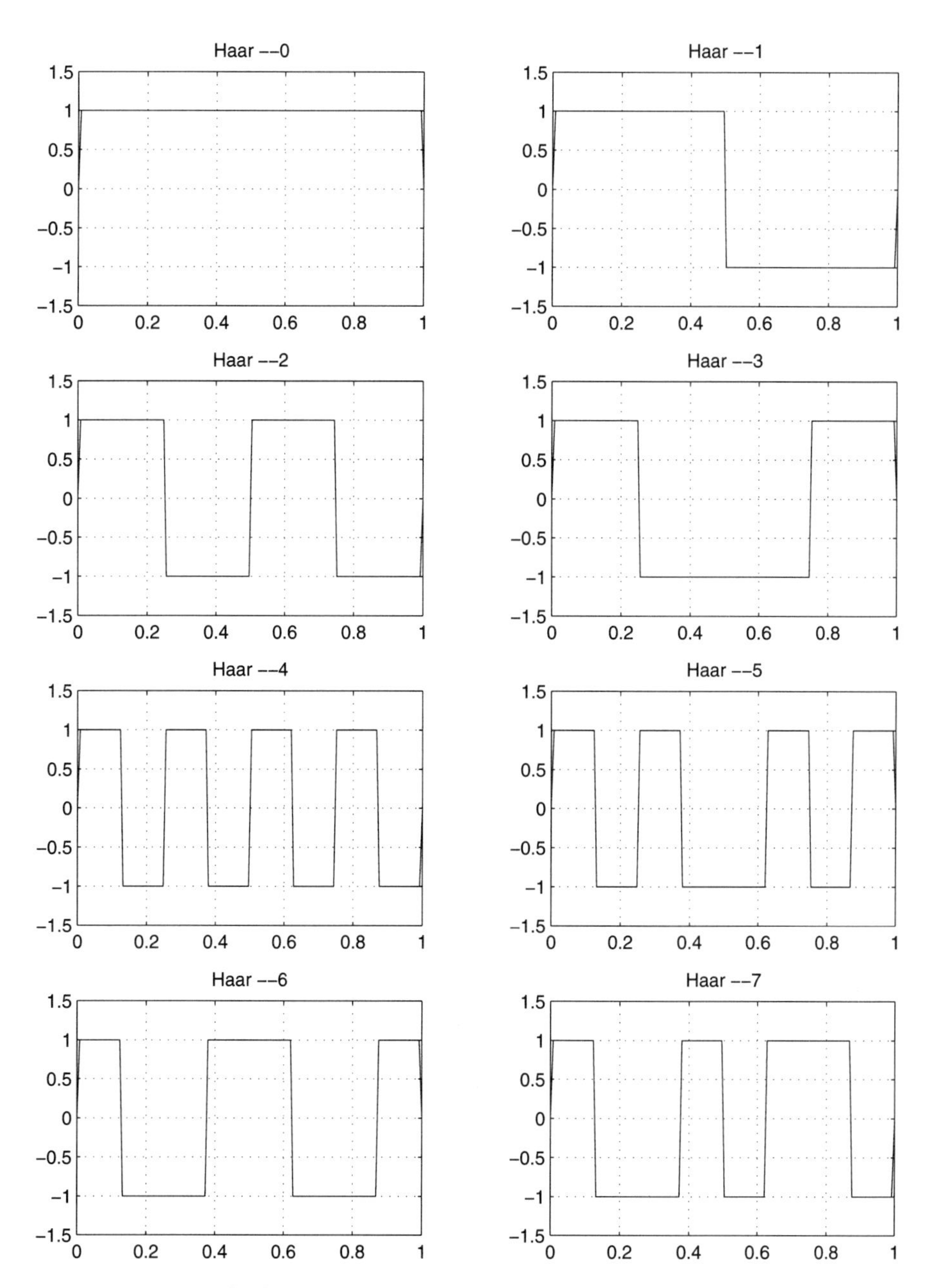

FIGURE 9.2 Wavelet packets of Haar scaling function.

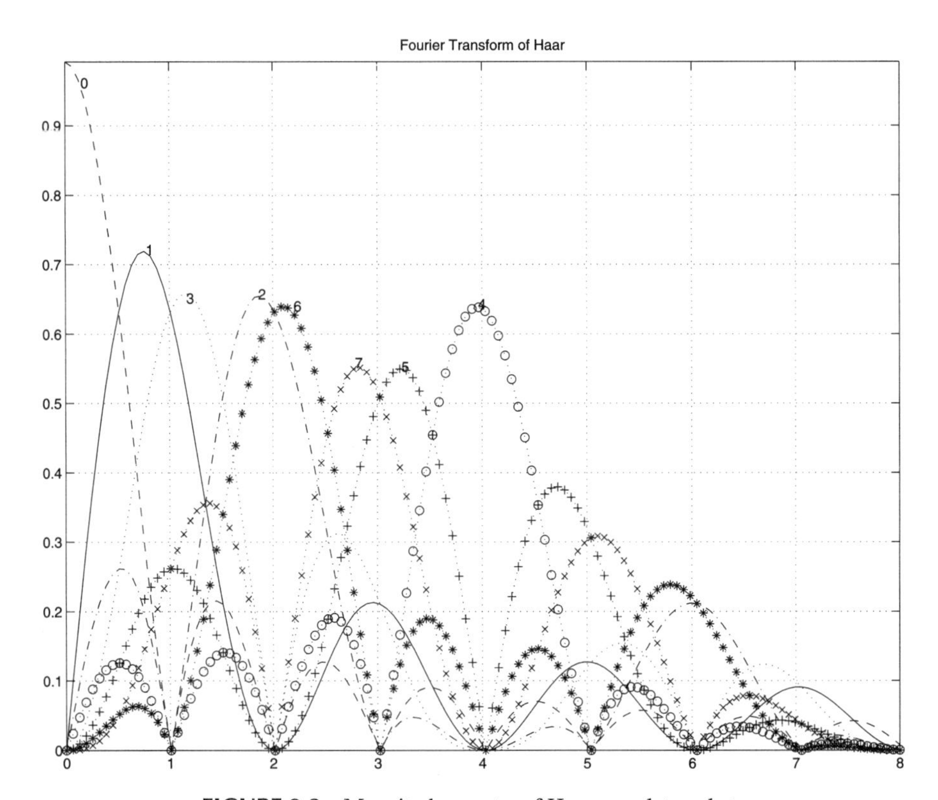

FIGURE 9.3 Magnitude spectra of Haar wavelet packets.

of coefficient sets is 2^m if the original coefficient set is processed for m resolutions. Figure 9.6 demonstrates the wavelet packet tree for $m = 2$.

It is important to keep track of the indices of the wavelet packet coefficients in the decomposition algorithm. To achieve perfect reconstruction, if a coefficient set has been processed by $h_0[n]$ and $(\downarrow 2)$, the result should be processed by $g_0[n]$ and $(\uparrow 2)$. The same order is applicable to $h_1[n]$ and $g_1[n]$. For example, if we process a set of data first by $h_0[n]$ and $(\downarrow 2)$ followed by $h_1[n]$ and $(\downarrow 2)$, the resulting signal must be processed by $(\uparrow 2)$ and $g_1[n]$ and then followed by $(\uparrow 2)$ and $g_0[n]$ to achieve perfect reconstruction. Thus signal processing using wavelet packets requires accurate bookkeeping of different orders of digital filtering and sampling rate changes.

9.3 THRESHOLDING

Thresholding is one of the most commonly used processing tools in wavelet signal processing. It is widely used in noise reduction, signal and image compression, and sometimes in signal recognition. We consider three simple thresholding methods [3]

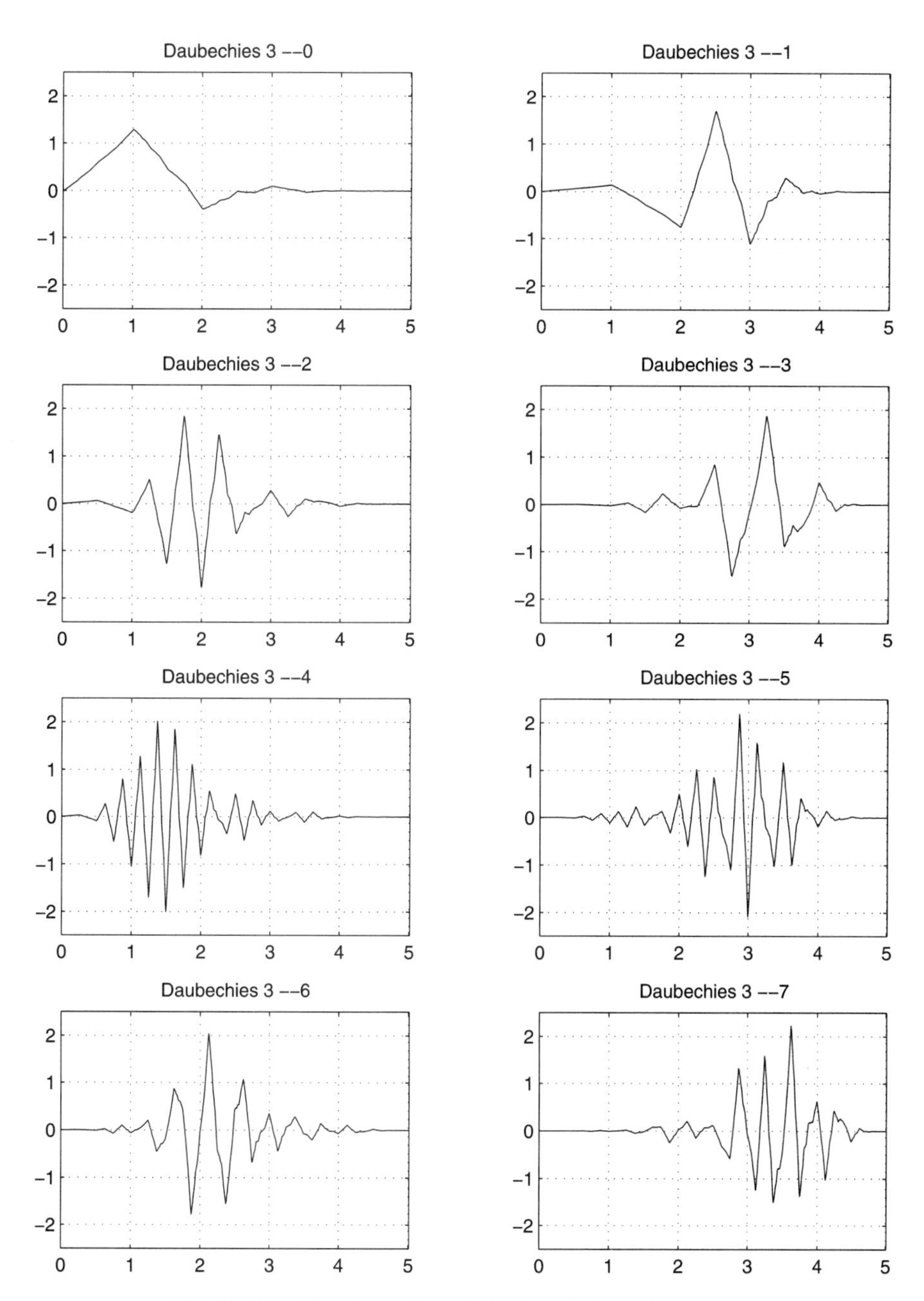

FIGURE 9.4 Wavelet packets of Daubechies 3 scaling function.

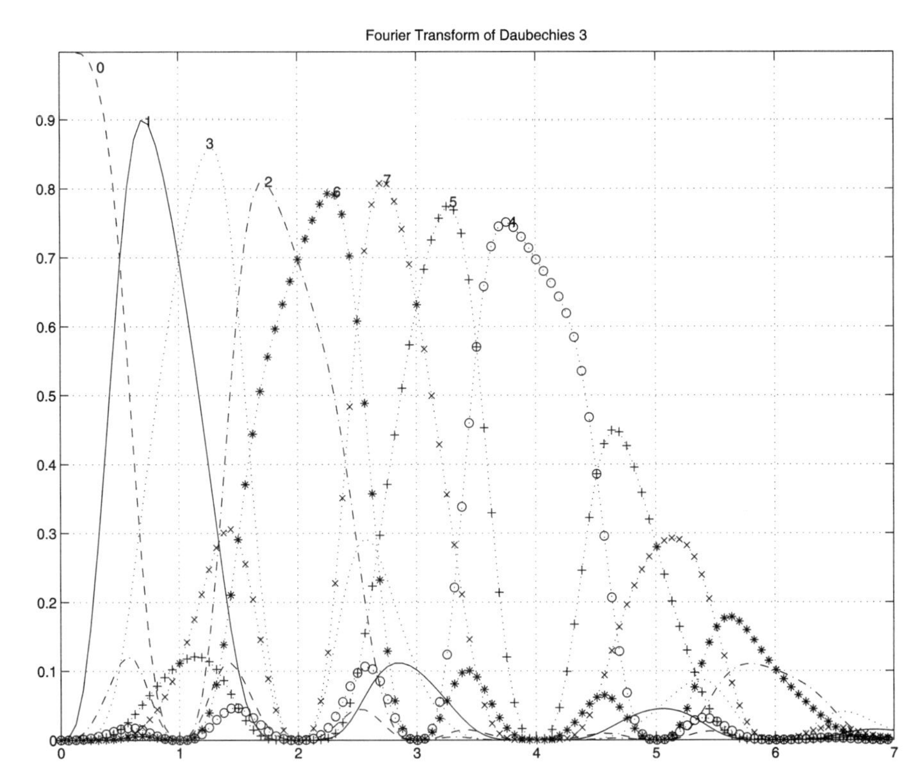

FIGURE 9.5 Magnitude spectra of Daubechies 3 wavelet packets.

here: (1) hard thresholding, (2) soft thresholding, and (3) percentage thresholding. The choice of thresholding method depends on the application. We discuss each type briefly.

9.3.1 Hard Thresholding

Hard thresholding is sometimes called *gating*. If a signal (or a coefficient) value is below a preset value, it is set to zero. That is,

$$y = \begin{cases} x & \text{for } |x| \geq \sigma \\ 0 & \text{for } |x| < \sigma, \end{cases} \tag{9.5}$$

where σ is the threshold value or the gate value. The graphical representation of the hard threshold is shown in Figure 9.7. Notice that the graph is nonlinear and discontinuous at $x = \sigma$.

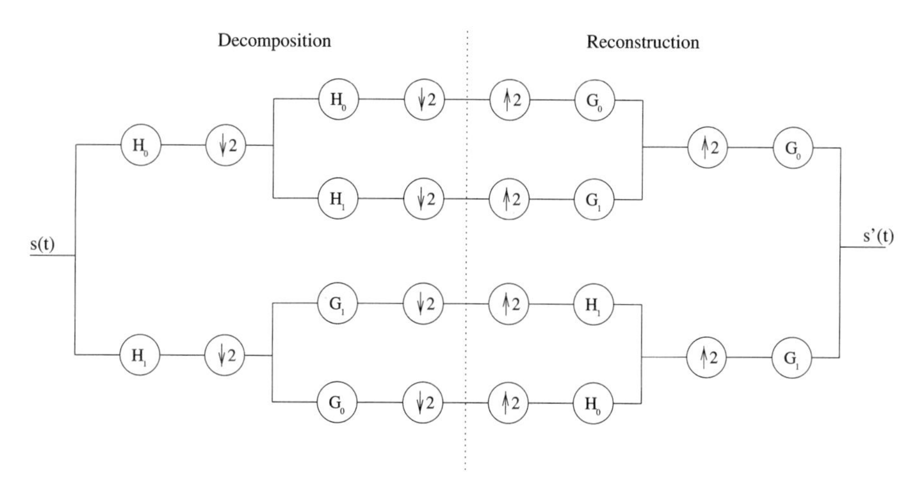

FIGURE 9.6 Block diagram for the decomposition and reconstruction algorithms for wavelet packets.

9.3.2 Soft Thresholding

Soft thresholding is defined as

$$y = \begin{cases} \operatorname{sgn}(x) f(|x| - \sigma) & \text{for } |x| \geq \sigma \\ 0, & \text{for } |x| < \sigma. \end{cases} \tag{9.6}$$

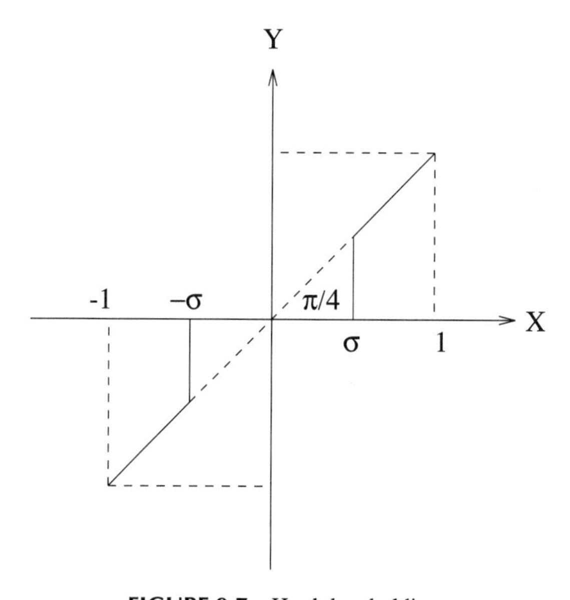

FIGURE 9.7 Hard thresholding.

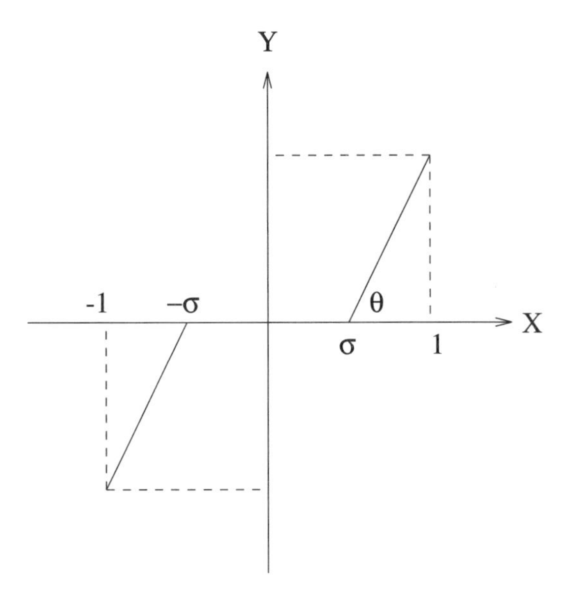

FIGURE 9.8 Soft thresholding.

The function $f(x)$ is generally a linear function (a straight line with slope to be chosen; see Figure 9.8). However, spline curves of third or fourth order may be used to effectively weight values greater than σ. In some signal compression applications, using a quadratic spline curves of order $m > 2$ may affect the compression ratio by a small amount.

9.3.3 Percentage Thresholding

In certain applications such as image compression, where a bit quota has been assigned to the compressed file, it is more advantageous to set a certain percentage of wavelet coefficients to zero to satisfy the quota requirement. In this case, the setting of the threshold value σ is based on the histogram of the coefficient set and the total number of coefficients. The thresholding rule is the same as hard thresholding once we have determined the threshold σ.

9.3.4 Implementation

Implementations of the hard, soft, and percentage thresholding methods are quite simple: One simply subtracts the threshold value from the magnitude of each coefficient. If the difference is negative, the coefficient is set to zero. If the difference is positive, no change is applied to the coefficient. To implement soft thresholding by using a linear function of unit slope, the thresholding rule is

$$y = \begin{cases} \operatorname{sgn}(x)(|x| - \sigma) & \text{if } |x| - \sigma \geq 0 \\ 0 & \text{if } |x| - \sigma < 0. \end{cases} \tag{9.7}$$

9.4 INTERFERENCE SUPPRESSION

The Wigner–Ville distribution and other nonlinear time–frequency distributions are often used in radar signal processing. Although they are not linear transformations, they have an advantage in that a linear chirp signal appears as a straight line on the time–frequency plane (see Chapter 4). However, the nonlinear distribution of a multicomponent signal produces interference that may have a high amplitude to cover up the signal. This example combines the Wigner–Ville distribution decomposition and the wavelet packets to suppress the interference [4]. We take the signal with interference and decompose it optimally into frequency bands by a best-basis selection [2]. We then apply the WVD decomposition to each of the wavelet packet signals. The cross terms are deleted in the distribution before reconstruction. This approach keeps the high resolution of WVD, yet reduces the cross-term interference to a minimum.

From the viewpoint of time–frequency analysis, an orthonormal wavelet ψ generates an orthonormal basis $\{\psi_{j,k}\}$, $j, k \in \mathbb{Z}$, of $L^2(R)$ in such a way that for each $j \in \mathbb{Z}$, the subfamily $\{\psi_{j,k} : k \in \mathbb{Z}\}$ is not only an orthonormal basis of W_s but is also a time–frequency window for extracting local information within the sth octave band $H_j := (2^{j+1}\Delta^{+}_{\hat{\psi}}, 2^{j+2}\Delta^{+}_{\hat{\psi}}]$, where $\Delta^{+}_{\hat{\psi}}$ is the RMS bandwidth of the wavelet. Unlike wavelets for which the width of the frequency band H_s increases with the frequency ranges, wavelet packets are capable of partitioning the higher-frequency octaves to yield better frequency resolution. Here $\Delta^{+}_{\hat{\psi}}$, as discussed in Chapter 4, is the standard deviation of $\hat{\psi}$ relative to the positive frequency range $(0, \infty)$. Let $\{\mu_n\}$ be a family of wavelet packets corresponding to some orthonormal scaling function $\mu_0 = \phi$, as defined in Section 8.1. Then the family of subspaces $U_0^n = \langle \mu_n(\cdot - k) : k \in \mathbb{Z}\rangle$, $n \in \mathbb{Z}^+$, is generated by $\{\mu_n\}$, and W_j can be expressed as

$$W_s = U_0^{2^j} \oplus U_0^{2^j+1} \oplus \cdots \oplus U_0^{2^{j+1}-1}. \tag{9.8}$$

In addition, for each $m = 0, \ldots, 2^j - 1$, and $j = 1, 2 \ldots$, the family

$$\{\mu_{2^j+m}(\cdot - k) : k \in Z\} \tag{9.9}$$

is an orthonormal basis of $U_0^{2^j+m}$. The jth frequency band H_j is therefore partitioned into 2^j subbands:

$$H_j^m, \quad m = 0, \ldots, 2^j - 1. \tag{9.10}$$

Of course, the orthonormal basis in (9.8) of $U_0^{2^j+m}$ provides time localization within the subband H_j^m. Any function $s(x) \in L^2(\mathbb{R})$ has a representation

$$s(t) = \sum_{j=1}^{\infty} \sum_{m=0}^{2^j-1} \sum_{n=-\infty}^{\infty} d_n^{j,m} \mu_{2^j+m}(t-n) = \sum_{j,m} s_{j,m}(t), \tag{9.11}$$

where $d_n^{j,m} = \langle s(t), \mu_{2^j+m}(t-n)\rangle$, and the component

$$s_{j,m}(t) = \sum_n d_n^{j,m} \mu_{2^j+m}(t-n) \tag{9.12}$$

represents the signal content of $s(t)$ within the mth subband of the jth band.

Let us rewrite the WVD for a multicomponent signal,

$$\begin{aligned} \mathrm{WVD}_s(t, f) &= \sum_{j,m} \mathrm{WVD}_{s_{j,m}}(t, f) \\ &\quad + 2 \sum_{j,m,k,n;m\neq n; j\neq k} \mathrm{WVD}_{s_{j,m},s_{k,n}}(t, f). \end{aligned} \tag{9.13}$$

Equation (9.13) partitions the traditional WVD into two subsets. The first summation in (9.13) represents the auto-term, whereas the second summation represents the cross-terms between components in each subband to be considered as interference. By removing this interference, we obtain the wavelet packet–based cross-term-deleted representation (WPCDR), given by

$$\mathrm{WPCDR}_s(t, f) = \sum_{j,m} WVD_{s_{j,m}}(t, f). \tag{9.14}$$

We remark here that the WPCDR actually gives the auto WVD of the signal components within each subband; therefore, it is quite effective and is perhaps the best choice for analyzing a multicomponent signal. In addition, the WPCDR is computationally advantageous, since both decomposition and representation can be implemented efficiently.

Best-Basis Selection Equation (9.8) is actually a special case of

$$W_j = U_{j-k}^{2^k} \oplus U_{j-k}^{2^k+1} \oplus \cdots \oplus U_{j-k}^{2^{k+1}-1} \tag{9.15}$$

when $k = j$. Equation (9.15) means that the jth frequency band H_j can be partitioned into 2^k, $k = 0, 1, ..., j$, subbands

$$H_j^{k,m}, \qquad m = 0, \ldots, 2^k - 1. \tag{9.16}$$

The uniform division of the frequency axis and the logarithmic division for wavelets are just two extreme cases, when k in (9.16) takes on the values of j and 0, respectively. In fact, k is allowed to vary among H_j, $j = 1, 2, \ldots,$ so that the subbands adapt to (or match) the local spectra of the signal and thereby yield the best representation or the best basis of the signal. The best basis can be obtained by minimizing the global cost functional or entropy. Specifically, the following algorithm is used in [2] to find the adapted frequency subband or the equivalent best basis:

$$\widetilde{H}_j^{k,m} = \begin{cases} H_j^{k,m} & \text{if } E\big(H_j^{k,m}\big) < E\big(H_j^{k+1,2m}\big) \\ & \quad + E\big(H_j^{k+1,2m+1}\big) \\ \widetilde{H}_j^{k+1,2m} \bigcup \widetilde{H}_j^{k+1,2m+1}, & \text{otherwise,} \end{cases} \tag{9.17}$$

where $\widetilde{H}_j^{k,m}$ represents the adapted frequency subband and $E(H_j^{k,m})$ denotes the entropy of the local spectrum of the signal restricted to the $H_j^{k,m}$.

Although the minimum entropy-based best basis is useful for applications to segmentation in speech processing, it is not effective in our case, since the resultant distribution yields interference. We modify the algorithm to yield

$$\widetilde{H}_j^{k,m} = \begin{cases} H_j^{k,m} & \text{if } \mathrm{Var}(H_j^{k,m}) < \sigma \\ \widetilde{H}_j^{k+1,2m} \bigcup \widetilde{H}_j^{k+1,2m+1} & \text{otherwise,} \end{cases}$$

where $\mathrm{Var}(H_j^{k,m})$ denotes the variance of the local spectrum and σ is a preset threshold. The idea behind this algorithm is that a narrow analysis band should be used when the local spectrum is well concentrated or when a small variance is obtained, while a wide band should be used when the local spectrum is spread or variance is large. We note in passing that a best basis is usually obtained between the third and fourth layers, since deeper layers may yield some adverse effects due to the amplitude increase in their spectral sidelobes.

Once we have chosen a best basis, the signal is readily expressed as

$$\begin{aligned} s(t) &= \sum_{j=1}^{\infty} \sum_{m=0}^{2^k-1} \sum_{n=-\infty}^{\infty} d_n^{j;k,m} \mu_{2^k+m}(2^{j-k}t - n) \\ &= \sum_{j,m} s'_{j,m}(t), \end{aligned} \tag{9.18}$$

where $d_n^{j,m} = \langle s(t), \mu_{2^k+m}(2^{j-k}t - n) \rangle$ and

$$s'_{j,m}(t) = \sum_n d_n^{j;k,m} \mu_{2^k+m}(2^{j-k}t - n). \tag{9.19}$$

The WPCDR with a best-basis selection is given by

$$\mathrm{WPCDR}_s(t, f) = \sum_{j,m} \mathrm{WVD}_{s'_{j,m}}(t, f). \tag{9.20}$$

We apply this algorithm to a bicomponent signal consisting of a sinusoid and a linear chirp signal. When compared with the WVD (see Figure 9.9), the WPCDR of the same signal shown in Figure 9.10, the interference is suppressed. Figure 9.11 shows the WPCDR with a best-basis selection produces the highest resolution on the time–frequency plane.

9.5 FAULTY BEARING SIGNATURE IDENTIFICATION

9.5.1 Pattern Recognition of Acoustic Signals

Acoustic signal recognition has gained much attention in recent years. It is applicable to the recognition of ships from sonar signatures, cardiopulmonary diagnostics from

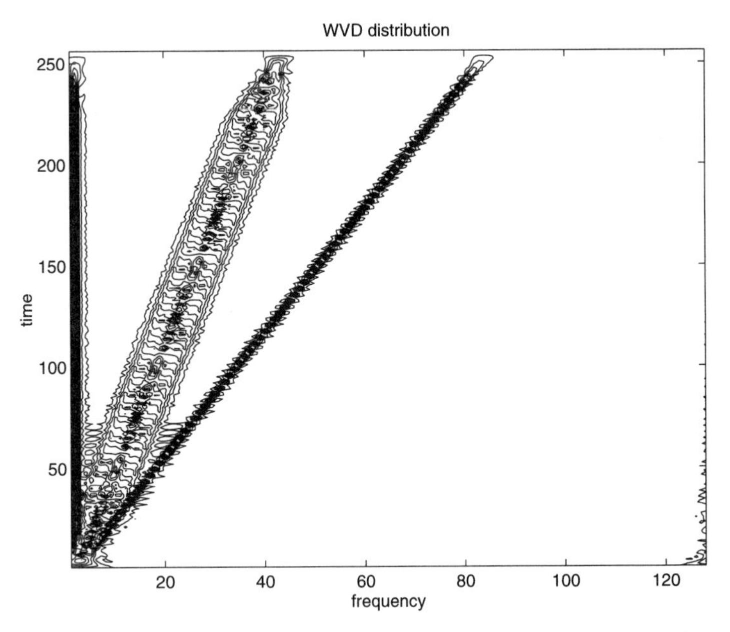

FIGURE 9.9 WVD of a bicomponent signal. (Reprinted with permission from [4], copyright © 1998 by Wiley.)

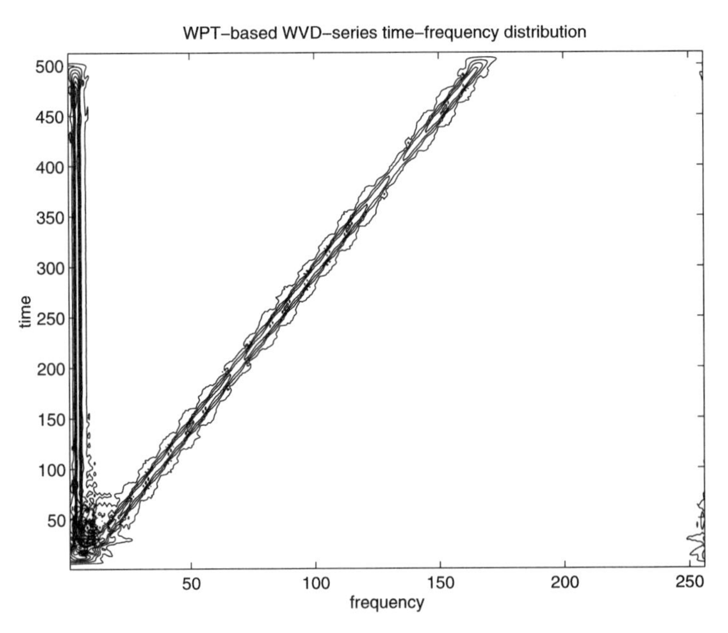

FIGURE 9.10 WPCDR of a bicomponent signal. (Reprinted with permission from [4], copyright © 1998 by Wiley.)

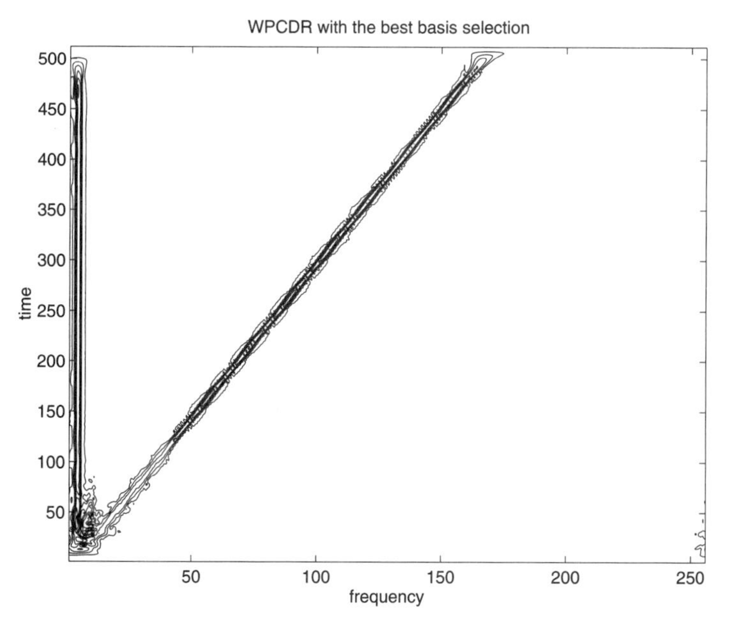

FIGURE 9.11 Best-basis WPCDR of a bicomponent signal. (Reprinted with permission from [4], copyright © 1998 by Wiley.)

heart sounds, safety warnings and noise suppression in factories, and recognition of different types of bearing faults in the wheels of railroad cars. The basic goal of acoustic signal recognition is to identify an acoustic signal pattern from a library of acoustic signatures. Practically all acoustical signature patterns are statistical in nature, and they are also highly nonstationary. Using wavelets to extract feature signals for recognition has great potential for success.

To recognize an acoustic pattern reliably, it is necessary to have a set of distinctive features forming a feature vector for each pattern. These feature vectors from different patterns are obtained by applying many data sets belonging to a particular event in order to train a recognition algorithm such as an artificial neural network (ANN). After we obtain the feature vectors through training, they are stored in a library and used to compare with feature vectors from unknown events.

In this example [5] we apply wavelet techniques and an ANN to identify several different types of faults in the wheel bearings of railroad cars. For information purposes, we show the acoustic signal and its spectrum of a data set with a known bearing defect (see Figure 9.12).

The American Association of Railroads (AAR) provides us with acoustic signals of 18 different types of bearing faults in two classes of sizes, the E-class and the F-class. Each data set is about 0.5 megabite sampling at greater than 260 kHz. Each bearing was tested under two different load conditions, and the wheel was rotating at

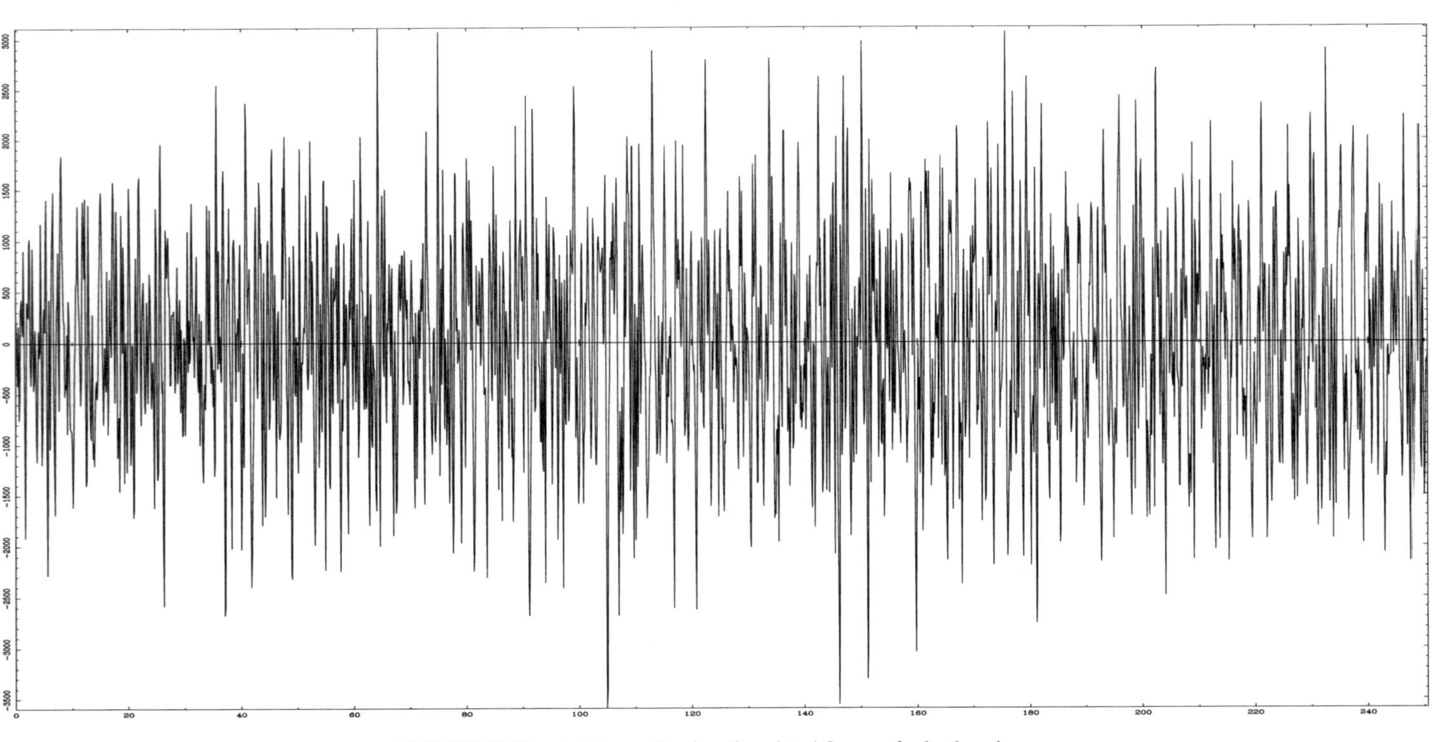

FIGURE 9.12 (*a*) Acoustic signal emitted from a faulty bearing.

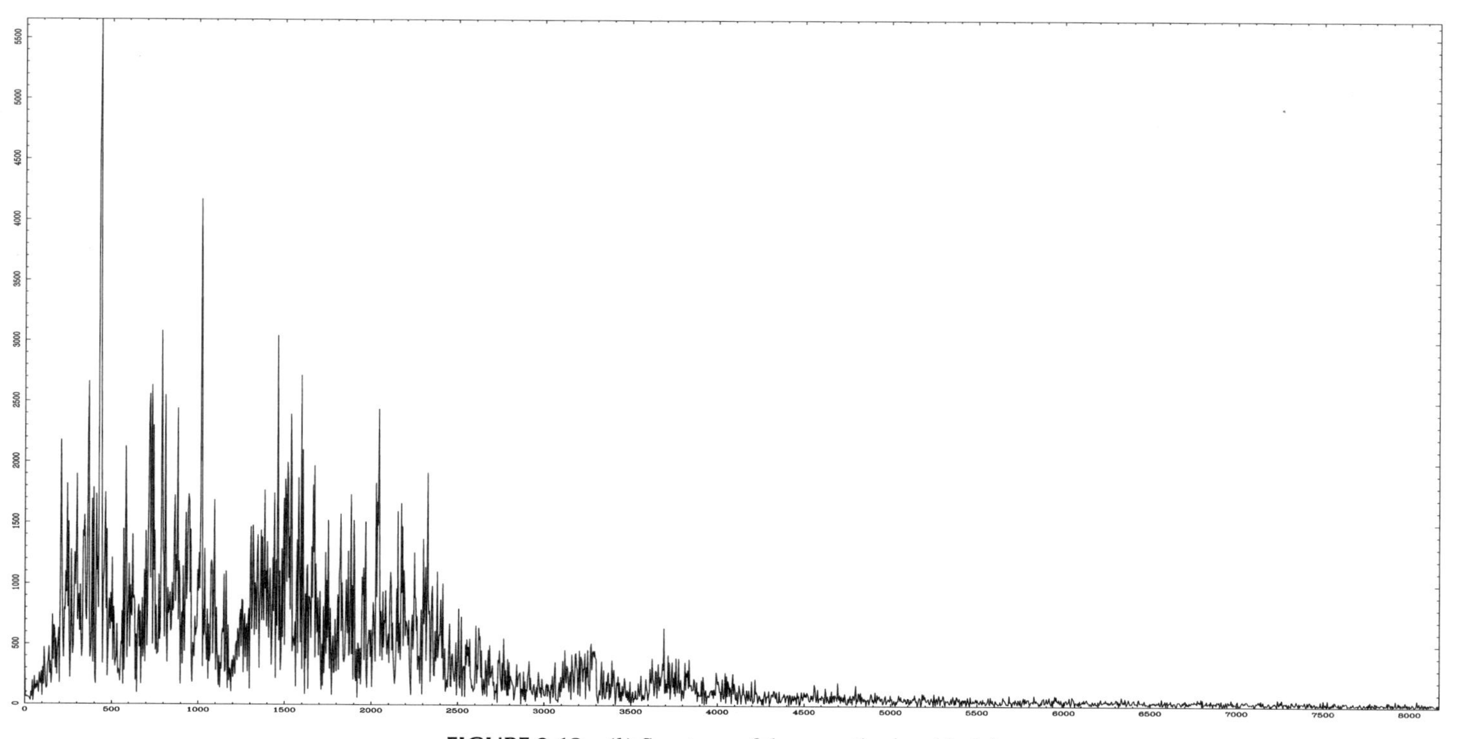

FIGURE 9.12 (*b*) Spectrum of the acoustic signal in (*a*).

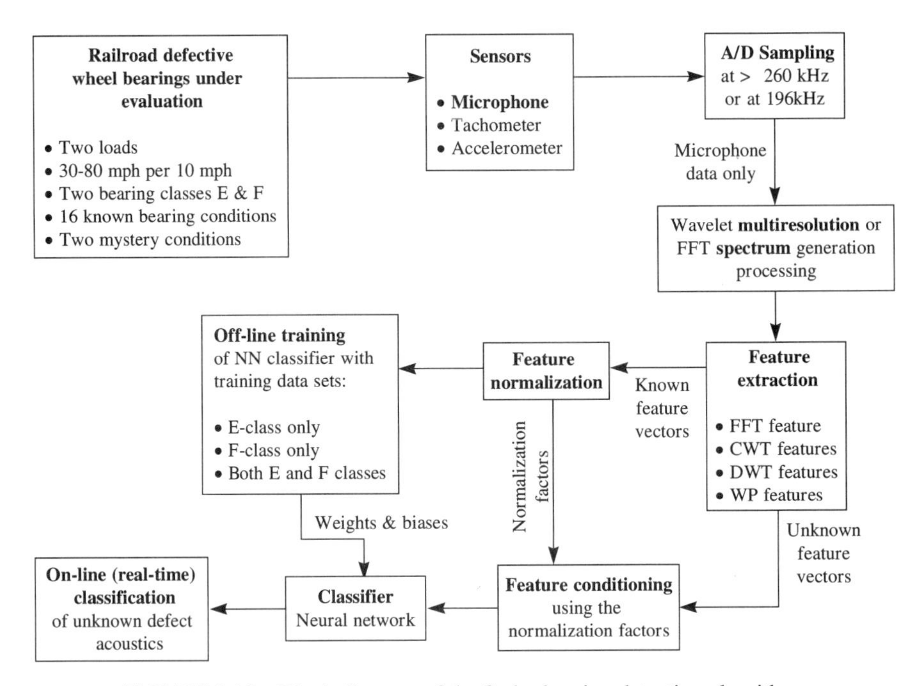

FIGURE 9.13 Block diagram of the faulty bearing detection algorithm.

equivalent train speeds of 30 to 80 mph. Only 25 percent of each data set was used for training the feature vector in every case. The recognition algorithm is given in Figure 9.13.

9.5.2 Wavelets, Wavelet Packets, and FFT Features

The number of samples in each training data is 2^{12}. The formation of the feature vectors for each technique is described below.

Wavelet Feature Extraction

1. Perform the discrete wavelet decomposition on the signal (DWT) to the twelfth level of resolution.
2. From the wavelet coefficients of each of the 12 resolution and approximation coefficients, compute the average energy content of the coefficients at each resolution. There are a total of 13 subbands (12 wavelet subbands and one approximation subband) from which features are extracted. The ith element of a feature vector is given by

$$v_i^{\text{dwt}} = \frac{1}{n_i} \sum_{j=1}^{n_i} w_{i,j}^2, \qquad i = 1, 2, ..., 13, \tag{9.21}$$

where $n_1 = 2^{11}$, $n_2 = 2^{10}$, $n_3 = 2^9, \ldots, n_{12} = 2^0$, $n_{13} = 2^0$; v_i^{dwt} is the ith feature element in a DWT feature vector; n_i is the number of samples in an individual subband; and $w_{i,j}^2$ is the jth coefficient of the ith subband. As a result, a DWT feature vector is formed as given by

$$\mathbf{v}^{\mathrm{dwt}} = \left\{v_1^{\mathrm{dwt}}, v_2^{\mathrm{dwt}}, \ldots, v_{13}^{\mathrm{dwt}}\right\}^t. \tag{9.22}$$

Wavelet Packet Feature Extraction

1. Perform the wavelet packet multiresolution analysis to the fifth level of resolution to obtain 32 subbands. Each subband contains a total of 128 wavelet packet coefficients.
2. From each subband at the fifth level of resolution, compute the average energy content in the wavelet packet coefficients such that

$$v_i^{\mathrm{wp}} = \frac{1}{n_i}\sum_{j=1}^{n_i} p_{i,j}^2, \qquad i = 1, 2, \ldots, 32 \quad \text{and} \quad n_i = 128 \qquad \forall\, i, \tag{9.23}$$

where v_i^{wp} is the ith feature in a wavelet packet feature vector, n_i is the number of sample in each subband, and $p_{i,j}$ is the jth wavelet packet coefficient in the ith subband. The WP feature vector is represented as follows:

$$\mathbf{v}^{\mathrm{wp}} = \left\{v_1^{\mathrm{wp}}, v_2^{\mathrm{wp}}, \ldots, v_{32}^{wp}\right\}^t. \tag{9.24}$$

Spectral Feature Extraction We also use the traditional FFT approach to solve this problem for the sake of comparison. The FFT feature vectors are constructed following the same pattern.

1. From the 2^{12} data points, we compute the FFT and take only the positive frequency information represented by 2^{12} spectral coefficients.
2. We divide the spectrum into 32 nonoverlapping bands with equal width. From each band we compute the average energy contained in the coefficients. The feature element becomes

$$v_i^{\mathrm{fft}} = \frac{1}{n_i}\sum_{j=1}^{n_i} s_{i,j}^2, \qquad i = 1, 2, \ldots, 32 \quad \text{and} \quad n_i = 128 \quad \forall\, i, \tag{9.25}$$

where $s_{i,j}^2$ is the jth FFT coefficient in the ith subband. Consequently, the feature vector become

$$\mathbf{v}^{\mathrm{fft}} = \left\{v_1^{\mathrm{fft}}, v_2^{\mathrm{fft}}, \ldots, v_{32}^{\mathrm{fft}}\right\}^t. \tag{9.26}$$

TABLE 9.1 Overall Performance of the Network for F, E, and F and E Class Bearings

Values (%)	FFT	CWT	DWT	WP
		F-Class Bearings		
Correct decision	96.06	95.70	92.37	92.87
Misclassification	1.67	1.69	4.56	2.49
Miss	2.26	2.61	3.07	4.64
		E-Class Bearings		
Correct decision	95.96	94.16	87.50	93.76
Misclassification	1.28	2.35	7.05	2.50
Miss	2.76	3.50	5.45	3.74
		F- and E-Class Bearings		
Correct decision	95.18	94.22	87.61	92.41
Misclassification	0.93	2.35	8.88	3.19
Miss	3.89	3.42	3.51	4.40

After the feature vectors have been obtained, we apply the feature vector normalization to separate the vectors farther apart to improve performance of recognition. These vectors are used to train an ANN. There are three hidden neurons in this ANN. Details of construction and training of the ANN are beyond the scope of this book and we refer the interested reader to [6].

Results The recognition results obtained using the wavelet techniques combined with the ANN are astounding. Every fault in every class is identified using the unused (not for training) portion of each data set. In fact, the two "mystery" (unknown) bearings containing more than one fault are all identified. Although the traditional FFT approach produces roughly the same results as the wavelet approach, it fails to recognize the unknown bearing by missing one of the two faults. We conclude that the new feature extraction methods using DWT and WP are comparable if not superior to the FFT approach. The FFT lacks the time-domain information and thus misses some of the more localized faults.

The feature vector normalization and conditioning play a key role in the convergence of the neural network while training. Without the normalization, the network does not converge to the desired network error. Convergence of the network produces the biases and the weights necessary for testing of the real data. Three hidden layers are used to improve the convergence. The results are given collectively in Table 9.1.

9.6 TWO-DIMENSIONAL WAVELETS AND WAVELET PACKETS

9.6.1 Two-Dimensional Wavelets

When the input signal is two-dimensional (2-D), it is necessary to represent the signal components by 2-D wavelets and a 2-D approximation function. For any scaling function ϕ with its corresponding wavelet ψ, we construct three different 2-D wavelets and one 2-D approximation function using the tensor-product approach. We write the 2-D wavelets as

$$\Psi_{i,j}^{[1]}(x, y) = \phi(x - i)\psi(y - j) \tag{9.27}$$

$$\Psi_{i,j}^{[2]}(x, y) = \psi(x - i)\phi(y - j) \tag{9.28}$$

$$\Psi_{i,j}^{[3]}(x, y) = \psi(x - i)\psi(y - j), \tag{9.29}$$

and the 2-D scaling function as

$$\Phi_{i,j}(x, y) = \phi(x - i)\phi(y - j). \tag{9.30}$$

$\Psi_{i,j}^{[1]}(x, y)$, $\Psi_{i,j}^{[1]}(x, y)$, and $\Psi_{i,j}^{[1]}(x, y)$ are all wavelets since they satisfy

$$\int_{-\infty}^{\infty}\int_{-\infty}^{\infty} \Psi_{i,j}^{[j]}(x, y)\, dx\, dy = 0 \qquad \text{for } j = 1, 2, 3.$$

The 2-D approximation function and wavelets of the cubic spline are shown in Figure 9.14. In the spectral domain, each of the wavelets and the scaling function oc-

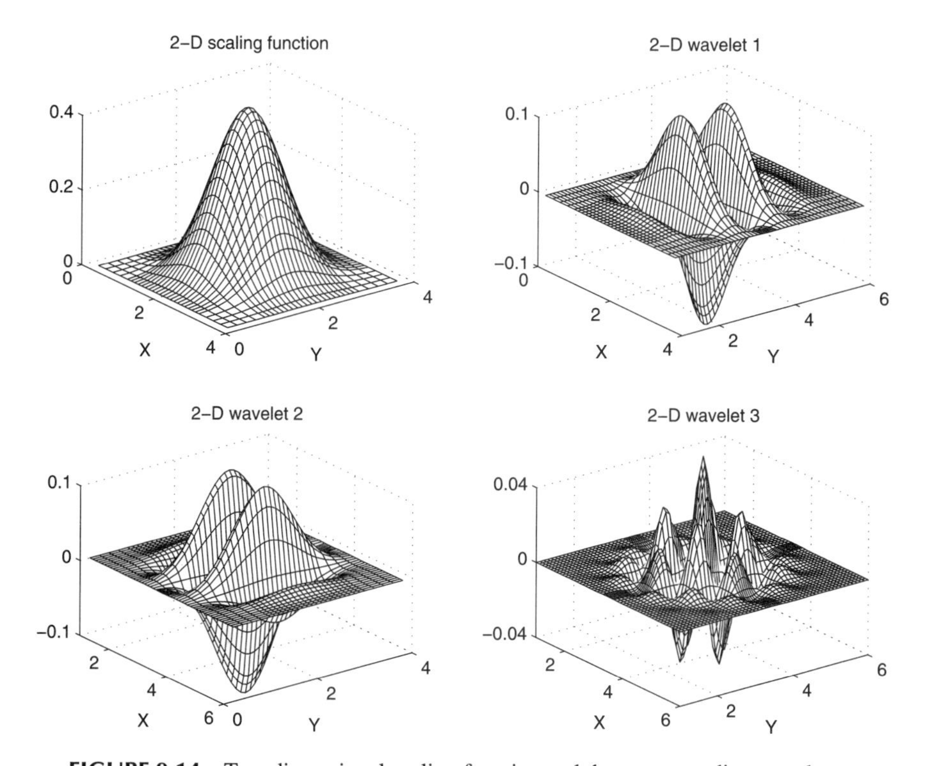

FIGURE 9.14 Two-dimensional scaling function and the corresponding wavelets.

cupy a different portion of the 2-D spectral plane. The spectral distributions of each of these four 2-D functions are shown in Figure 9.15. The spectral bands that are labeled low–high (LH), high–low (HL) and high–high (HH) correspond to the spectra of the wavelets $\Psi_{i,j}^{[M]}(x, y)$, $M = 1, 2, 3$. The low–low (LL) band corresponds to the 2-D approximation function. The terms *low-* and *high-* refer to whether the processing filter is lowpass or highpass. The decomposition of a 2-D signal results in the well-known hierarchical pyramid. Due to the downsampling operation, each image is decomposed into four subimages. The size of each subimage is only a quarter of the original image. An example of hierarchical decomposition of a gray-scale image is given in Figure 9.16.

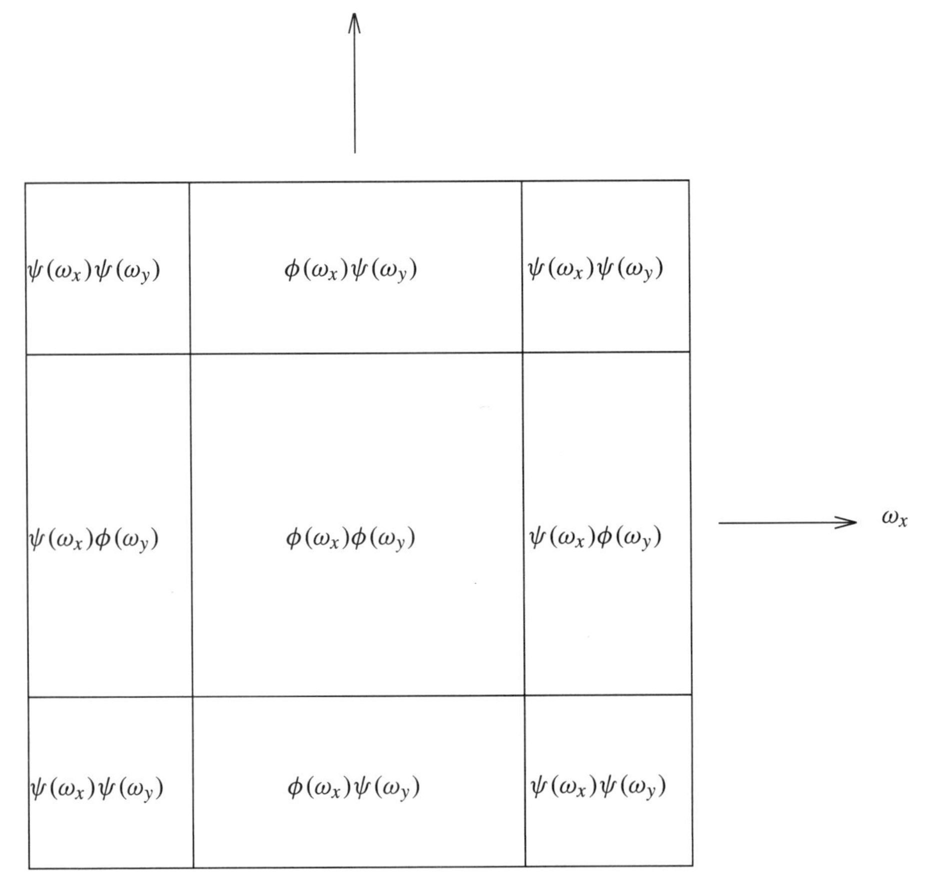

FIGURE 9.15 Regions on the 2-D spectral plane occupied by the 2-D scaling function and wavelets.

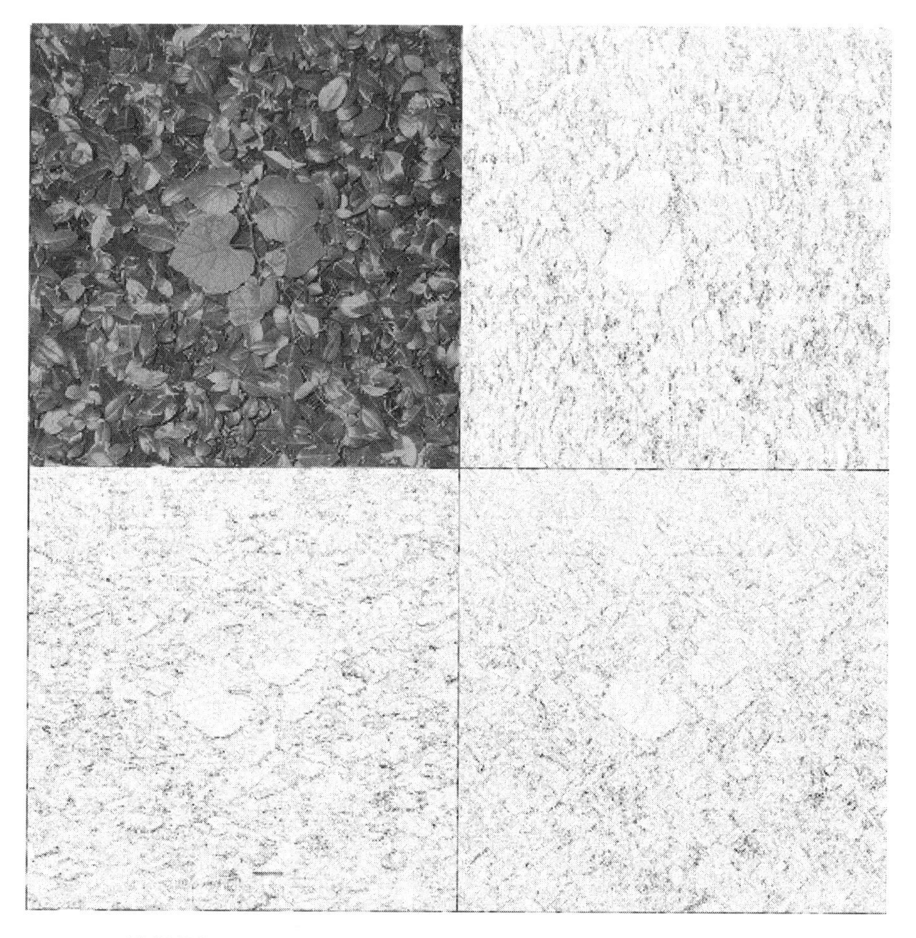

FIGURE 9.16 Two-dimensional wavelet decomposition of an image.

9.6.2 Two-Dimensional Wavelet Packets

Two-dimensional wavelet packets are refinements of the 2-D wavelets similar 1-D case. Using the notation $\mu_k(x)$ to represent the kth wavelet packet belonging to the approximation function $\mu_0(x) = \phi(x)$, a tensor product of any two wavelet packets generates a 2-D wavelet packet. That is,

$$\mu_{k,\ell}(x, y) = \mu_k(x)\mu_\ell(y). \tag{9.31}$$

Consequently, there are many 2-D wavelet packets that can be chosen to form bases in L^2 for signal representation. For example, in the 1-D case we use the two-scale relations for three levels, resulting in $2^3 = 8$ wavelet packets, including the LLL components of the approximation function. Taking the tensor product of any two packets, we obtain 64 different 2-D wavelet packets, including the 2-D approximation functions

$$\mu_{0,0}(x, y) = \mu_0(x)\mu_0(y). \tag{9.32}$$

There are too many 2-D wavelet packets to be shown individually. Two examples of 2-D wavelet packets are shown in Figure 9.17.

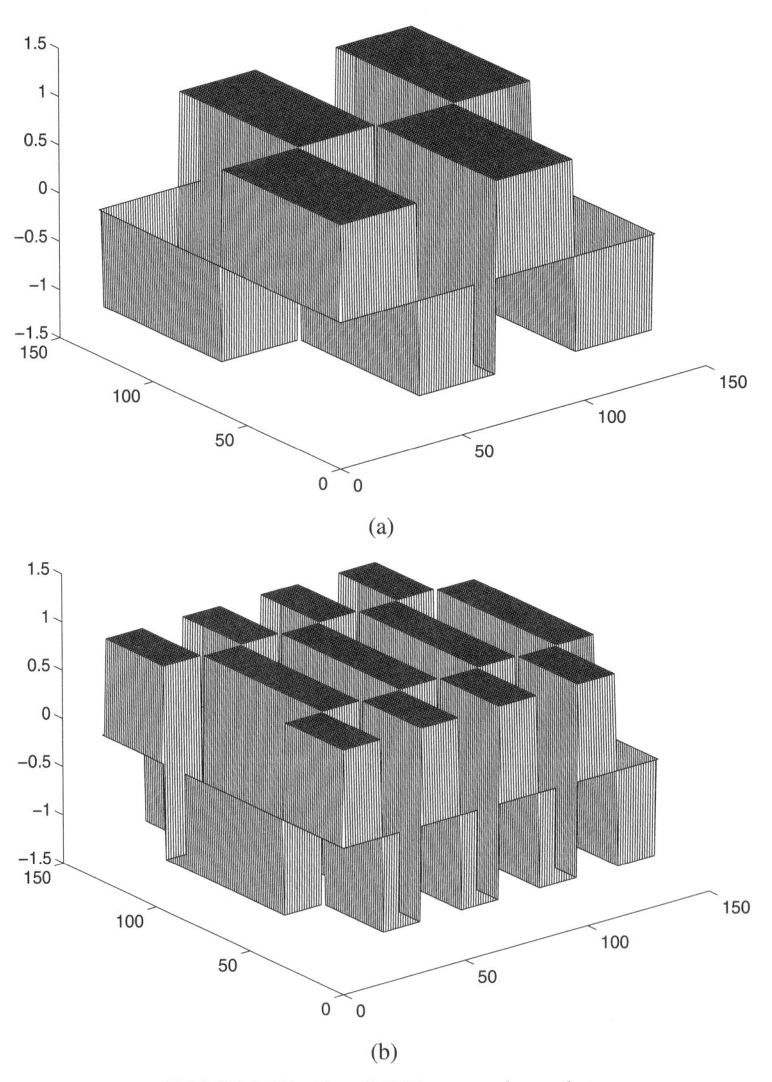

FIGURE 9.17 Two 2-D Haar wavelet packets.

9.7 WAVELET AND WAVELET PACKET ALGORITHMS FOR TWO-DIMENSIONAL SIGNALS

9.7.1 Two-Dimensional Wavelet Algorithm

We have discussed in previous sections that the 2-D wavelets are tensor products of the 1-D scaling function and the wavelet. Corresponding to the scaling function ϕ and the wavelet ψ in one dimension are three 2-D wavelets and one 2-D scaling function at each level of resolution. As a result, the 2-D extension of the wavelet algorithms is the 1-D algorithm applied to both the x- and y-directions of the 2-D signal. Let us consider a 2-D signal as a rectangular matrix of signal values. In the case where the 2-D signal is an image, we call these signal values PIXEL values corresponding to the intensity of the optical reflection. Consider the input signal $c^j(m,n)$ as an $N \times N$ square matrix. We may process the signal along the x-direction first. That is, we decompose the signal row-wise for every row using the 1-D decomposition algorithm. Because of the downsampling operation, the two resultant matrices are rectangular of size $N \times N/2$. These matrices are then transposed, and they are processed row-wise again to obtain four $N/2 \times N/2$ square matrices, namely, $c^{j-1}(m,n)$, $d_1^{j-1}(m,n)$, $d_2^{j-1}(m,n)$, and $d_3^{j-1}(m,n)$. The subscripts of the d matrices correspond to the three different wavelets. The algorithm for 2-D decomposition is shown in Figure 9.18. This procedure can be repeated for an arbitrary number of times to the $c^{\ell}(m,n)$ matrix (or the LL component), and the total number of coefficients after the decomposition is always equal to the initial input coefficient N^2. An example of the decomposition is shown in Figure 9.19.

If the coefficients are not processed, the original data can be recovered exactly through the reconstruction algorithm. The procedure is simply the reverse of the decomposition except that the sequences are $\{g_0[k], g_1[k]\}$ instead of $\{h_0[k], h_1[k]\}$. Care should be taken to remember upsampling before convolution with the input

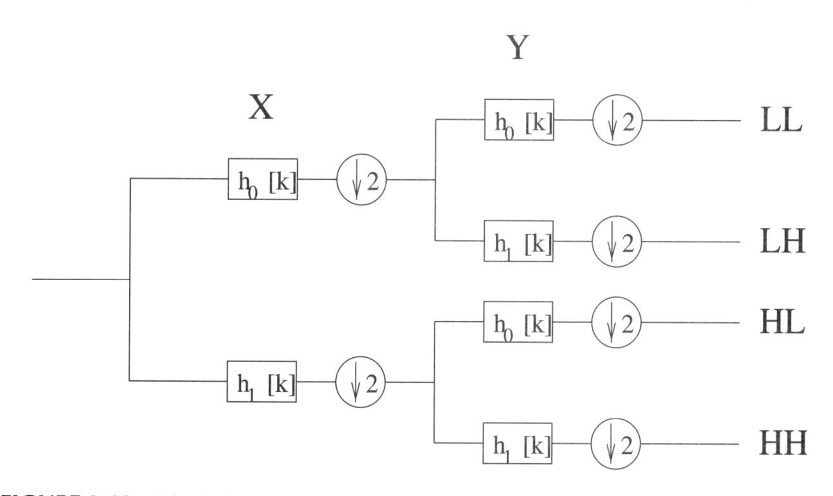

FIGURE 9.18 Block diagram of the two-dimensional wavelet decomposition algorithm.

FIGURE 9.19 A two-dimensional hierarchichal decomposition of an image.

sequences. The perfectly reconstructed image is identical to the original image in Figure 9.20.

9.7.2 Wavelet Packet Algorithm

The 2-D wavelet packet algorithm mimics the 1-D case. It simply repeats the algorithms first along the x-direction and then along the y-direction. Not only is the LL component (the approximation function component) decomposed to obtain further details of the image, the other wavelet components (LH, HL, HH) are also further decomposed. For example, starting with an original image with size 256×256, a 2-D wavelet decomposition of this image will result in four subimages of size 128×128. Continuing the decomposition, one gets 16 2-D wavelet packet subimages of size 64×64. The computational algorithm for 2-D wavelet packets is no more difficult

FIGURE 9.20 Perfect reconstruction from components shown in Figure 9.19.

than that for the 2-D wavelets. It requires orderly bookkeeping to keep track of the directions (x or y) and the filters that have been used in processing. It is necessary to reverse the order to reconstruct the image from its wavelet packet components. An example of 2-D wavelet packet decomposition of an image and its reconstruction is shown in Figures 9.21 and 9.22.

9.8 IMAGE COMPRESSION

9.8.1 Image Coding

There exist in the literature many coding schemes for image compression. The basic idea of compression is to try to reduce the average number of bits per pixel to adequately represent the image. Image compression is urgently needed for very large medical or satellite images, both for reducing the storage requirements and for improving transmission efficiency. Coding schemes are classified as *lossless* (recoverable) and *lossy* (nonrecoverable) types. Compression ratios are usually very low (approximately 10 or less) for lossless coding. Schemes like the Huffman code, run-length code, arithmetic code, predictive coding, and bit plane coding belong to the lossless class. Lossy coding methods may be constructed through a variety of ways.

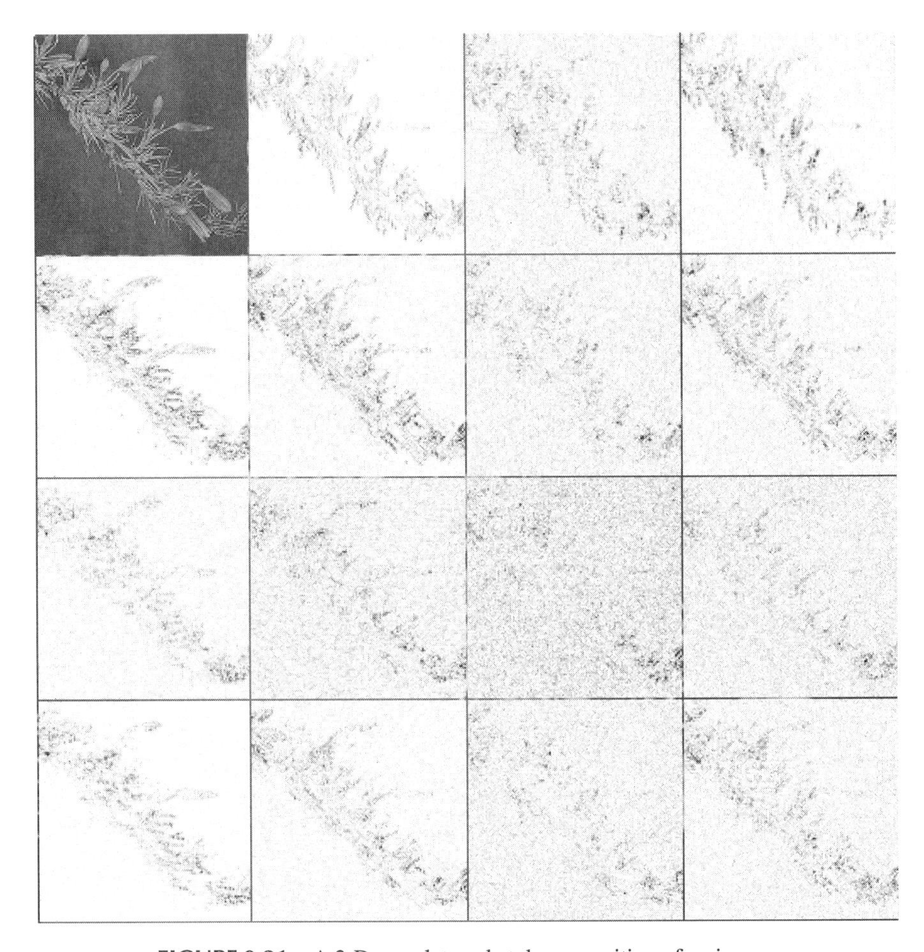

FIGURE 9.21 A 2-D wavelet packet decomposition of an image.

We will use the zero-tree algorithm to demonstrate the application of wavelets to image compression.

9.8.2 Wavelet Tree Coder

In general, tree coders [7] use a tree structure which takes advantage of the correlation between the discrete wavelet coefficients (DWCs) in each of the three spatial directions (i.e., HL, LH, and HH), as shown in Figure 9.23. That means that if a DWC at a higher decomposition level is smaller than a specified threshold, there is a great possibility that all of its children and grandchildren are smaller than the threshold. Thus, all of these *insignificant* DWCs can be encoded with one symbol. The encoding of a *significant* DWC may need more bits. Many tree structures have been

FIGURE 9.22 Perfect reconstruction from wavelet packet components shown in Figure 9.21.

developed for improving the efficiency of encoding the *locations* of the correlated DWCs, such as EZW, SOT, and GST.

A brief description of a generic tree coder is as follows:

1. The DWCs are selected in groups with decreasing thresholds such that larger DWCs are encoded earlier.
2. The first threshold is selected to be an integer $T_0 = 2^j$, where j is the nearest integer $\leq \log_2 \max |\text{DWC}|$ and the kth threshold is $T_k = T_0/2^k$, that is, the uniform quantization.
3. Choosing the threshold T_k, all the *locations* of this group of DWCs, $C_{i,j}$ with $T_k \leq |C_{i,j}| < T_{k-1}$, are encoded with a tree structure, and signs of these encoded DWCs are also appended. This process is called the *dominant pass*.
4. Those $C_{i,j}$ encoded in previous higher thresholds are refined to a better accuracy by appending the bit corresponding to T_k. This process is called the *subordinate pass*.
5. With decreasing thresholds, the leading *zero* bits of encoded DWCs are saved to achieve compression.
6. The number of bits in the coded bit stream from a tree coder can be further reduced using a lossless entropy coder (e.g., an arithmetic coder [8]).

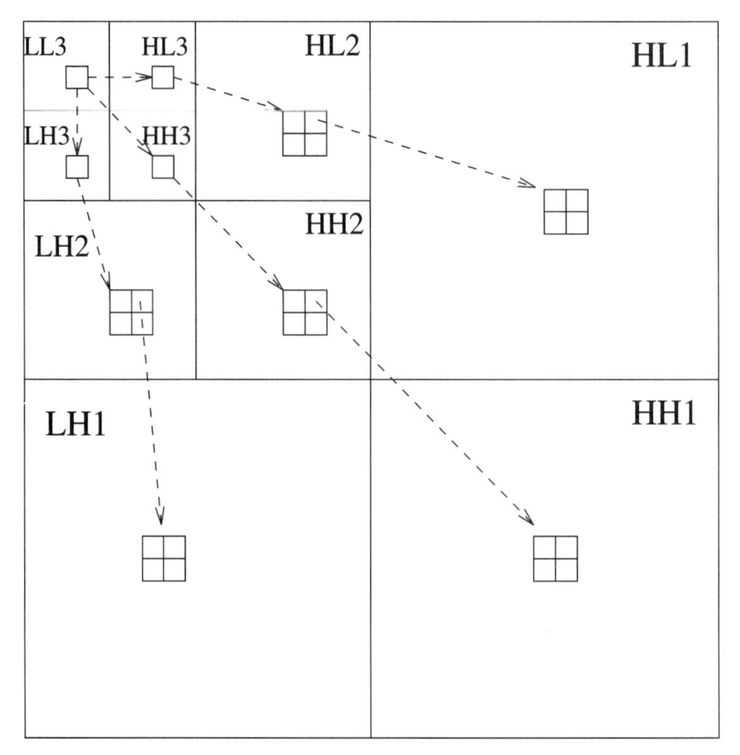

FIGURE 9.23 Spatial correlation and parent-child relationship among wavelet coefficients at different resolutions.

9.8.3 EZW Code

Initiated by Shapiro [9], the zero-tree structure combined with bit plane coding is an efficient compression scheme for the discrete wavelet transformation. The *embedded zero-tree wavelet* (EZW) coding scheme has proven its efficiency and flexibility in still image coding in terms of image quality and computation simplicity. Also, the EZW image coding algorithm generates an embedded bit stream in which information is sent to the decoder in the order of its importance; importance is judged by how much the information reduces the distortion of the reconstructed image. This embedded technique has two important advantages. First, the bit rate control allows one to stop the coding process at any point. Second, the image can be reconstructed from a point where the encoded bit stream has been disrupted, even with reduced quality.

As an entropy coder, the zero-tree coder takes advantage of the correlation between interlevel subbands of DWCs. Four symbols, *ZTR*, *POS*, *NEG*, and *IZ*, are used in the zero-tree. A *ZTR*, zero-tree root, represents a DWC and all of its descendants if they are insignificant, and it is the symbol that gets most of the compression. A *POS* or *NEG* symbol stands for a significant DWC with a positive or negative sign.

An *IZ* represents an insignificant DWC with at least one significant descendent. It is this symbol that reduces the compression since more symbols are needed for encoding its descendants. With the definitions above, two bits per symbol on average are needed for each of the four symbols. Without a good follow-up entropy coder for those symbols, zero-tree cannot get good compression results. Also, all of the encoded symbols must be *reordered* so that the entropy coder can achieve compression on them. Both of these procedures increase the computation overhead.

9.8.4 EZW Example

We use a well-designed example similar to the one appearing in Shapiro's original paper [9] to demonstrate the procedure. The EZW is a successive approximation algorithm. It uses the *dominant pass* and the *subordinant pass* recursively to achieve the approximation. The coefficient map of this example is shown as follows:

58	41	−44	−17	−8	13	5	8
29	−47	42	−13	3	4	−1	10
22	−14	25	−9	35	−11	6	7
−9	−11	−16	12	−3	6	14	−1
−13	2	0	−11	0	−30	15	7
10	−5	9	4	−7	−1	−9	0
−22	4	−13	3	6	−8	11	5
9	−1	9	−12	2	5	9	−3

Following the steps outlined in Section 9.8.3, we find the largest value to be coded is $M = 58$. The nearest power-of-2 integer $2^j \geq 58/2 = 29$ is $2^5 = 32$. We set the initial threshold value $T_0 = 32 = 2^5$. We then use this threshold and compare it with all of the coefficients in a dominant pass, described below.

In wavelet-tree coding, not only do the values of the coefficients need to be coded, the location of the coefficients must also be known to the decoder. The EZW makes use of the parent–children relationship between coefficients in adjacent coefficient maps to record the location of a given coefficient. This relationship is embedded in the EZW code to reduce the coding overhead. This relationship is best shown in Figure 9.23. The coding procedure is listed as follows.

1. For the first dominant pass, the threshold is set at 32 and the results of the code assignment are given in the following table. We assign a symbol to each of the codeable values:

Value	Symbol	Reconstructed Value	Value	Symbol	Reconstructed Value
58	P	48	−16	ZTR	0
41	P	48	12	ZTR	0
29	ZTR	0	−8	Z	0
−47	N	−48	13	Z	0
−44	N	−48	3	Z	0
−17	ZTR	0	4	Z	0
42	P	48	35	P	48
−13	ZTR	0	−11	Z	0
25	ZTR	0	−3	Z	0
−9	ZTR	0	6	Z	0

2. After the first dominant pass has been completed, the first subordinate pass refines the coded values. Only those significant values (P and N) are coded in this pass. The symbol P in the first subordinate pass states that the value lies in the interval (64, 32], and the symbol N, in the interval (−64, −32]. The subordinate pass refines these values by narrowing the interval from (64, 32] to (64, 48] and (48, 32]. If the value is in the upper interval, the subordinate pass appends a "1" to the code, a "0" if the value lies in the lower interval. The six significant values now have the following code file:

Coefficient Magnitude	Symbol	Reconstructed Magnitude	Binary Representation
58	1	56	**11**1010
41	0	40	**10**1001
47	0	40	**10**1111
44	0	40	**10**1100
42	0	40	**10**1010
35	0	40	**10**0011

This completes the first iteration of both passes. The user has to remember that in the subordinate pass, if a "1" is appended to the code, one has to subtract the refinement amount, namely 16 in this case, from the value that is over the threshold for this pass. For example, the coefficient 58 has $58 - 32 - 16 = 10$ yet to be refined by the next iteration. In addition, the user should also remember that the coded values are now replaced by zero in the coefficient map and will not be coded from later iterations.

3. We repeat step 1 with the second dominant pass with a threshold $T_1 = 16$. We have the following codes from this pass:

Coefficient Value	Symbol	Reconstructed Value	Coefficient Value	Symbol	Reconstructed Value
29	P	24	−11	Z	0
−17	N	−24	−3	Z	0
−13	ZTR	0	6	Z	0
22	P	24	−13	Z	0
−14	ZTR	0	2	Z	0
−9	IZ	0	10	Z	0
−11	ZTR	0	−5	Z	0
25	P	24	−22	N	−24
−9	ZTR	0	4	Z	0
−16	N	−24	9	Z	0
12	ZTR	0	−1	Z	0
−8	Z	0	0	Z	0
13	Z	0	−30	N	−24
3	Z	0	−7	Z	0
4	Z	0	−1	Z	0
5	Z	0	6	Z	0
−8	Z	0	−8	Z	0
−1	Z	0	2	Z	0
10	Z	0	5	Z	0

4. The second subordinate pass will separate the intervals more finely by dividing all intervals at their midpoints. Hence we have intervals (64,56], (56,48], (48,40], (40,32], (32,24], and (24,16]. This pass also updates all previous code files. The value to be compared with in this pass is 8. The updated code file becomes:

Coefficient Magnitude	Symbol	Reconstructed Magnitude	Binary Representation
58	1	60	11**1**010
41	1	44	10**1**001
47	1	44	10**1**111
44	1	44	10**1**100
42	1	44	10**1**010
35	0	36	10**0**011
29	1	28	01**1**101
17	0	20	01**0**001
22	0	20	01**0**110
25	1	28	01**1**001
16	0	20	01**0**000
22	0	20	01**0**110
30	1	28	01**1**110

FIGURE 9.24 Original image for EZW image coding.

The passes are repeated by cutting the threshold by half each time. If all the coefficients are coded, we have a lossless code. The compression ratio achieved in this manner is limited. High compression ratio is achieved using EZW followed by a lossless entropy coder. The user may stop coding at any time or when the bit budget is exhausted. We have a lossy compression scheme wherein users may control the bit budget but cannot control the compression ratio. An original image and the image recovered from EZW coding are shown in Figures 9.24 and 9.25.

9.8.5 Spatial-Oriented Tree

Said and Pearlman [10,11] discovered set partitioning principles to improve the performance up to 1.3 dB over that of the zero-tree method. They observed that there is a spatial self-similarity between subbands, and the discrete wavelet coefficients (DWCs) are expected to be better magnitude ordered if one moves downward in the pyramid following the same spatial orientation. Based on this observation, a tree structure called a *spatial orientation tree* (SOT) is used to define the spatial relationship of the DWCs in the hierarchical structure. Three main concepts are proposed by Said and Pearlman to adapt the SOT to obtain better performance in image coding: (1) partial ordering of the transformed image by magnitude and transmission of coordinates via a subset partitioning algorithm, (2) ordered bit plane transmission of

FIGURE 9.25 Decoded image at compression of 30:1.

refinement bits, and (3) exploitation of the self-similarity of the DWCs across the various scales.

For the SOT [10–12], only two symbols, *zero* and *one*, are used, and each symbol has a different meaning at a different part of the tree. The symbol *one* may represent (1) a significant DWC, (2) the negative sign of a significant DWC, (3) the case that any one of the four children is significant, or (4) the case that any of the grandchildren is significant. The symbol *zero* could indicate (1) an insignificant DWC, (2) the positive sign of a significant DWC, (3) the case that all four children are insignificant, or (4) the case that all grandchildren are insignificant. To maintain the SOT for the DWCs along with the different scales, three lists are used as follows:

1. LIS, *list of insignificant sets*, is a list of the roots of a tree for further tracing, and types A and B of the roots of the tree are used interchangeably to obtain better adaptability.
2. LIP, *list of insignificant pixels*, is a list of the DWCs that are not roots of the tree currently but are the candidates to be placed into the LSP.
3. LSP, *list of significant pixels*, is a list of the DWCs that have been encoded and are to be further refined.

9.8.6 Generalized Self-Similarity Tree

Based on the SOT, a *generalized self-similarity tree* (GST) coding algorithm has been constructed that can handle images of any size and any gray level [13]. In the GST, the wavelet decomposition/rcconstruction algorithm with boundary reflection techniques is used so that perfect reconstruction can be achieved. Analysis of the GST coder shows results comparable to the original SOT coder for images of dyadic size, and it even outperforms the SOT for images of nondyadic size.

9.9 MICROCALCIFICATION CLUSTER DETECTION

The majority of early breast cancers are indicated by the presence of one or more clusters of microcalcifications on a mammogram. Although breast cancer can be fatal, women have one of the highest chances of survival among cancer types if the tumors can be detected and removed in an early stage. Thus the detection of microcalcifications with minimal false-positive rates is critical to screening mammograms. Microcalcifications are small deposits of calcium phosphate hydroxide in breast tissue with sizes raging from 0.05 to 1.0 mm in diameter which appear as bright specks on photonegative x-ray film [14]. They are difficult to detect because they vary in size and shape, and they are embedded in parenchymal tissue structures of varying density [15].

Screening mammograms have been one of the main thrusts in the health care program of the United States. However, even partial compliance with the rule set by the ACR would produce a huge volume of data to be read by a limited number of radiologists. Consequently, human error can run the percentages of false negatives (a true target missed) up to 20 percent [16]. If a computer-aided-diagnostic (CAD) algorithm were designed and constructed, it could serve as a second opinion to help the radiologist by pointing out suspicious regions in the mammogram needing a more detailed diagnostic screening. This application example attempts to show how a two-dimensional wavelet pyramid algorithm working in conjunction with other image processing techniques can identify the microcalcifications in mammograms and localize the suspicious regions.

9.9.1 CAD Algorithm Structure

Success in signature recognition depends strongly on the features one can extract from a signature. The more distinct the features, the higher the success rate for making a positive identification. The most important objective in the detection and recognition of microcalcifications is to remove the background noise and enhance the object to be identified. We use several traditional image processing techniques to work with the wavelet decomposition algorithm to achieve this objective. Decision-making rules in some of these algorithms are goal oriented and therefore are problem dependent. Parameter choices often depend on the data to be analyzed. The CAD algorithm for microcalcification cluster detection in a highly textured and cluttered

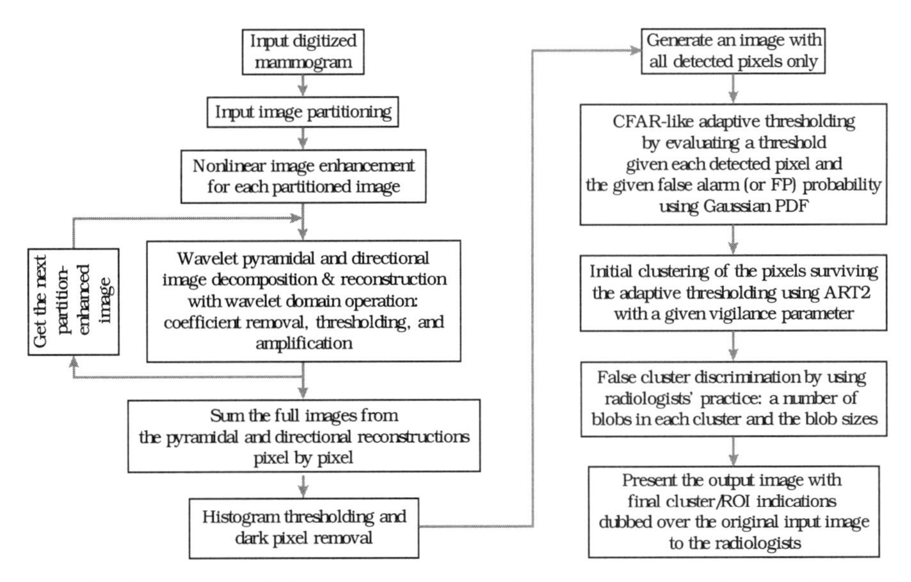

FIGURE 9.26 Block diagram of the microcalcification cluster detection algorithm.

background is illustrated in Figure 9.26. The image processing techniques used in this CAD algorithm include nonlinear image enhancement, wavelet pyramidal and directional image decomposition and reconstruction, wavelet coefficient domain operations, dark pixel removal, constant-false-alarm-rate (C-FAR) adaptive thresholding, adaptive resonance theory clustering, and false cluster discrimination.

9.9.2 Partitioning of Image and Nonlinear Contrast Enhancement

We partition the mammogram to be analyzed simply by dividing the image into a number of equal-sized subimages. In this case the size of the mammogram is 1024 $\times$ 1024 and we divide it up into 64 subimages of size 128 $\times$ 128. Each partitioned subimage is processed separately to bring out locally significant details of the input image with image contrast enhancement. This step provides better localization for detection of the targets. Since wavelet processing is known to handle the image boundary better than DCT, we are assured that information is not lost in the partitioning. We use cubic mapping to suppress pixels with low gray-scale values and to enhance pixels with large gray-scale values.

9.9.3 Wavelet Decomposition of the Subimages

We decompose each subimage using the wavelet decomposition algorithm so that the high-frequency components of the subimage are singled out. There are many choices of wavelets in this applications. We chose the Haar wavelet after examining all of Daubechies' orthogonal and biorthogonal wavelets and the coiflets, because the spatial-domain window of the Haar wavelet is very small for better spatial localization. Higher-order wavelets tend to average and blur the high-frequency information to produce a low-amplitude wavelet coefficient.

Two types of wavelet MRA tree decompositions, pyramidal and directional decompositions, are applied simultaneously to the same subimage. The pyramidal MRA decomposes only the subband image obtained through the LL-subband in the column and row direction at each level of resolution (LOR). The directional MRA, on the other hand, decomposes images in only one direction. The decomposition wavelet coefficient maps of these two MRA trees are shown in Figures 9.27 and 9.28.

9.9.4 Wavelet Coefficient Domain Processing

Once we have the wavelet coefficients computed as shown in the preceding section, the goal of processing these coefficients is to retain only the significant wavelet coefficients that pertain to microcalcifications and other high-frequency information. Processing these coefficients includes removal, thresholding, and amplification. These are operations without a user interface; we must generate rules and parameters to guide these operations.

To retain high-frequency information that contains microcalcifications and other high-frequency noise, the wavelet coefficients in the lower-resolution subbands are removed. For each partitioned subimage, we compute the global (all wavelet coefficients in the subimage) standard deviation-to-mean ratio (GSMR) and local

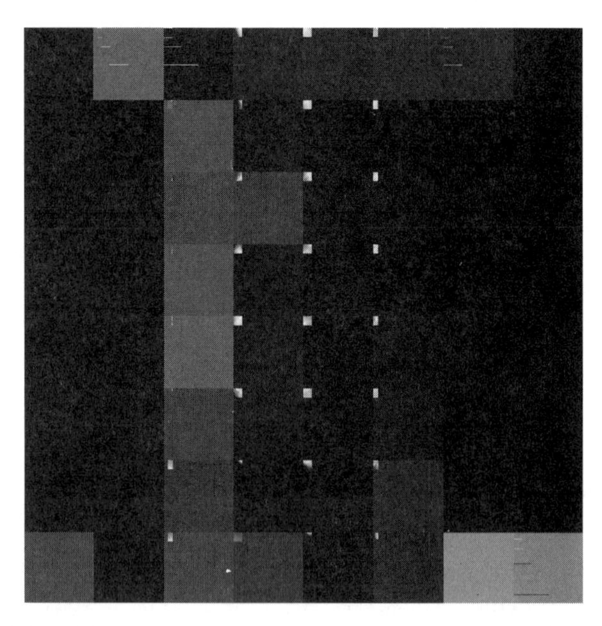

FIGURE 9.27 Hierarchical wavelet decomposition of a segmented mammogram.

FIGURE 9.28 Directional wavelet decomposition of a segmented mammogram.

(wavelet coefficients in one subband of the subimage) standard deviation-to-mean ratio (LSMR) for the removal of coefficients containing low-frequency or insignificant high-frequency information. Let us denote $\gamma_{g,k}$ and $\gamma_{j,k}$ as the GSMR and LSMR computed from the wavelet coefficients of the kth subimage. The wavelet coefficients $w_{j,k}(\cdot)$ are set to zero according to the following rules:

$$\begin{array}{lll} w_{j,k}(\cdot) = 0 & \text{for coefficients in subband } j & \text{if } \gamma_{g,k} > \gamma_{j,k} \\ \text{retain in subband } j \text{ for further processing} & & \text{if } \gamma_{g,k} \leq \gamma_{j,k}. \end{array} \tag{9.33}$$

We use the mean and standard deviation, $\mu_{j,k}$ and $\sigma_{j,k}$, from each subband to set the thresholding and amplification criterion. The rule is stated as follows:

$$w_{j,k}(\cdot) = \begin{cases} 0 & \text{if } w_{j,k}(\cdot) \leq \mu_{j,k} + 2.5 \times \sigma_{j,k} \\ \text{abs}\left[\dfrac{\sigma_{j,k}}{\mu_{j,k}}\right] \times w_{j,k}(\cdot) & \text{otherwise.} \end{cases} \tag{9.34}$$

After the operations in the wavelet coefficient maps have been completed, we reconstruct the image using the remaining subband coefficients. The reconstructed images have a dark background with white spots representing the microcalcifications and high-frequency speckle noise. To differentiate the microcalcifications and noise, we use histogram thresholding and dark pixel removal.

9.9.5 Histogram Thresholding and Dark Pixel Removal

Since the reconstructed images also contain information that is not relevant to the microcalcifications, we need to filter out this erroneous information. The histogram threshold requires the peak value of the histogram of a given gray scale g_{peak}. We formulate the following thresholding rule:

$$\begin{aligned} &v_r(x, y) = 0, \ \text{if } v_r(x, y) \leq (g_{\text{peak}} + 1) + 0.5 \times \sigma_{nz}^r \\ &v_r(x, y) \text{ remain unchanged for further processing,} \end{aligned} \tag{9.35}$$

where σ_{nz}^r and μ_{nz}^r are the standard deviation and mean obtained from all nonzero (nz) pixels in the reconstructed images, and $v_r(x, y)$ is the value of the (x, y)th pixel in the reconstructed image. After this step, the CAD algorithm adds the two images together in a spatially coherent fashion to form a composite image in which all microcalcification information is contained.

We now refer back to the original image. Since the pixel intensity from the microcalcification is greater than 137 in an 8-bit (256 levels) linear gray scale, we make use of this information to formulate a dark pixel removal threshold as follows:

$$\begin{aligned} v_r(x, y) &= 0 && \text{if } v_r(x, y) = 0 \\ v_r(x, y) &= 0 && \text{if } v_r(x, y) \neq 0 \text{ and } u(x, y) \leq \mu_{\text{org}} + 0.5 \times \sigma_{\text{org}} \\ v_r(x_{nz}, y_{nz}) &= v_r(x, y) && \text{if } v_r(x, y) \neq 0 \text{ and } u(x, y) > \mu_{\text{org}} + 0.5 \times \sigma_{\text{org}}, \end{aligned} \tag{9.36}$$

where μ_{org} and σ_{org} are the mean and standard deviations obtained using nonzero pixels in the original input mammogram and $u(x, y)$ is the pixel value of the (x, y)th pixel in the original image.

After the dark pixels are set to zero, potential microcalcification regions (PMRs) are identified in the enhanced image. The nonzero pixel locations indicate potential sites of microcalcifications. These sites are then made the centers of 5×5 pixel PMRs. Each of these PMRs must go through a CFAR-like detector to reduce the number of PMRs to a manageable level. The CFAR acts like a probabilistic discriminator. The PMRs with high probabilities are retained for further analysis. Hence the 5×5 pixel region acts like a window through which an adaptive rule is set up to determine its probability as a microcalification. To evaluate the CFAR threshold, one needs the mean and standard deviation from the PMR, an a priori probability distribution, and a desired false alarm rate. The detailed theory of CFAR is beyond the scope of this book; interested readers are referred to [17].

9.9.6 Parametric ART2 Clustering

The suspicious regions are formed by using ART (adaptive resonance theory [5]) with a vigilance factor, ρ_v, or 25 pixels. In this example application, we choose the search region to be an area corresponding to 1 cm $\times$ 1 cm, which has approximately 50×50 pixels of the image. Once an initial clustering is completed, each cluster must be tested for false alarm discrimination. Each cluster must have at least three

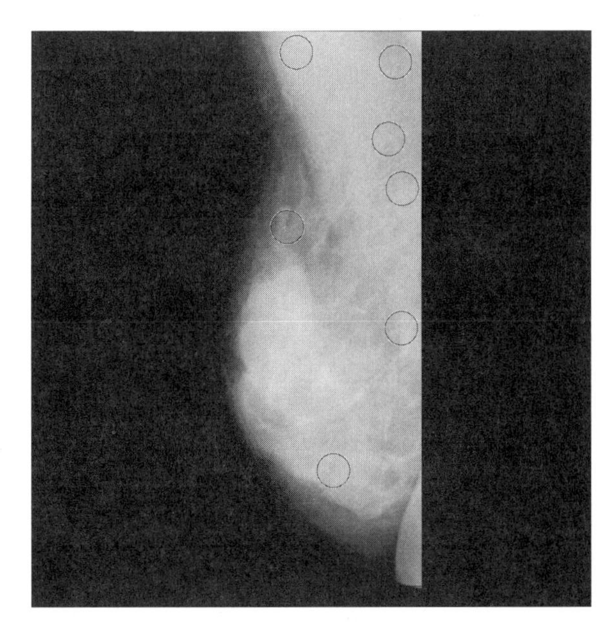

FIGURE 9.29 Clusters detected in a segmented mammogram.

microcalcifications whose individual size must not exceed 5 × 5 pixels. If an initial cluster does not meet this criterion, it is declared a false positive (FP) cluster and removed from the list of suspicious regions.

9.9.7 Results

We have applied this CAD algorithm to 322 mammograms obtained from the MIAS MiniMammographic Database of England. We found 150 truly suspicious regions and 1834 false alarms with 37 undeterminable regions. When we compare the results from the algorithm with the biopsy results (which came with the data set), all 31 true positives (TPs) were correctly recognized, with one false negative (FN). There were 119 false positives (FPs). In terms of sensitivity, the CAD algorithm achieved between 87 to 97 percent accuracy. In terms of the number of FPs, it attains 0.35 to 5 per image, with 0.04 to 0.26 FN per image. These results compare favorably with respect to results from other CAD algorithms as well as statistics from the radiological community. An original mammogram and the algorithm output of the same mammogram are shown in Figures 9.29 and 9.30.

9.10 MULTICARRIER COMMUNICATION SYSTEMS

Multicarrier modulation is the principle of transmitting data by dividing the data stream into several parallel bit streams, each of which has a much lower bit rate.

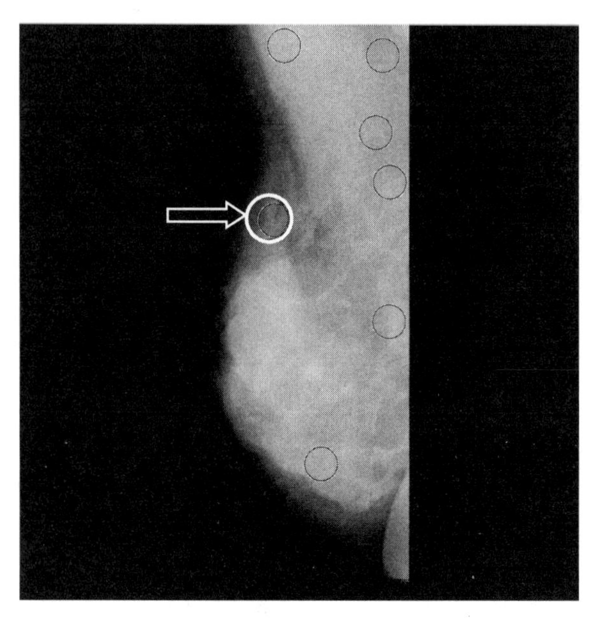

FIGURE 9.30 Comparison of clusters detected with a true positive.

Each of these substreams modulates an individual carrier. Figure 9.31 shows the block diagram of the transmitter of a multicarrier communication system (MCCS). A serial-to-parallel buffer segments the information sequences into frames of N_f bits. The N_f bits in each frame are parsed into M groups, where the ith group is assigned n_i bits so that

$$\sum_{i=0}^{M-1} n_i = N_f. \tag{9.37}$$

It is convenient to view the multicarrier modulation as having M independent channels, each operating at the same symbol rate $1/T$. The data in each channel are modulated by a different subcarrier. We denote the signal input to the subchannels by $S_i, i = 0, \ldots, M-1$. To modulate the M subcarriers, we use an orthogonal basis $\Phi = \{\phi\}_{k=0}^{M-1}$ such that

$$\langle \phi_m, \phi_\ell \rangle = \epsilon \delta_{m,\ell}.$$

9.10.1 OFDM Multicarrier Communication Systems

Orthogonal frequency-division multiplexing (OFDM) is a special form of MCCS with densely spaced subcarriers and overlapping spectra. It abandons the use of steep bandpass filters that completely separate the spectra of individual subcarriers. Instead, OFDM time-domain waveforms are chosen such that mutual orthogonal-

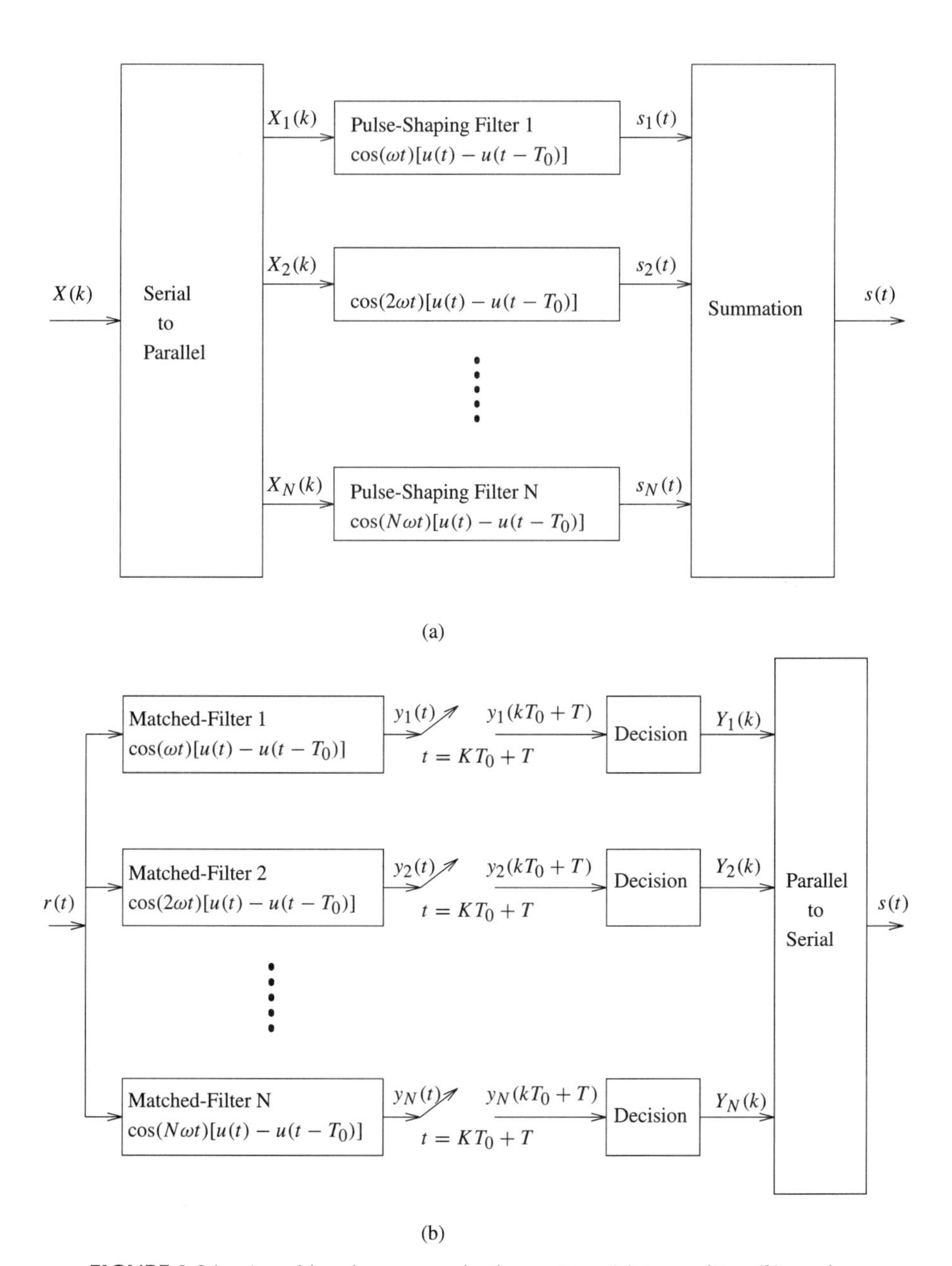

FIGURE 9.31 A multicarrier communication system: (*a*) transmitter, (*b*) receiver.

ity is ensured even though subcarrier spectra may overlap. OFDM is more robust against time-domain impulse interference, due to its long symbol time, which tends to average out the effects. OFDM subcarriers may lose their mutual orthogonality if high-frequency errors occur in the channel. As shown in Figure 9.31, the operating principle is simple. The data are transmitted on several subcarriers. The spectra of the subcarriers may overlap, but the mutual orthogonality is ensured. These subcarriers are summed together and transmitted over the channel. On the receiver end of the channel, the signal received is sent in parallel to the matched filters in each subchannel. The output of the matched filter is sampled before the decision is made on the signal. In general, each subchannel uses the binary phase-shift key (BPSK) scheme [18] to represent the signal.

When the channel behaves well and does not introduce frequency dispersion, the bit error rate (P_e) is very small. The imperfection may be due to noise in the channel. On the other hand, when frequency dispersion is present due to time variation of the channel parameter, P_e increases. Phase jitters and receiver frequency offsets introduce interchannel interferences that degrade the P_e.

9.10.2 Wavelet Packet-Based MCCS

Instead of the sine or cosine functions used in the OFDM, WP-based MCCS uses different wavelet packets as the time-domain waveforms. If the approximation function ϕ generates an orthonormal set in the L^2 space, the corresponding wavelet packets are guaranteed to be orthogonal. The subcarriers are now wavelet packets, and the matched filters in the receiver are designed accordingly (see Figure 9.32). Since there are a large number of wavelet packets to be chosen for the subcarriers, our experiment chooses those whose spectra are very close to those of the OFDM. Under this condition, we can make a fair comparison between the results of these two systems.

The curves in Figure 9.33 represent the P_e verse symbols per second. Without any frequency offsets, all system performances are very close to being the same. When the frequency offset is 10 percent, the wavelet packet system performs slightly better than the OFDM. When we stress the system by allowing 25 percent offset, the WP system works far better than the OFDM. In particular, the Daubechies D_3 orthogonal WP system seems to be the best. Comparing the spectra of the subcarriers in both systems, they are very similar. However, there appears to be an optimal set of wavelet packets through which the system produces the best performance under a highly stressed system. An iterative and an analytical approach are being investigated.

9.11 THREE-DIMENSIONAL MEDICAL IMAGE VISUALIZATION

Medical image visualization is becoming increasingly popular for planning treatment and surgery. Medical images captured by various instruments are two-dimensional gray-level signals. A 3-D image reconstructed from 2-D slices provides much more information about surfaces and localization of objects in a 3-D space. The medical

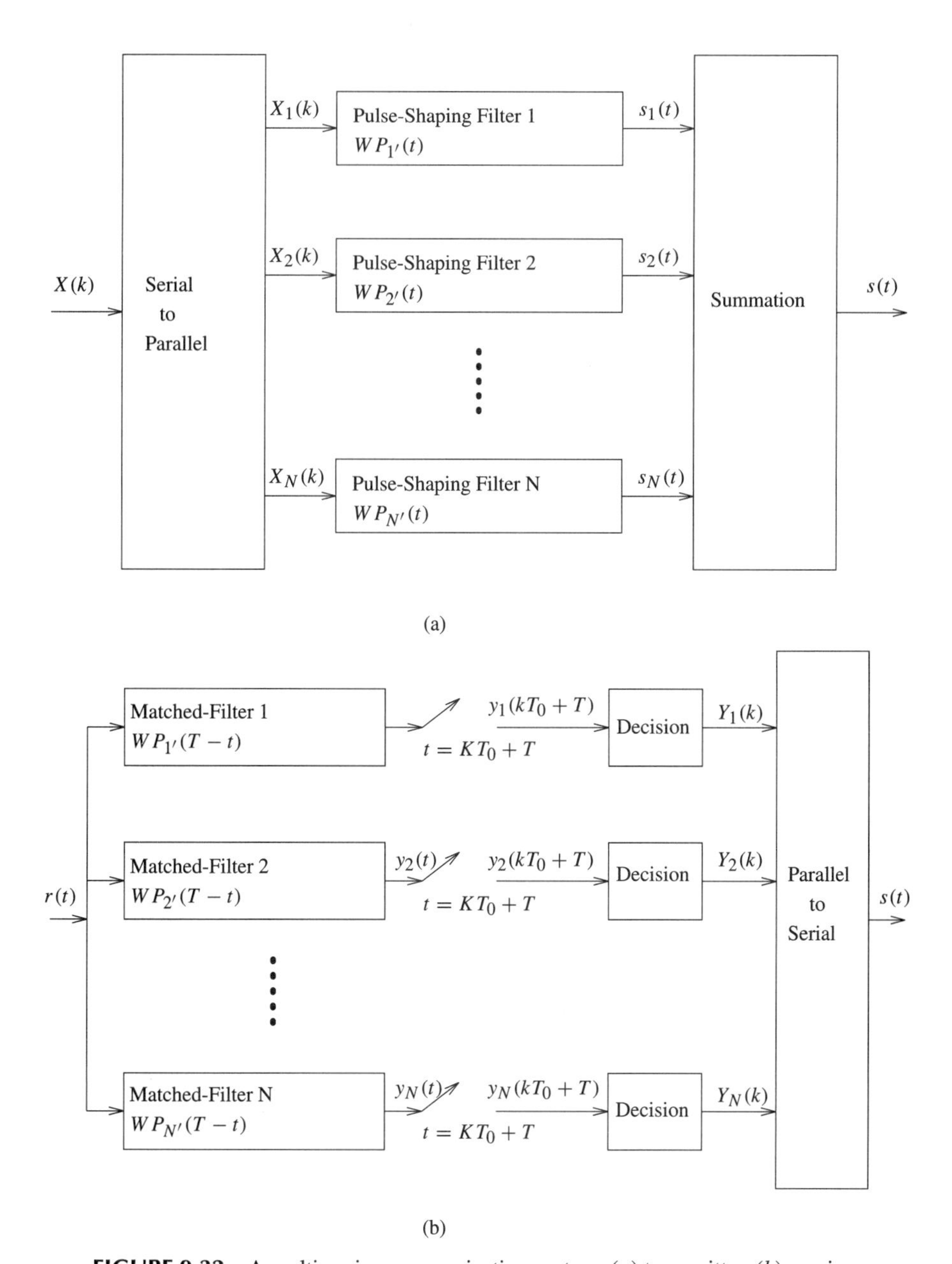

FIGURE 9.32 A multicarrier communication system: (*a*) transmitter, (*b*) receiver.

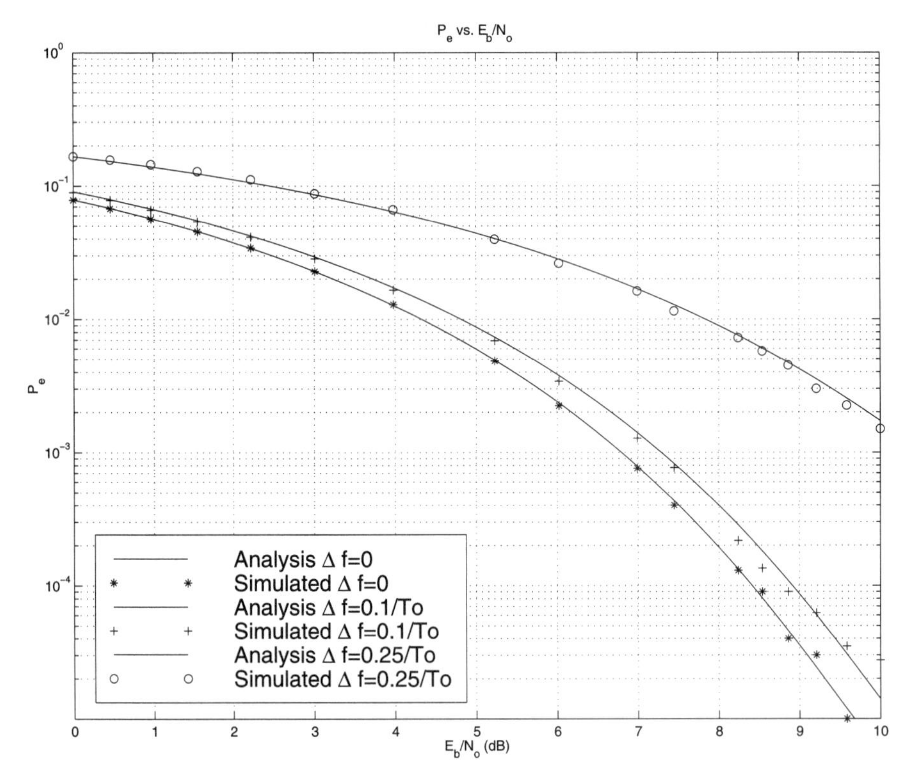

FIGURE 9.33 Probability of error vs. signal/noise ratio for the D-3 wavelet packet multi-carrier communication system for various frequency offsets.

community has increasingly taken advantage of recent advances in communication and signal processing technology to improve diagnostic accuracy and treatment planning. Through teleradiology, it is now possible to have surgeons making diagnoses and plan treatments for a patient who lives at a remote location. This is possible by transmitting 2-D images of the infected region of the patient and reconstructing the image in 3-D at a place where a group of experts can make an accurate diagnosis of the disease. Several problems that have hindered the progress of this work include:

1. *Large storage requirements.* 3-D data sets occupy huge memory space, and storing them for easy retrieval is an important issue.
2. *Low transmission rate.* Channel bandwidths for telephone lines or ISDN lines are small, thus slowing down the transmission speed.
3. *Low-speed image reconstruction.* Rendering algorithms is complex, and it takes time to maneuver these huge sets of data.

We use a 3-D wavelet decomposition and reconstruction algorithm for compression of 3-D data sets. For region of interest (ROI) volume compression, the advantages of using wavelets are that (1) upon reconstruction from a spectral-spatial localized representation of highly correlated 3-D image data, a more natural and artifact-free 3-D visualization is produced, even at high compression rates; and (2) the localized nature of the transform in the space and frequency domains allows for ROI transmission of data.

9.11.1 Three-Dimensional Wavelets and Algorithms

Similar to the 2-D wavelet, 3-D wavelet decomposition can be performed for discrete volume data by a filtering operation, as shown in Figure 9.34.

After a single 3-D level wavelet transform, the volume data would be decomposed into eight blocks, as shown in Figure 9.35. The 3-D volume can be approximated by

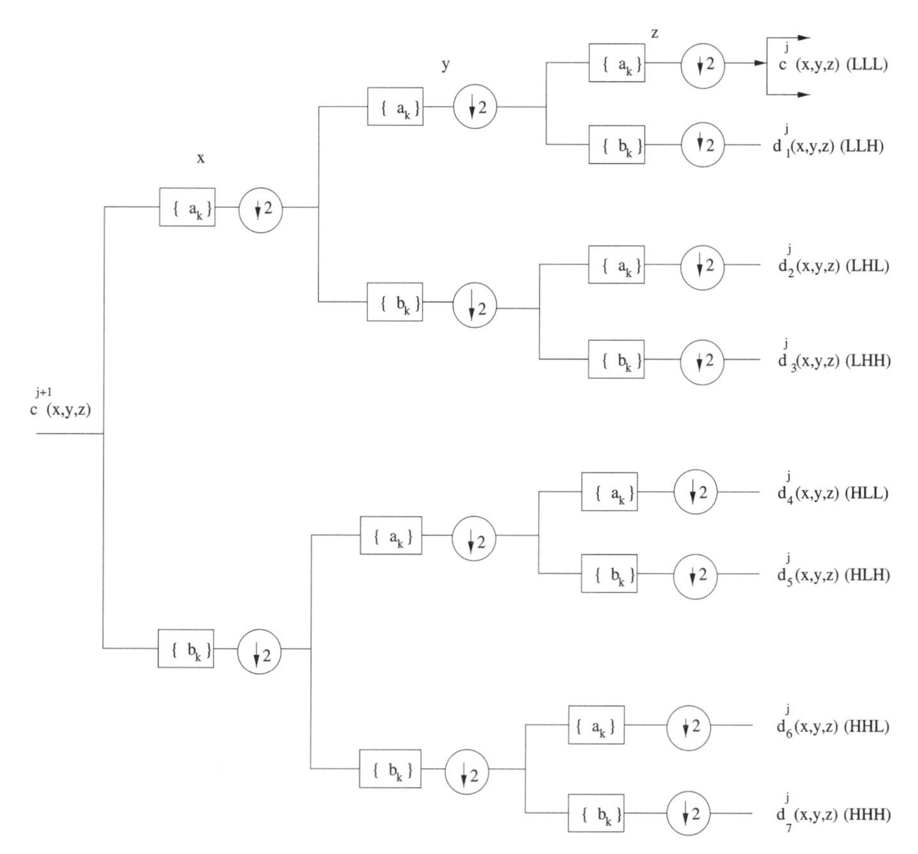

FIGURE 9.34 Block diagram of a three-dimensional hierarchical wavelet decomposition algorithm.

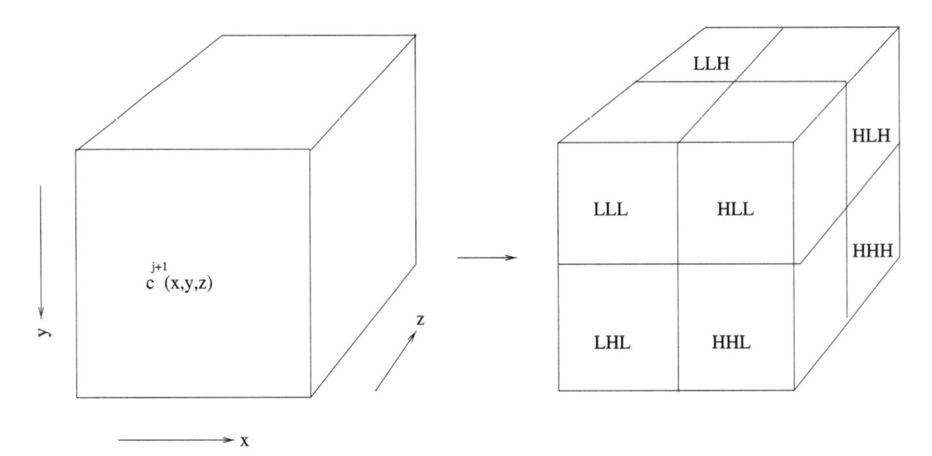

FIGURE 9.35 Labeling wavelet coefficient sets in eight different octants.

using

$$a^{j+1}(x, y, z) = \sum_{n,m,l} a^j_{n,m,l}\phi(2^j x - n, 2^j - m, 2^j z - l),$$

where $\phi(x, y, z) = \phi(x)\phi(y)\phi(z)$ and $a^j_{n,m,l}$ is the scaling function coefficient. We can add the details by adding the 3-D wavelet functions at the resolution 2^j

$$\begin{aligned}
\sum_{n,m,l} [(w_1^j)_{n,m,l}\psi^1(2^j x - n, 2^y - m, 2^j - l)& \\
+ (w_2^j)_{n,m,l}\psi^2(2^j x - n, 2^y - m, 2^j - l)& \\
+ (w_3^j)_{n,m,l}\psi^3(2^j x - n, 2^y - m, 2^j - l)& \\
+ (w_4^j)_{n,m,l}\psi^4(2^j x - n, 2^y - m, 2^j - l)& \\
+ (w_5^j)_{n,m,l}\psi^5(2^j x - n, 2^y - m, 2^j - l)& \\
+ (w_6^j)_{n,m,l}\psi^6(2^j x - n, 2^y - m, 2^j - l)& \\
+ (w_7^j)_{n,m,l}\psi^7(2^j x - n, 2^y - m, 2^j - l)],& \qquad (9.38)
\end{aligned}$$

where w_1^j through w_7^j are the wavelet coefficients. We can reconstruct the original 3-D function volume to any refinement by adding some of the details listed above.

9.11.2 Rendering Techniques

Rendering is the process of generating images using computers. In data visualization, our goal is to transform numerical data into graphical data, or *graphical primitives,* for rendering. Traditional techniques assumed that when an object was rendered, the surfaces and their interactions with light were viewed. However, common objects such as clouds and fog are translucent and scatter light that passes through them. Therefore, for proper rendering, we need to consider the changing properties inside the object.

When we render an object using surface rendering techniques, we model the object mathematically with a surface description such as points, lines, triangles, polygons, or surface splines. The interior of the object is not described or is represented only implicitly by the surface representation.

One of the key developments in volume visualization of scalar data was the marching cubes algorithm of Lorenson and Cline [19]. The basic assumption of this technique and its higher-dimension counterparts is that a contour can pass through a cell in only a finite number of ways. A case table is constructed that enumerates all possible topological states of a cell given combinations of scalar values at the cell points. The number of topological states depends on the number of cell vertices and the number of inside–outside relationships a vertex can have with respect to the contour value. A vertex is considered to be inside a contour if its scalar value is larger than the scalar value of the contour line. Vertices with scalar values less than the contour value are said to be *outside* the contour. For example, if a cell has four vertices and each vertex can be either outside or inside the contour, there are $2^4 = 16$ possible ways that contour lines can pass through the cell. There are 16 combinations for a square cell, but these can be reduced to four cases by utilizing symmetry (see Figure 9.36). Once the proper case is selected, the location of the contour–cell edge intersection can be calculated using interpolation. The algorithm processes a cell and then moves or *marches* to the next cell. After all cells are visited, the contour will be complete. (*Note:* The dashed line in Figure 9.36 indicates a contouring ambiguity.)

In summary, the marching algorithm proceeds as follows:

1. Select a cell.
2. Calculate the inside–outside state of each vertex of the cell.
3. Create an index by storing the binary state of each vertex in a separate bit.

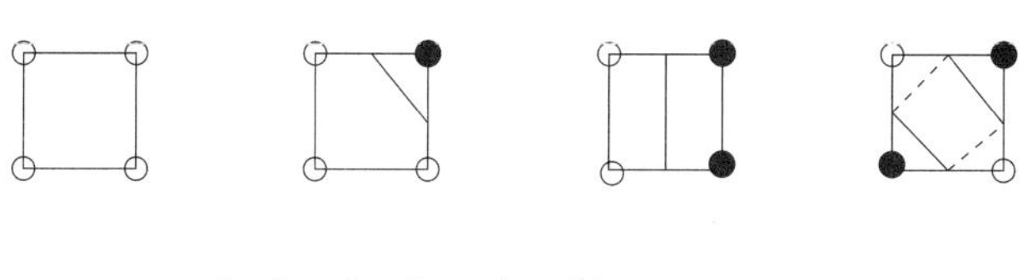

○ = less than iso-value of interest

● = greater than iso-value of interest

FIGURE 9.36 Inside–outside relationship of a square whose vertices have numerical value either higher or lower than a set threshold.

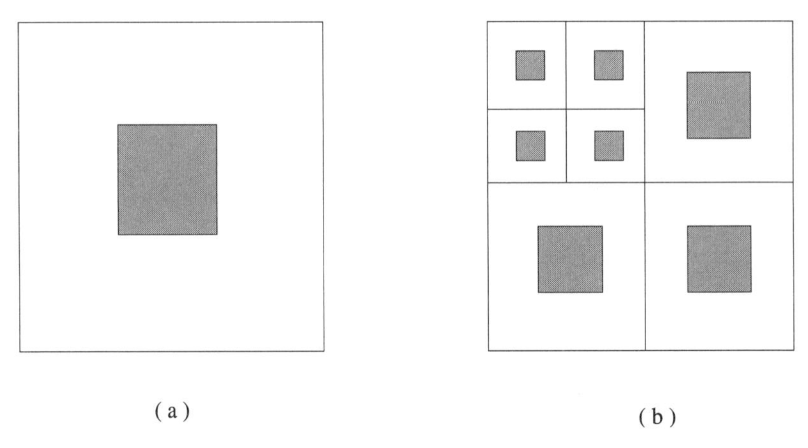

FIGURE 9.37 Region of interest: (*a*) in original image, (*b*) in subimages.

4. Use the index to look up the topological state of the cell in a case table.
5. Calculate the contour locations for each edge in the case table.

9.11.3 Region of Interest

Due to the localized nature of wavelets in frequency and space domains, ROI refinement can be achieved by adding details only in the regions required. Figure 9.37 shows the ROI in the original image and in the wavelet domain for two levels of decomposition. Thus wavelets can be a useful tool for compression, as the image can be approximated by first reconstructing the lowpass coefficients, and the detail can be restored to the ROI solely by transmitting the appropriate highpass coefficients in the ROI. The results of this volume rendering using a 3-D wavelet algorithm are shown in Figures 9.38 to 9.40.

9.11.4 Summary

The 3-D wavelet decomposition and reconstruction algorithm is useful for 3-D image visualization. It improves the speed of the rendering algorithm and achieves data compression by using an ROI approach.

9.12 COMPUTER PROGRAMS

9.12.1 Two-Dimensional Wavelet Algorithms

```
%
% PROGRAM algorithm2D.m
%
```

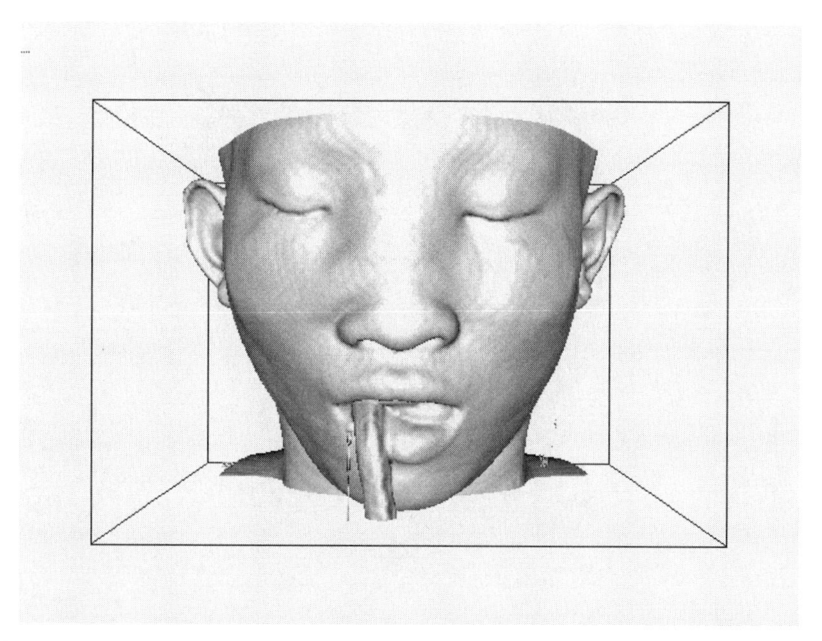

FIGURE 9.38 Multiresolution rendering of 93 slices of 64 $\times$ 64, 8-bit image using isosurfacing with the marching cubes algorithm.

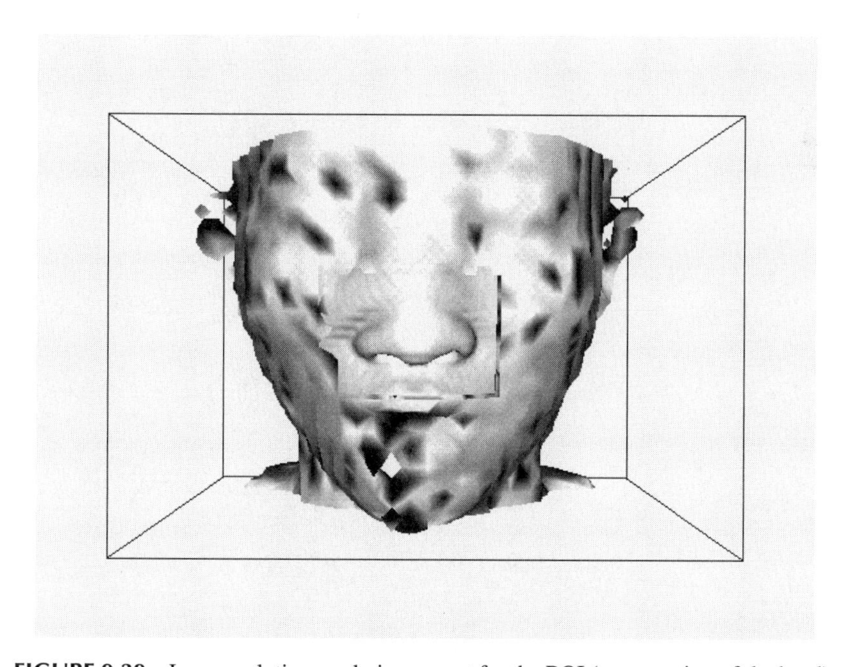

FIGURE 9.39 Low-resolution rendering except for the ROI (nose portion of the head).

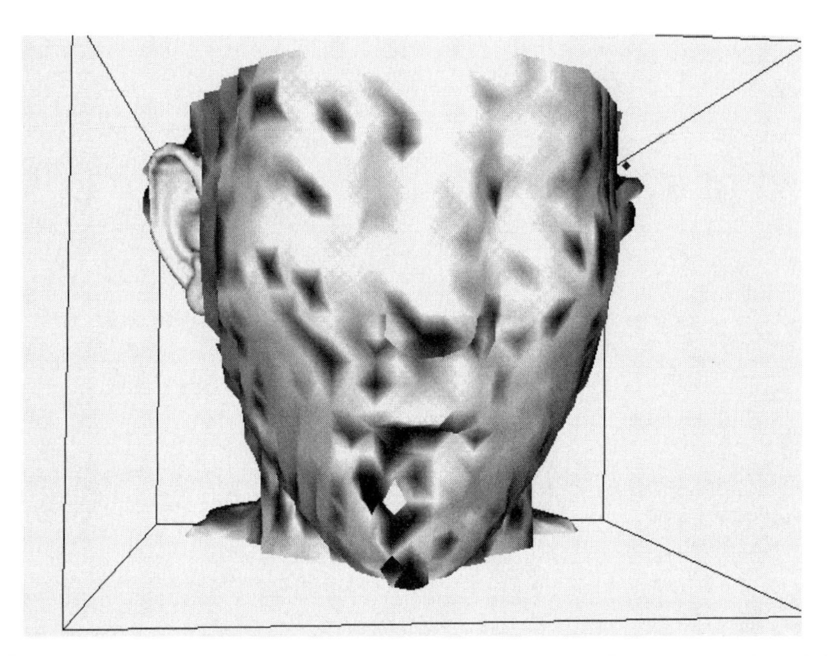

FIGURE 9.40 Low-resolution rendering except for the ROI (left-side ear portion of the head.)

```
% Decomposes and reconstructs a 256x256 image using
% Daubechies' wavelet (m = 2). The initial coefficients
% are taken as the function values themselves.

% test image

picture = 256 * ones(256);
for i = 1:256
  picture(i,i) = 0;
  picture(i,257-i) = 0;
end

image(picture)
title('Original Image')

% Decomposition and reconstruction filters

g0 = [0.68301; 1.18301; 0.31699; -0.18301];
k = [0; 1; 2; 3];
g1 = flipud(g0).*(-1).^k;
h0 = flipud(g0) / 2;
h1 = flipud(g1) / 2;
```

```
% Decomposition process

% First level decomposition

for k=1:256
   s=[0; 0; picture(:,k); 0; 0];
   x=conv(s,h0);
   a=x(1:2:length(x));   %downsampling
   x=conv(s,h1);
   w=x(1:2:length(x));   %downsmapling
   C(:,k)=[a; w];
end

for k=1:256+8
   s=rot90([0 0 C(k,:) 0 0],3);
   x=conv(s,h0);
   a=x(1:2:length(x));  %downsampling
   x=conv(s,h1);
   w=x(1:2:length(x));  %downsmapling
   CC(k,:)=rot90([a; w]);
end

LL=CC(1:132,1:132);
HL=CC(133:264,1:132);
LH=CC(1:132,133:264);
HH=CC(133:264,133:264);


figure(2)

axes('position',[0.1 0.5 0.3 0.3])
image(LL)
title('LL')
axes('position',[0.5 0.5 0.3 0.3])
image(LH)
title('LH')
axes('position',[0.1 0.1 0.3 0.3])
image(HL)
title('HL')
axes('position',[0.5 0.1 0.3 0.3])
image(HH)
title('HH')

clear C
clear CC

% Second level decompostion

for k=1:132
```

```
   s=LL(:,k);
   x=conv(s,h0);
   a=x(1:2:length(x));  %downsampling
   x=conv(s,h1);
   w=x(1:2:length(x));  %downsmapling
   C(:,k)=[a; w];
end

for k=1:128+8
   s=rot90(C(k,:),3);
   x=conv(s,h0);
   a=x(1:2:length(x));  %downsampling
   x=conv(s,h1);
   w=x(1:2:length(x));  %downsmapling
   CC(k,:)=rot90([a; w]);
end;

LL_LL=CC(1:68,1:68);
HL_LL=CC(69:136,1:68);
LH_LL=CC(1:68,69:136);
HH_LL=CC(69:136,69:136);
clear C
clear CC

% Reconstruction Process

% Second level reconstruction

s=[LL_LL LH_LL; HL_LL HH_LL];
for k=1:136
   x=zeros(136,1);
   x(1:2:136)=rot90(s(k,1:68),3);
   y=zeros(136,1);
   y(1:2:136)=rot90(s(k,69:136),3);

   x=conv(x,g0)+conv(y,g1);
   C(k,:)=rot90(x(4:length(x)-4));
end

s=C;
clear C

for k=1:132
   x=zeros(136,1);
   x(1:2:136)=s(1:68,k);
   y=zeros(136,1);
   y(1:2:136)=s(69:136,k);

   x=conv(x,g0)+conv(y,g1);
```

```
   C(:,k)=x(4:length(x)-4);
end
LL_rec=C;
clear C

% First level reconstruction

s=[LL_rec LH; HL HH];
for k=1:264
   x=zeros(264,1);
   x(1:2:264)=rot90(s(k,1:132),3);
   y=zeros(264,1);
   y(1:2:264)=rot90(s(k,133:264),3);
   x=conv(x,g0)+conv(y,g1);
   C(k,:)=rot90(x(4:length(x)-4));
end

s=C;
clear C

for k=1:260
   x=zeros(264,1);
   x(1:2:264)=s(1:132,k);
   y=zeros(264,1);
   y(1:2:264)=s(133:264,k);

   x=conv(x,g0)+conv(y,g1);
   C(:,k)=x(4:length(x)-4);
end

picture_rec=C(3:258,3:258);

figure(3)
image(picture_rec)
title('Reconstructed Image')
```

9.12.2 Wavelet Packets Algorithms

```
%
% PROGRAM waveletpacket.m
%
% Wavelet packet decomposition and reconstruction of a
% function using Daubechies' wavelet (m = 2). The initial
% coefficients are taken as the function values themselves.
%

% Signal

v1 = 100;          % frequency
```

```
v2 = 200;
v3 = 400;
r = 1000;          %sampling rate

k = 1:100;
t = (k-1) / r;
s = sin(2*pi*v1*t) + sin(2*pi*v2*t) + sin(2*pi*v3*t);

% Decomposition and reconstruction filters

g0 = [0.68301; 1.18301; 0.31699; -0.18301];
k = [0; 1; 2; 3];
g1 = flipud(g0).*(-1).^k;
h0 = flipud(g0) / 2;
h1 = flipud(g1) / 2;

% Decomposition process

% First level decomposition

x=conv(s,h0);
a=x(1:2:length(x)); %downsampling
x=conv(s,h1);
w=x(1:2:length(x)); %downsmapling

%second level decomposition
x=conv(a,h0);
aa=x(1:2:length(x));
x=conv(a,h1);
aw=x(1:2:length(x));

x=conv(w, g0);
wa=x(1:2:length(x));
x=conv(w, g1);
ww=x(1:2:length(x));

% Reconstruction process

% Second level reconstruction

x=zeros(2*length(aa),1);
x(1:2:2*length(aa))=aa(1:length(aa));
y=zeros(2*length(aw),1);
y(1:2:2*length(aw))=aw(1:length(aw));
x=conv(x,g0)+conv(y,g1);
a_rec=x(4:length(x)-4);

x=zeros(2*length(wa),1);
x(1:2:2*length(aw))=wa(1:length(wa));
```

```
y=zeros(2*length(ww),1);
y(1:2:2*length(ww))=ww(1:length(ww));
x=conv(x, h0)+conv(y,h1);
w_rec=x(4:length(x)-4);

% First level reconstruction

y=zeros(2*length(w_rec), 1);
y(1:2:2*length(w_rec))=w_rec(1:length(w_rec));
x=zeros(2*length(a_rec), 1);
x(1:2:2*length(a_rec))=a_rec;

x=conv(x,g0);
y=conv(y,g1);
y=x(1:length(y))+y;
s_rec=y(4:length(y)-4);
```

REFERENCES

1. I. Daubechies, *Ten Lectures on Wavelets*, CBMS-NSF Ser. Appl. Math. **61**. Philadelphia: SIAM, 1992, 326 pp.
2. R. R. Coifman and M. V. Wickerhauser, "Entropy-based algorithms for best basis selection," *IEEE Trans. Inform. Theory,* **38**(2), pp. 713–718, 1992.
3. B. Vidaković and P. Müller, *Wavelets for Kids*. Durham, N.C.: Institute of Statistics and Decision Sciences, Duke University, 1991.
4. M. Wang and A. K. Chan, "Wavelet-packet-based time–frequency distribution and its application to radar imaging," *Int. J. Numer. Model.: Electron. Networks Devices Fields*, **11**, pp. 21–40, 1998.
5. H. C. Choe, Ph.D. dissertation, Texas A&M University, College Station, Texas, 1997.
6. D. K. Kil and F. B. Shin, *Pattern Recognition and Prediction with Applications to Signal Characterization*. Woodbury, N.Y.: American Institute of Physics Press, 1996.
7. T. F. Yu, Ph.D. dissertation, Texas A&M University, College Station, Texas, 1997.
8. M. Rabbani and P. W. Jones, *Digital Image Compression Techniques*, Vol TT7. Bellingham, Wash.: SPIE Press, 1991.
9. J. M. Shapiro, "Embedded image coding using zerotrees of wavelet coefficients," *IEEE Trans. Signal Process.,* **41**(12), pp. 3445–3462, 1993.
10. A. Said and W. A. Pearlman, "Image compression using a spatial-orientation tree," *IEEE Int. Symp. on Circuit and Systems*, pp. 279–282, 1993.
11. A. Said and W. A. Pearlman, "A new, fast, and efficient image codec based on set partitioning in hierarchical trees," *IEEE Trans. Circuits Syst. Video Technol.*, **6**, pp. 243–250, 1996.
12. A. Said, "An image multiresolution representation for lossless and lossy compression," *IEEE Trans. Image Process.,* **5**, pp. 1303–1310, 1996.
13. N. W. Lin, D. K. Shin, T. F. Yu, J. S. Liu, and A. K. Chan, "The generalized self-similarity tree coder," *Technical Report.* College Station, Texas: Wavelet Research Laboratory, Texas A&M University, 1996.

14. M. E. Peters, D. R. Voegeli, and K. A. Scanlan, in *Breast Imaging*, R. L. Eisenberg (Ed.), New York: Churchill Livingstone, 1989.
15. R. N. Strickland and H. I. Hahn, "Wavelet transforms for detecting microcalcification in mammograms," *IEEE Trans. Med. Imaging*, **15**, pp. 218–229, 1996.
16. D. A. McCandless, S. K. Rogers, J. W. Hoffmeister, D. W. Ruck, R. A. Raines, and B. W. Suter, "Wavelet detection of clustered microcalcification," *Proc. SPIE Wavelet Appl.*, **2762**, pp. 388–399, 1996.
17. P. P. Gandhi and S. A. Kassam, "Analysis of CFAR processors in nonhomogeneous background," *IEEE Trans. Aerosp. Electron. Syst.*, **24**(4), pp. 427–445, 1988.
18. J. G. Proakis and M. Salehi, *Communication Systems Engineering*, Upper Saddle River, N.J.: Prentice Hall, 1994.
19. W. E. Lorenson and H. E. Cline, "Marching cubes: A high resolution 3-D surface construction algorithm," *Comput. Graphics*, **21**(4), pp. 163–169, 1987.

CHAPTER TEN

Wavelets in Boundary Value Problems

All of the applications discussed so far deal with processing a given function (signal, image, etc.) in the time, frequency, and time–frequency domains. We have seen that wavelet-based time-scale analysis of a function can provide important additional information that cannot be obtained by either time- or frequency-domain analyses. There is another class of problems that we quite often come across that involve solving boundary value problems (BVP). In BVPs functions are not known explicitly; some of their properties along with function values are known at a set of certain points in the domain of interest. In this chapter we discuss the applications of wavelets in solving such problems.

Much of the phenomena studied in electrical engineering can be described mathematically by second order partial differential equations (PDE). Some examples of PDEs are the Laplace, Poisson, Helmholtz, and the Schrödinger equations. Each of these equations may be solved analytically for some, but not for all cases of interest. These PDEs can often be converted to integral equations. One of the attractive features of integral equations is that boundary conditions are built-in and therefore do not have to be applied externally [1]. Mathematical questions of existence and uniqueness of a solution may be handled with more ease with the integral form.

Either approach—differential or integral equations—to represent a physical phenomenon can be viewed in terms of an operator operating on an unknown function in order to produce a known function. In this chapter we deal with the linear operators. The linear operator equation is converted to a system of linear equations with the help of a set of complete bases which is then solved for the unknown coefficients. The finite element and finite difference techniques used to solve PDEs result in sparse and banded matrices, whereas integral equations almost always lead to a dense matrix; an exception being the case when the basis functions, chosen to represent the unknown functions, happen to be the eigen functions of the operator.

Two of the main properties of wavelets vis-a-vis boundary value problems are their hierarchical nature and the vanishing moments properties. Because of their hierarchical (multiresolution) nature, wavelets at different resolutions are interrelated, a property that makes them suitable candidates for multigrid-type methods in solving PDEs. On the other hand, the vanishing moment property by virtue of which wavelets, when integrated against a function of certain order, make the integral zero, is attractive in sparsifying a dense matrix generated by an integral equation.

A complete exposition of the application of wavelets to integral and differential equations is beyond the scope of this chapter. Our objective is to provide the readers with some preliminary theory and results on the application of wavelets to boundary value problems and give references where more details may be found. Since most often, in electrical engineering problems, we encounter integral equations, we will emphasize their solutions using wavelets. We give a few examples of commonly occurring integral equations. The first and the most important step in solving integral equations is to transform them into a set of linear equations. Both conventional and wavelet-based methods in generating matrix equations are discussed. Both the methods fall under the general categories of method of moments (MoM). We will call the method with conventional bases (pulse, triangular, piecewise sinusoid, etc.) *conventional MoM* while the method with wavelet bases will be referred to as *wavelet MoM*. Some numerical results are presented which illustrate the advantages of the wavelet-based technique. We also discuss wavelets on a bounded interval. Some of the techniques applied to solving integral equations are useful for differential equations as well. At the end of the chapter we describe briefly the applications of wavelets in PDEs and provide references where readers can find further information.

10.1 INTEGRAL EQUATIONS

Consider the following first-kind integral equation

$$\int_a^b f(x')K(x,x')\,dx' = g(x), \tag{10.1}$$

where f is the unknown function and the kernel K and the function g are known. This equation, depending on the kernel and the limits of integration, is referred to by different names, such as Fredholm, Volterra, convolution, and the Weiner–Hopf integral equation. Such integral equations appear frequently in practice [2]; for instance, in inverse problems in which the objective is to reconstruct the function f from a set of known data represented in the functional form of g, one encounters first-kind integral equations. In some electromagnetic scattering problems, discussed next, the current distribution on the metallic surface is related to the incident field in the form of an integral equation of type (10.1) with Green's function as the kernel. Observe that solving for f is equivalent to finding the inverse transform of g with respect to the kernel K; in particular, if $K(x,x') = e^{-jxx'}$, then f is nothing but the inverse Fourier transform of g. We assume that (10.1) has a unique solution. Although we discuss solutions of first-kind integral equations only, the method can be extended to second-kind [3, 4] and higher-dimension integral equations [5] with little additional work.

As an example of (10.1), consider that an infinitely long metallic cylinder is illuminated by a TM (transverse magnetic) plane wave as shown in Figure 10.1. An integral equation relating the surface current distribution and the incident field can be formulated by enforcing the boundary condition

$$\hat{\boldsymbol{n}} \times \boldsymbol{E}(\boldsymbol{\rho}) = \hat{\boldsymbol{n}} \times [\boldsymbol{E}_i(\boldsymbol{\rho}) + \boldsymbol{E}_s(\boldsymbol{\rho})] = 0, \qquad \boldsymbol{\rho} \in \boldsymbol{S}, \tag{10.2}$$

where $\boldsymbol{E}$, $\boldsymbol{E}_i$, and $\boldsymbol{E}_s$ are the total, incident, and scattered electric fields, respectively. The surface of the cylinder is represented by $\boldsymbol{S}$. For the TM plane wave incident field,

$$\boldsymbol{E}_i = \hat{z}\, E_z^i, \quad \boldsymbol{H}_i = \hat{\boldsymbol{\ell}}\, H_\ell^i, \quad \text{and} \quad \boldsymbol{J} = \hat{z}\, J_{sz}, \tag{10.3}$$

where, as usual, $\boldsymbol{H}_i$ is the incident magnetic field and $\boldsymbol{J}$ is the induced electric current on the surface of the cylinder. This electric current is related to the incident field and the Green's function by an integral equation,

$$j\omega\mu_0 \int_C J_{sz}(\ell')G(\ell, \ell')\, d\ell' = E_z^i(\ell), \tag{10.4}$$

where

$$G(\ell, \ell') = \frac{1}{4j}\, H_0^{(2)} \left[k_0\, |\boldsymbol{\rho}(\ell) - \boldsymbol{\rho}(\ell')|\right], \tag{10.5}$$

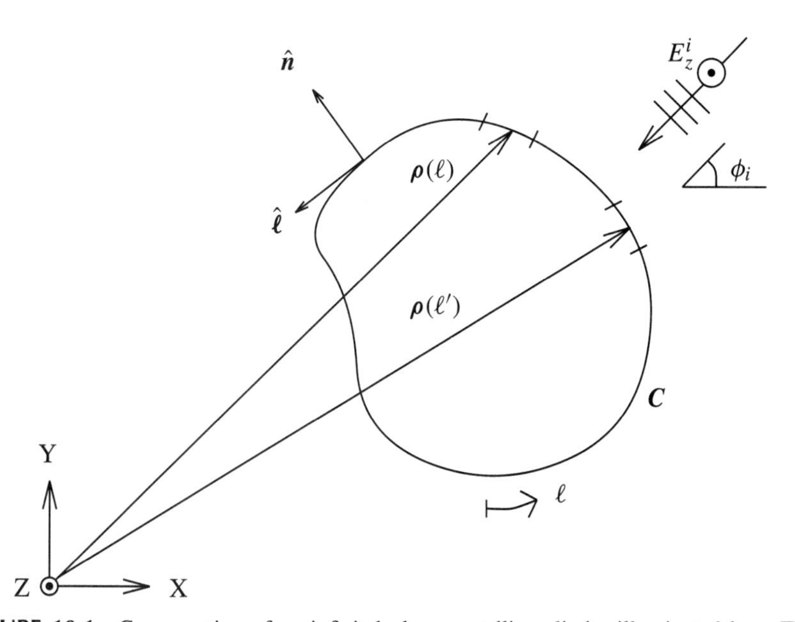

FIGURE 10.1 Cross section of an infinitely long metallic cylinder illuminated by a TM plane wave.

with $k_0 = 2\pi/\lambda_0$, and λ_0 denoting the wavelength. E_z^i is the z-component of the incident electric field and $H_0^{(2)}$ is the second-kind Hankel function of order 0. Here, the contour of integration has been parameterized with respect to the chord length. The field component E_z^i can be expressed as

$$E_z^i(\ell) = E_0 \exp[jk_0(x(\ell)\cos\phi_i + y(\ell)\sin\phi_i)], \tag{10.6}$$

where ϕ_i is the angle of incidence.

It is clear that (10.4) is of the form of equation (10.1). Our objective is to solve (10.4) for the unknown current distribution J_{sz} and compute the radar cross section (RCS); the latter being given by

$$\frac{\text{RCS}}{\lambda_0} = k_0\rho\frac{|E_z^s|^2}{|E_z^i|^2} = \frac{\eta_0^2 k_0^2}{8\pi}|F_\phi|^2, \tag{10.7}$$

where $\eta_0 = \sqrt{\mu_0/\epsilon_0}$ is a known constant and

$$F_\phi = \int_C \exp[jk_0(x(\ell')\cos\phi + y(\ell')\sin\phi)]J_{sz}(\ell')\,d\ell'. \tag{10.8}$$

Scattering from a thin perfectly conducting strip, as shown in Figure 10.2a, gives rise to an equation similar to (10.4). For this case we have

$$\int_{-h}^{h} J_{sy}(z')G(z,z')dz' = E_y^i(z), \tag{10.9}$$

where $G(z, z')$ is given by (10.5).

As a final example, consider scattering from a thin wire as shown in Figure 10.2b. Here the current on the wire and the incident field are related to each other as

$$\int_{-\ell}^{\ell} I(z')K_w(z,z')\,dz' = -E^i(z), \tag{10.10}$$

where the kernel K_w is given by

$$K_w(z,z') = \frac{1}{4\pi j\omega\epsilon_0}\frac{\exp(-jk_0R)}{R^5}\left[(1 + jk_0R) \times (2R^2 - 3a^2) + k_0^2a^2R^2\right] \tag{10.11}$$

$$E^i(z) = E_0 \sin\theta \exp(jk_0 z\cos\theta). \tag{10.12}$$

This kernel is obtained by interchanging integration and differentiation in the integrodifferential form of Pocklington's equation and by using the reduced kernel distance $R = [a^2 + (z - z')^2]^{1/2}$, where a is the radius of the wire [6].

The first step in solving any integral or differential equation is to convert these into a matrix equation which is then solved for the unknown coefficients. Let us rewrite

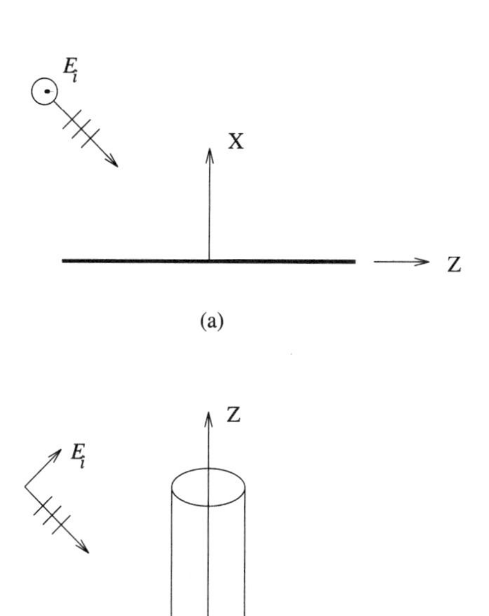

FIGURE 10.2 (*a*) Thin half-wavelength-long metallic strip illuminated by a TM wave; (*b*) Thin wire of length $\lambda/2$ and thickness $\lambda/1000$ illuminated by a plane wave.

(10.1) as $L_K f = g$, where

$$L_K f = \int_a^b f(x')K(x, x')\, dx'. \tag{10.13}$$

The goal is to transform equation (10.1) to a matrix equation

$$Ac = b, \tag{10.14}$$

where A is a two-dimensional matrix, sometimes referred to as impedance matrix, c is the column vector of unknown coefficients, and b is another column vector related to g. Computation time depends largely upon the way we obtain and solve (10.14). In the following sections we describe conventional and wavelet basis functions that are used to represent the unknown function.

10.2 METHOD OF MOMENTS

Method of moments [7] is probably the most widely used technique for solving integral equations in electromagnetics. In conventional MoM, the boundary of integration is approximated by discretizing it into many segments. Then the unknown function is expanded in terms of known basis functions with unknown coefficients. These bases may be global (entire-domain), extending the entire length $[a, b]$, or they may be local (subdomain), covering only a small segment of the interval, or a combination of both. Finally, the resultant equation is tested with the same or different functions, resulting in a set of linear equations whose solution gives the unknown coefficients.

The unknown function $f(x)$ can be written as

$$f(x) = \sum_n c_n \Lambda_n(x), \tag{10.15}$$

where the $\{\Lambda_n\}$ form a complete set of basis functions. For an exact representation of $f(x)$ we may need an infinite number of terms in the series above. However, in practice, a finite number of terms suffices for a given acceptable error. Substituting the series representation of $f(x)$ into the original equation (10.1), we get

$$\sum_{n=1}^{N} c_n L_K \Lambda_n \approx g. \tag{10.16}$$

For the present discussion we assume N to be large enough so that the representation above is exact. Now by taking the inner product of (10.16) with a set of *weighting functions* or *testing functions* $\{\xi_m : m = 1, \ldots, M\}$ we get a set of linear equations

$$\sum_{n=1}^{N} c_n \langle \xi_m, L_K \Lambda_n \rangle = \langle \xi_m, g \rangle, \qquad m = 1, \ldots, M, \tag{10.17}$$

which can be written in the matrix form as

$$[A_{mn}][c_n] = [b_m], \tag{10.18}$$

where

$$\begin{aligned} A_{mn} &= \langle \xi_m, L_K \Lambda_n \rangle, \qquad m = 1, \ldots, M, \qquad n = 1, \ldots, N \\ b_m &= \langle \xi_m, g \rangle, \qquad m = 1, \ldots, M. \end{aligned}$$

Solution of the matrix equation gives the coefficients $\{c_n\}$ and thereby the solution of the integral equations. Two main choices of the testing functions are: (1) $\xi_m(x) = \delta(x - x_m)$, where x_m is a discretization point in the domain, and (2) $\xi_m(x) = \Lambda_m(x)$. In the former case the method is called *point matching*, whereas the latter method is known as *Galerkin method*. Observe that the operator L_K in the preceding paragraphs could be any linear operator—a differential as well as an integral operator.

10.3 WAVELET TECHNIQUES

Conventional bases (local or global), when applied directly to the integral equations, generally lead to a dense (fully populated) matrix A. As a result, the inversion and final solution of such a system of linear equations are very time consuming. In later sections it will be clear why conventional bases give a dense matrix while wavelet bases produce sparse matrices. Observe that conventional MoM is a single-level approximation of the unknown function in the sense that the domain of the function ($[a, b]$, for instance), are discretized only once, even if we use nonuniform discretization of the domain. On the other hand, wavelet-MoM is inherently multilevel in nature.

Beylkin et al. [8] first proposed the use of wavelets in sparsifying an integral equation. Alpert et al. [3] used waveletlike basis functions to solve second-kind integral equations. In electrical engineering, wavelets have been used to solve integral equations arising from electromagnetic scattering and transmission line problems [5,9–23]. In what follows we briefly describe four different ways in which wavelets have been used in solving integral equations.

10.3.1 Use of Fast Wavelet Algorithm

In this method, the impedance matrix A is obtained via the conventional method of moments using basis functions such as triangular functions, and then wavelets are used to transform this matrix into a sparse matrix [9,10]. Consider a matrix W formed by wavelets. The transformation of the original MoM impedance matrix into the new wavelet basis is obtained as

$$WAW^T \cdot (W^T)^{-1}c = Wb, \tag{10.19}$$

which can be written as

$$A_w \cdot c_w = b_w, \tag{10.20}$$

where W^T represents the transpose of the matrix W. The new set of wavelet transformed linear equations are

$$\begin{aligned} A_w &= WAW^T \\ c_w &= (W^T)^{-1}c \\ b_w &= Wb. \end{aligned} \tag{10.21}$$

The solution vector c is then given by

$$c = W^T(WAW^T)^{-1}Wb.$$

For orthonormal wavelets $W^T = W^{-1}$ and the transformation (10.19) is unitary similar. It has been shown in [9, 10] that the impedance matrix A_w is sparse, which reduces the inversion time significantly. Discrete wavelet transform (DWT) algorithms can be used to obtain A_w and finally the solution vector c.

10.3.2 Direct Application of Wavelets

In another method of applying wavelets to integral equations, wavelets are directly applied; that is, first the unknown function is represented as a superposition of wavelets at several levels (scales) along with the scaling function at the lowest level, before using Galerkin's method described previously.

Let us expand the unknown function f in (10.1) in terms of the scaling functions and wavelets as

$$f(x) = \sum_{s=s_0}^{s_u} \sum_{k=K_1}^{K(s)} w_{k,s} \psi_{k,s}(x) + \sum_{k=K_1}^{K(s_0)} a_{k,s_0} \phi_{k,s_0}(x). \tag{10.22}$$

It should be pointed out here that the wavelets $\{\psi_{k,s}\}$ by themselves form a complete set; therefore, the unknown function could be expanded entirely in terms of the wavelets. However, to retain only a finite number of terms in the expansion, the scaling function part of (10.22) must be included. In other words, $\{\psi_{k,s}\}$, because of their bandpass filter characteristics, extract successively lower and lower frequency components of the unknown function with decreasing values of the scale parameter s, while ϕ_{k,s_0}, because of its lowpass filter characteristics, retains the lowest-frequency components or the coarsest approximation of the original function.

In (10.22), the choice of s_0 is restricted by the order of wavelet, while the choice of s_u is governed by the physics of the problem. In applications involving electromagnetic scattering, as a rule of thumb, the highest scale s_u should be chosen such that $1/2^{s_u+1}$ does not exceed $0.1\lambda_0$.

The expansion of f given by (10.22) is substituted in (10.1), and the resulting equation is tested with the same set of expansion functions. This result gives a set of linear equations as

$$\begin{bmatrix} [A_{\phi,\phi}] & [A_{\phi,\psi}] \\ [A_{\psi,\phi}] & [A_{\psi,\psi}] \end{bmatrix} \begin{bmatrix} [a_{k,s_0}]_k \\ [w_{n,s}]_{n,s} \end{bmatrix} = \begin{bmatrix} \langle E_z^i, \phi_{k',s_0} \rangle_{k'} \\ \langle E_z^i, \psi_{n',s'} \rangle_{s',n'} \end{bmatrix}, \tag{10.23}$$

where

$$[A_{\phi,\phi}] := \langle \phi_{k',s_0}, (L_K \phi_{k,s_0}) \rangle_{k,k'} \tag{10.24}$$

$$[A_{\phi,\psi}] := \langle \phi_{k',s_0}, (L_K \psi_{n,s}) \rangle_{k',n,s} \tag{10.25}$$

$$[A_{\psi,\phi}] := \langle \psi_{n',s'}, (L_K \phi_{k,s_0}) \rangle_{k,n',s'} \tag{10.26}$$

$$[A_{\psi,\psi}] := \langle \psi_{n',s'}, (L_K \psi_{n,s}) \rangle_{n,s,n',s'} \tag{10.27}$$

$$\langle f, g \rangle := \int_a^b f(x) g(x)\, dx, \tag{10.28}$$

$$(L_K f)(x) := \int_a^b f(x') K(x, x')\, dx'. \tag{10.29}$$

In (10.23), $[w_{n,s}]_{n,s}$ is a one-dimensional vector and should not be confused with a two-dimensional matrix. Here the index n is varied first for a fixed value of s.

We can explain the denseness of the conventional MoM and the sparseness of the wavelet MoM by recalling the fact that unlike wavelets, the scaling functions discussed in this book do not have vanishing moments properties. Consequently, for two pulse or triangular functions ϕ_1 and ϕ_2 (usual bases for the conventional MoM and suitable candidates for the scaling functions), even though $\langle \phi_1, \phi_2 \rangle = 0$ for nonoverlapping supports, $\langle \phi_1, L_K \phi_2 \rangle$ is not very small since $L_k \phi_2$ is not small. On the other hand, as is clear from the vanishing moment property,

$$\int_{-\infty}^{\infty} t^p \psi_m(t)\, dt = 0, \quad p = 0, \ldots, m-1, \tag{10.30}$$

the integral vanishes if the function against which the wavelet is being integrated behaves as a polynomial of a certain order locally. Away from the singular points, the kernel usually has a locally polynomial behavior. Consequently, integrals such as $(L_K \psi_{n,s})$ and the inner products involving the wavelets are very small for nonoverlapping supports.

Because of its total positivity property, the scaling function has a smoothing or variation diminishing effect on a function against which it is integrated. The smoothing effect can be understood as follows. If we convolve two pulse functions, both of which are discontinuous but totally positive, the resultant function is a linear B-spline that is continuous. Similarly, if we convolve two linear B-splines, we get a cubic B-spline that is twice continuously differentiable. Analogous to these, the function $L_K \phi_{k,s_0}$ is smoother than the kernel K itself. Furthermore, because of the MRA properties that give

$$\langle \phi_{k,s}, \psi_{\ell,s'} \rangle = 0, \quad s \le s', \tag{10.31}$$

the integrals $\langle \phi_{k',s_0}, (L_K \psi_{n,s}) \rangle$ and $\langle \psi_{n',s'}, (L_K \phi_{k,s_0}) \rangle$ are quite small.

Although diagonally dominant, the $\left[A_{\phi,\phi}\right]$ portion of the matrix usually does not have entries that are very small compared to the diagonal entries. In the case of conventional MoM, all the elements of the matrix are of the form $\langle \phi_{k',s}, (L_K \phi_{k,s}) \rangle$. Consequently, we cannot threshold such a matrix to sparsify it. In the case of the wavelet MoM, the entries of $\left[A_{\phi,\phi}\right]$ occupy a very small portion (5×5 for linear and 11×11 for cubic spline cases) of the matrix, while the rest contain entries whose magnitudes are very small compared to the largest entry; hence a significant number of entries can be set to zero without affecting the solution appreciably.

10.3.3 Wavelets in Spectral Domain

In all of the preceding chapters, we have used wavelets in the time (space) domain. In previous sections, the local support and vanishing moment properties of wavelet bases were used to obtain a sparse matrix representation of an integral equation. In some applications, particularly in the spectral domain methods in the electromagnetic problems, wavelets in the spectral domain may be quite useful. Whenever we have a problem in which the unknown function is expanded in terms of the basis function in the time domain while the numerical computation takes place in the spectral domain, we should look at the time–frequency window product to determine the efficiency of using the particular basis function. Because of the nearly optimal time (space)–frequency (wavenumber) window product of the cubic spline and the corresponding semiorthogonal wavelet, the double integral appearing in the transmission line discontinuity problems can be evaluated efficiently. In this section we consider an example from a transmission line discontinuity to illustrate the usefulness of wavelets in the spectral domain.

Transmission-line configurations are shown in Figure 10.3. Formulation of the integral equation for these configurations is not the purpose of this section. Readers may refer to [5, 24, 25] for details on such formulation. The integral equation obtained is

$$\int \Lambda(k_x, k_y) \hat{f}_y(k_y) \hat{\phi}_k(-k_y)\, dk_x\, dk_y = 0 \tag{10.32}$$

with

$$\Lambda(k_x, k_y) = \hat{G}_{yy}(k_x, k_y) J_0^2(k_x d) \begin{Bmatrix} \cos^2 k_x p \\ \sin^2 k_x p \end{Bmatrix}, \tag{10.33}$$

where G_{yy} is the appropriate Green's function and $\cos^2 k_x p$ and $\sin^2 k_x p$ refer to even and odd modes, respectively. The functions $\hat{f}_y(k_y)\hat{\phi}_k(k_y)$ are the Fourier transforms of the basis functions representing the y-dependence of the magnetic current. To find the propagation constant k_{ye} of any infinite transmission line, we assume that all the field and current distributions have their y-dependence as $e^{-jk_{ye}y}$. It is easy, for this case, to arrive at

$$\int_{-\infty}^{\infty} \Lambda^{\mathrm{HM}}(k_x, k_{ye})\, dk_x = 0. \tag{10.34}$$

Since the transverse dimensions of slots (strips) are very small compared with the wavelength, we assume that the current distribution, sufficiently away from the discontinuity, is due to the fundamental mode only. Near the discontinuity, in addition to the entire domain basis functions resulting from the fundamental mode, subdomain basis functions are used to expand the unknown current (magnetic or electric) to account for the deviation caused by the presence of higher-order modes generated by the discontinuity. We use three different sets of subdomain basis functions: (1) piecewise sinusoids (PWSs), (2) cubic B-splines, and (3) a combination of cubic B-splines and the corresponding semiorthogonal wavelets.

For cases (1) and (2), the longitudinal variation of current is given as

$$f_y(y) = s_i(y) \pm \Gamma s_r(y) + \sum_{k=0}^{K} a_k \phi_k(y), \tag{10.35}$$

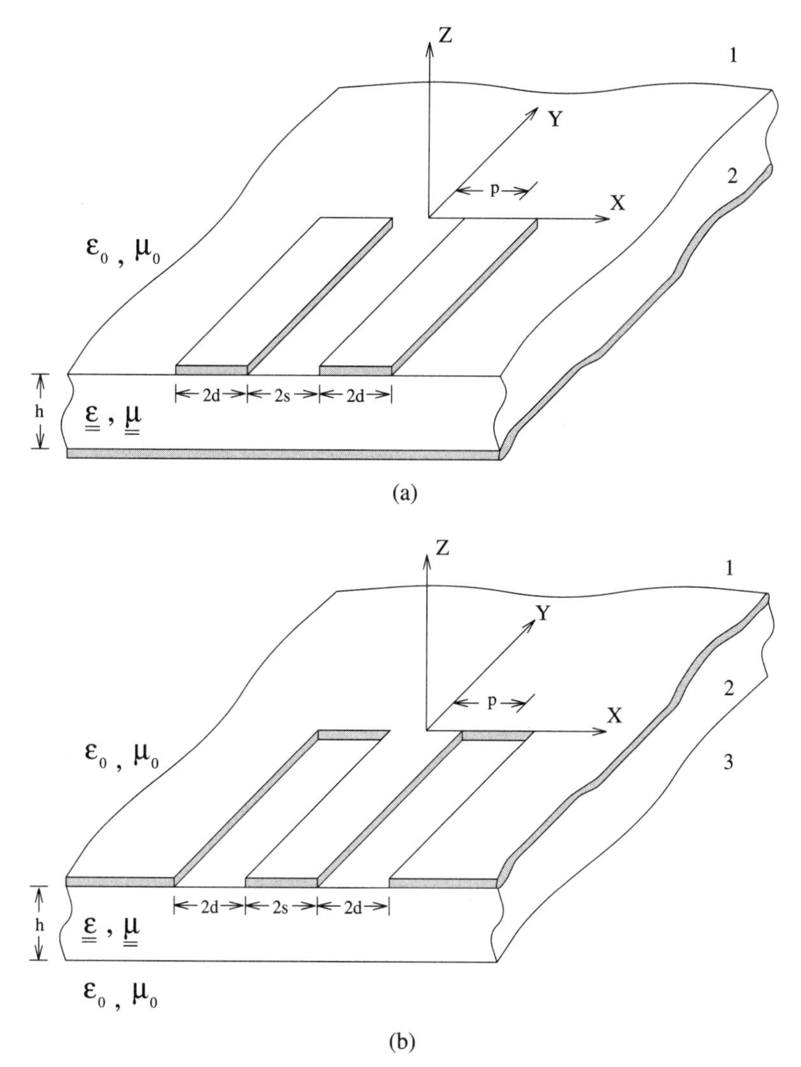

FIGURE 10.3 (*a*) Open coupled-microstrip; (*b*) Short-circuited coupled slot line with uniaxial substrate.

where $K > 0$ and plus and minus signs apply to transmission line configurations with strips (Figure 10.3a), and with slots (Figure 10.3b), respectively. Ideally, the magnitude of the reflection coefficient Γ should be 1; we will see, however, that $|\Gamma| < 1$, indicating the pseudo nature of such terminations, as shown in Figure 10.3. The entire domain functions s_i and s_r, representing incident and reflected waves, respectively, are given below.

$$s_i(y) := \cos k_{ye}y\, U\left(-y - \frac{\pi}{2k_{ye}}\right) - j \sin k_{ye}y\, U(-y) \tag{10.36}$$

$$\hat{s}_i(k_y) = \left[\exp\left(-j\frac{\pi\, k_y}{2\, k_{ye}}\right) - j\right] \times \left\{\frac{k_{ye}}{k_y^2 - k_{ye}^2} + j\,\frac{\pi}{2}\,[\delta(k_y - k_{ye}) - \delta(k_y + k_{ye})]\right\} \tag{10.37}$$

$$s_r(y) := \cos k_{ye}y\, U\left(-y - \frac{\pi}{2k_{ye}}\right) + j \sin k_{ye}y\, U(-y) \tag{10.38}$$

$$\hat{s}_r(k_y) = \left[\exp\left(-j\frac{\pi\, k_y}{2\, k_{ye}}\right) + j\right] \times \left\{\frac{k_{ye}}{k_y^2 - k_{ye}^2} + j\,\frac{\pi}{2}\,[\delta(k_y - k_{ye}) - \delta(k_y + k_{ye})]\right\}, \tag{10.39}$$

where k_{ye} is the propagation constant for the fundamental mode and $U(y)$ is the Heaviside function, defined in the usual way as

$$U(y) = \begin{cases} 1, & y \geq 0 \\ 0, & y < 0. \end{cases} \tag{10.40}$$

The subdomain basis function ϕ_k for the piecewise sinusoid case is

$$\phi_k(y) := \cap\,[y + \tau(k+1)] \tag{10.41}$$

$$\cap(y) := \begin{cases} \dfrac{\sin k_{ye}(\tau - |y|)}{\sin k_{ye}\tau}, & |y| \leq \tau \\ 0, & \text{elsewhere} \end{cases} \tag{10.42}$$

$$\hat{\phi}_k(k_y) = e^{-j\tau(k+1)k_y}\,\frac{2k_{ye}}{k_y^2 - k_{ye}^2} \times \frac{\cos k_{ye}\tau - \cos k_y\tau}{\sin k_{ye}\tau}, \tag{10.43}$$

with $0 < \tau < \pi/2k_{ye}$. For the cubic B-spline case,

$$\phi_k(y) := N_4\left(\frac{y}{\tau} + k + 4\right) \tag{10.44}$$

$$\hat{\phi}(k_y) = \tau e^{-j(k+2)k_y\tau}\left[\frac{\sin(k_y\tau/2)}{k_y\tau/2}\right]^4. \tag{10.45}$$

For the third choice of basis function, we have the following representation of $f_y(y)$:

$$f_y(y) = s_i(y) \pm \Gamma s_r(y) + \sum_{s=s_0}^{0}\sum_{n=0}^{N(s)} w_{n,s}\psi_{n,s}(y) + \sum_{k=0}^{K(s_0)} a_k\phi_{k,s_0}(y), \tag{10.46}$$

where $N(s)$, $K(s_0) \geq 0$, $s_0 \leq 0$, and

$$\phi_{k,s} := N_4\left(\frac{2^s\, y}{\tau} + k + 4\right) \tag{10.47}$$

$$\psi_{k,s} := \psi_4\left(\frac{2^s\, y}{\tau} + k + 7\right) \tag{10.48}$$

$$\hat{\psi}_{k,s}(k_y) = \frac{\tau}{2^s}\, \exp\left[-j\, \frac{(2k+7)k_y\tau}{2^{s+1}}\right] Q\left(\frac{k_y\tau}{2^s}\right) \left[\frac{\sin(k_y\tau/2^{s+2})}{k_y\tau/2^{s+2}}\right]^4 \tag{10.49}$$

$$Q(k_y) := \frac{1}{2520}\left(\cos\frac{3k_y}{2} - 120\cos k_y + 1191\cos\frac{k_y}{2} - 1208\right) \times \sin^4\frac{k_y}{4}. \tag{10.50}$$

Observe that the definitions of $\phi_{k,s}$ and $\psi_{k,s}$ are slightly different from those used in previous chapters. The time–frequency window products of PWS, cubic spline, and the cubic spline wavelet are 0.545, 0.501, and 0.505, respectively. Observe that the product for the linear spline is 0.548; therefore, the double integral as discussed before will take about the same time to compute in both cases.

Application of the Galerkin method with the basis functions described previously leads to a set of linear equations, the solution of which gives the desired value of the reflection coefficient Γ. For the first two we have

$$\left[\ [A_{p,1}]\quad [A_{p,q}]\ \right]\left[\begin{array}{c}\Gamma\\ [c_k]\end{array}\right] = \left[\ B_p\ \right], \tag{10.51}$$

where in the case of cubic splines, the matrix elements take the form

$$A_{p,1} = 2\tau\, k_{ye} \int_0^\infty \int_0^\infty \Lambda^{\text{HM}}(k_x, k_y) \left[\frac{\sin(k_y\tau/2)}{k_y\tau/2}\right]^4 \times \frac{\cos[(p+1-\pi/2\, k_{ye}\, \tau)\, k_y\tau] + j\, \cos[(p+1)\, k_y\tau]}{k_y^2 - k_{ye}^2}\, dk_x\, dk_y, \tag{10.52}$$

$$A_{p,q} = 2\tau^2 \int_0^\infty \int_0^\infty \Lambda^{\text{HM}}(k_x, k_y) \left[\frac{\sin(k_y\tau/2)}{k_y\tau/2}\right]^8 \times \cos[(p-q+1)\, k_y\, \tau]\, dk_x\, dk_y\ , \tag{10.53}$$

with $p = 1, \ldots, M+2$ and $q = 2, \ldots, M+2$. Matrix elements for the PWS can be written in a similar way. In both cases, we observe that

$$A_{p,q} = A_{q-1,p+1},\quad p = 1, \ldots, M+1 \tag{10.54}$$

$$= A_{p-1,q-1},\quad p = 2, \ldots, M+2;\quad q = 3, \ldots, M+2, \tag{10.55}$$

indicating the symmetry and Toeplitz nature of the major portion of the matrix.

For the discontinuity problem, we find that the third representation, (10.46), does not give much advantage over the second one. Unlike the scattering problem in which the domain of the unknown function may be several wavelengths long, for most of the discontinuity problems, the domain of unknown is approximately one wavelength, since the effect of discontinuity on the current distribution is localized. The size of the matrix in the case of the discontinuity problems is usually small compared with the scattering problem. Consequently, achieving sparsity of the matrix may not be a major concern. On the other hand, the spectral integrals associated with each matrix element in the case of the discontinuity problems usually takes a considerably large amount of CPU time. Faster computations of these integrals are achieved using cubic splines due to their decay property which is better than that of PWS [see (10.43) and (10.45)]. For further details on the numerical results for the reflection coefficients, readers are referred to [5, 24, 25].

10.3.4 Wavelet Packets

Recently, discrete wavelet packet (DWP) similarity transformations has been used to obtain a higher degree of sparsification of the matrix than is achievable using the standard wavelets [21]. It has also been shown that DWP method gives faster matrix-vector multiplication than some of the fast multipole methods.

In the standard wavelet decomposition process, first we map the given function to a sufficiently high resolution subspace (V_M) and obtain the approximation coefficients $\{a_{k,M}\}$ (see Chapter 7). The approximation coefficients $\{a_{k,M-1}\}$ and wavelet coefficients $\{w_{k,M-1}\}$ are computed from $\{a_{k,M}\}$. This process continues, that is, the coefficients for the next lower level $M-2$ are obtained from $\{a_{k,M-1}\}$, and so on. Observe that in this scheme, only approximation coeffiecients $\{a_{k,j}\}$ are processed at any scale j; the wavelet coefficients are merely the outputs and remain untouched. In a wavelet packet, the wavelet coefficients are also processed which, heuristically, should result in higher degree of sparsity since in this scheme, the frequency bands are further divided compared with the standard decomposition scheme.

10.4 WAVELETS ON THE BOUNDED INTERVAL

In previous chapters we described wavelets and scaling functions defined on the real line. If we use these functions directly to expand the unknown function of an integral equation, some of the scaling functions and wavelets will have to be placed outside the domain of integration, thus necessitating the explicit enforcement of the boundary conditions. In signal processing, uses of these wavelets lead to undesirable jumps near the boundaries (see Figures 8.8–8.10). We can avoid this difficulty by periodizing the scaling function as [26, Sec. 9.3]

$$\phi^{p}_{k,s} := \sum_{\ell} \phi_{k,s}(x+\ell); \tag{10.56}$$

where the superscript p implies the periodic case. Periodic wavelets are obtained similarly. It is easy to show that if $\hat{\phi}(2\pi k) = \delta_{k,0}$, which is generally true for the scaling functions, then $\sum_k \phi(x-k) \equiv 1$. If we apply to (10.56) the last relation, which as discussed in Chapter 5 is also known as the partition of unity, we can show that $\{\phi^p_{0,0}\} \cup \{\psi^p_{k,s};\ s \in \mathbb{Z}^+ := \{0,1,2,\ldots\}, k = 0,\ldots,2^s-1\}$ generates $L^2([0,1])$.

The idea of periodic wavelets has been used by [18–20]. However, as mentioned in [26, Sec. 10.7], unless the function that is being approximated by the periodized scaling functions and wavelets is already periodic, we still have edge problems at the boundaries. Therefore, we follow a different approach to account for the boundary effects. We apply the compactly supported semiorthogonal spline wavelets of [23, 27, 28], which are specially constructed for the bounded interval [0, 1]. Other ways of obtaining intervallic wavelets are described in [29, 30].

As we discussed in Chapter 5, splines for a given simple knot sequence can be constructed by taking piecewise polynomials between the knots and joining them together at the knots in such a way as to obtain a certain order of overall smoothness. For instance, consider a knot sequence $\{0,1,2,3,4\}$. With this sequence we can construct the cubic spline ($m = 4$) by taking polynomials of order 4 between knots, such as $[0,1), [1,2), \ldots,$ and joining them together at 1, 2, and 3 so that the resultant function (cubic spline) is in C^2, that is, up to its second derivative is continuous in $[0,4)$. In general, cardinal B-splines of order m are in C^{m-2}. However, if we have multiple knots, say for example $\{0,0,1,2,3\}$, the smoothness at the point with multiple knots decreases. It is easy to verify that the smoothness decreases by $r-1$ at a point with r-tuple knots. Observe that at the boundary points 0 and 1, the knots coalesce and form multiple knots. Inside the interval, though, the knots are simple, and hence the smoothness remains unaffected.

For $s \in \mathbb{Z}^+$, let $\{t^s_k\}_{k=-m+1}^{2^s+m-1}$ be a knot sequence with m-tuple knots at 0 and 1 and simple knots inside the unit interval, that is,

$$\begin{aligned} t^s_{-m+1} &= t^s_{-m+2} = \cdots = t^s_0 = 0 \\ t^s_k &= k2^{-s}, k = 1,\ldots,2^s-1 \\ t^s_{2^s} &= t^s_{2^s+1} = \cdots = t^s_{2^s+m-1} = 0. \end{aligned} \tag{10.57}$$

For the knot sequence (10.57) we define the B-spline ($m \geq 2$) as [31, p. 108]

$$B_{m,s,k}(x) := (t^s_{k+m} - t^s_k)[t^s_k, t^s_{k+1}, \ldots, t^s_{k+m}]_t (t-x)^{m-1}_+, \tag{10.58}$$

where $[t^s_k,\ldots,t^s_{k+m}]_t$ is the mth-order divided difference of $(t-x)^{m-1}_+$ with respect to t and $(x)_+ := \max(0,x)$. Wavelets can be obtained from the corresponding spline scaling functions. Instead of going into the details of constructing scaling functions and wavelets on a bounded interval, we provide their explicit formulas in Section 10.9. Interested readers will find details in [23, 27, 28].

The support of the inner (without multiple knots) B-spline occupies m segments, and that of the corresponding semiorthogonal wavelet occupies $2m-1$ segments. At any scale s the discretization step is $1/2^s$, which for $s > 0$ gives 2^s segments in

[0, 1]. Therefore, to have at least one inner wavelet, the following condition must be satisfied:

$$2^s \geq 2m - 1. \tag{10.59}$$

Let s_0 be the scale for which the condition (10.59) is satisfied. Then for each $s \geq s_0$, let us define the scaling functions $\phi_{m,k,s}$ of order m as

$$\phi_{m,k,s}(x) := \begin{cases} B_{m,s_0,k}(2^{s-s_0}x), & k = -m+1, \ldots, -1 \\ B_{m,s_0,2^s-m-k}(1 - 2^{s-s_0}x), & k = 2^s - m + 1, \ldots, 2^s - 1 \\ B_{m,s_0,0}(2^{s-s_0}x - 2^{-s_0}k), & k = 0, \ldots, 2^s - m \end{cases} \tag{10.60}$$

and the wavelets $\psi_{m,k,s}$ as

$$\psi_{m,k,s}(x) := \begin{cases} \psi_{m,k,s_0}(2^{s-s_0}x), & k = -m+1, \ldots, -1 \\ \psi_{m,2^s-2m+1-k,s_0}(1 - 2^{s-s_0}x), & k = 2^s - 2m + 2 \ldots, 2^s - m \\ \psi_{m,0,s_0}(2^{s-s_0}x - 2^{-s_0}k), & k = 0, \ldots, 2^s - 2m + 1. \end{cases} \tag{10.61}$$

Observe that the inner scaling functions ($k = 0, \ldots, 2^s - m$) and the wavelets ($k = 0, \ldots, 2^s - 2m + 1$) are the same as those for the nonboundary case. There are $m - 1$ boundary scaling functions and wavelets at 0 and 1, and $2^s - m + 1$ inner scaling functions and $2^s - 2m + 2$ inner wavelets. Figure 10.4 shows all the scaling functions and wavelets for $m = 2$ at the scale $s = 2$. All the scaling functions for $m = 4$ and $s = 3$ are shown in Figure 10.5a, while Figure 10.5b gives only the corresponding boundary wavelets near $x = 0$ and one inner wavelet. The rest of the inner wavelets can be obtained simply by translating the first one, whereas the boundary wavelets near $x = 1$ are the mirror images of wavelets near $x = 0$.

10.5 SPARSITY AND ERROR CONSIDERATIONS

The study of the effects of thresholding the matrix elements on the sparsity and error in the solution is the objective of this subsection. By *thresholding* we mean setting those elements of the matrix to zero that are smaller (in magnitude) than some positive number δ ($0 \leq \delta < 1$), called the *threshold parameter*, times the largest element of the matrix.

Let $A_{\max}$ and $A_{\min}$ be the largest and the smallest elements of the matrix in (10.23). For a fixed value of the threshold parameter δ, define percent relative error (ϵ_δ) as

$$\epsilon_\delta := \frac{\| f_0 - f_\delta \|_2}{\| f_0 \|_2} \times 100 \tag{10.62}$$

and percent sparsity (S_δ) as

$$S_\delta := \frac{N_0 - N_\delta}{N_0} \times 100 \tag{10.63}$$

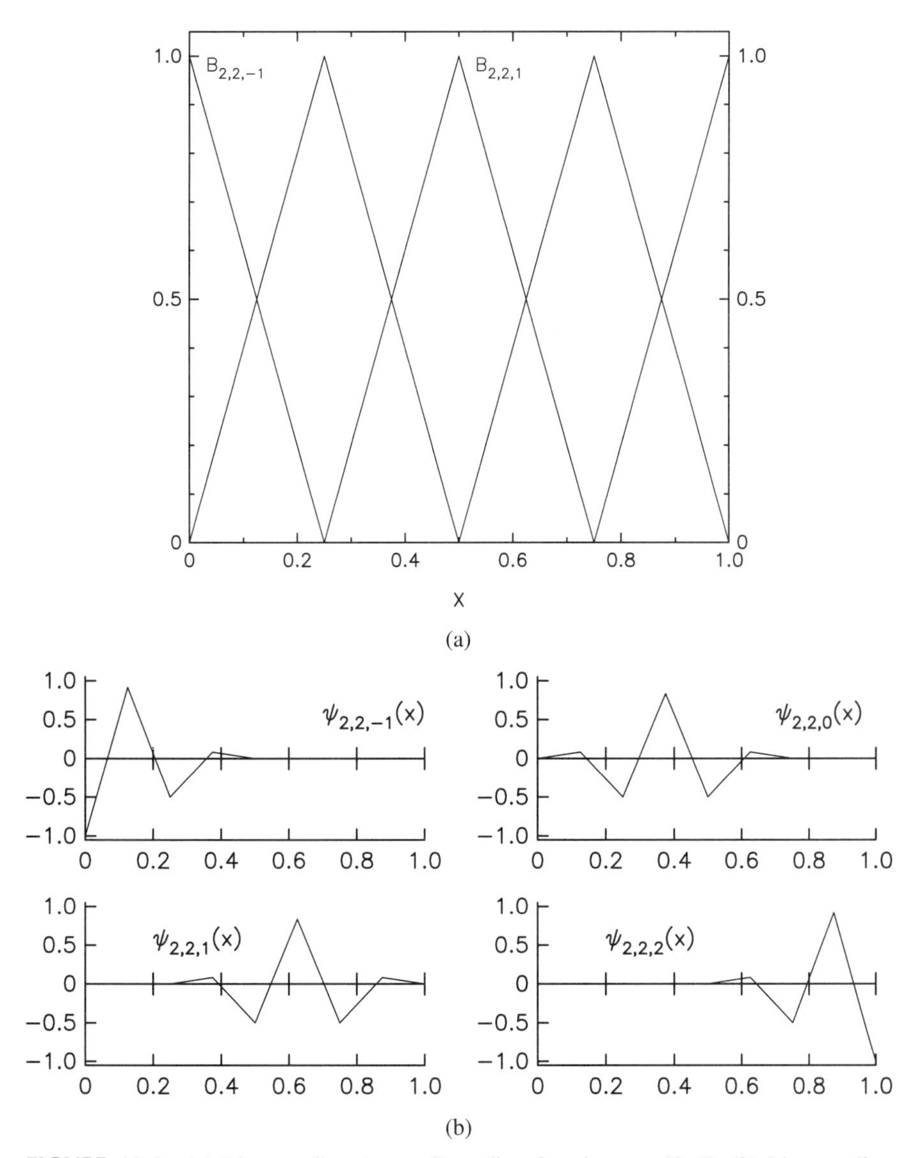

FIGURE 10.4 (*a*) Linear spline ($m = 2$) scaling functions on [0, 1]; (*b*) Linear spline wavelets on [0, 1]. The subscripts indicate the order of spline (m), scale (s), and the position (k), respectively. (Reprinted with permission from [23], copyright © 1995 by IEEE.)

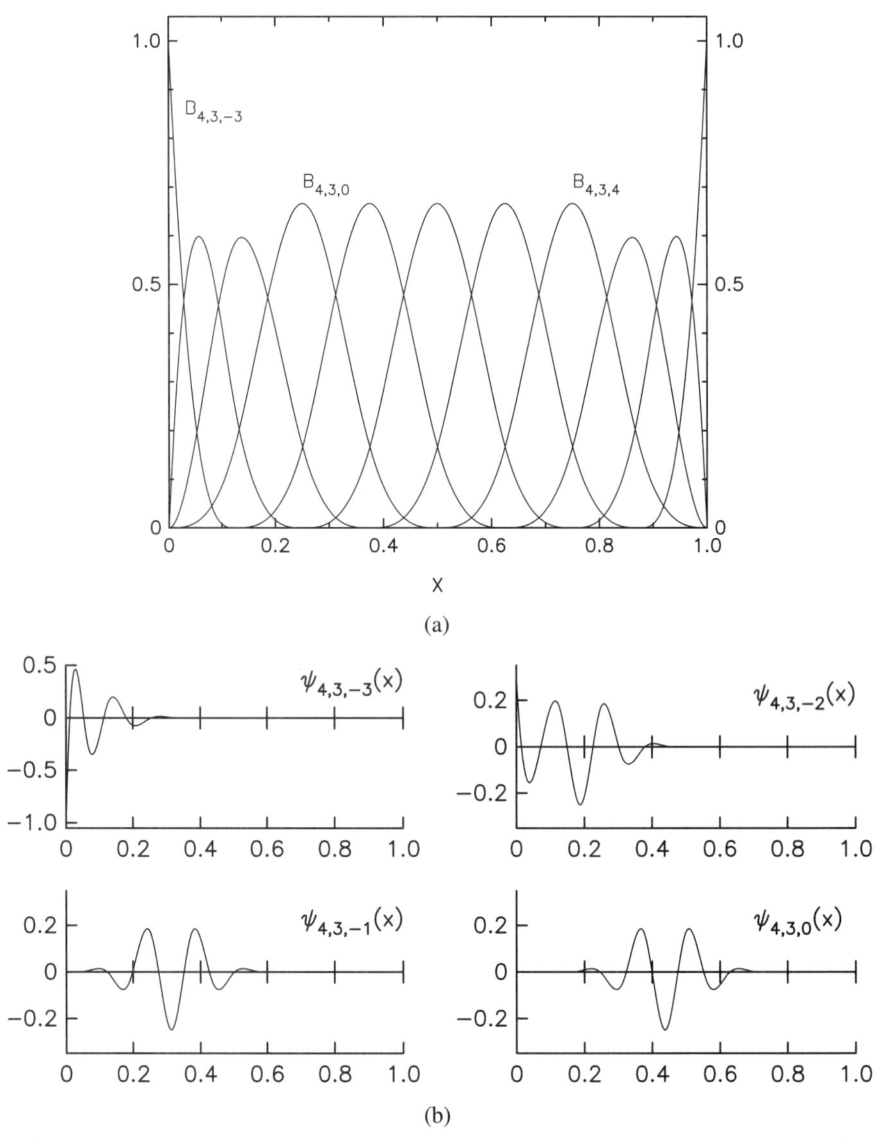

FIGURE 10.5 (*a*) Cubic spline ($m = 4$) scaling functions on [0, 1]; (*b*) Cubic spline wavelets on [0, 1]. The subscripts indicate the order of spline (m), scale (s), and the position (k), respectively. (Reprinted with permission from [23], copyright © 1995 by IEEE.)

TABLE 10.1 Relative Magnitudes of the Largest and Smallest Elements of the Matrix for Conventional and Wavelet MoM ($a = 0.1\lambda_0$)

	Conventional MoM	Wavelet MoM ($m = 2$)	Wavelet MoM ($m = 4$)
A_{max}	5.377	0.750	0.216
A_{min}	1.682	7.684×10^{-8}	8.585×10^{-13}
Ratio	3.400	9.761×10^{6}	2.516×10^{11}

Source: Reprinted with permission from [23], copyright © 1995 by IEEE.

In the above, f_δ represents the solution obtained from (10.23) when the elements whose magnitudes are smaller than δA_{max} have been set to zero. Similarly, N_δ is the total number of elements left after thresholding. Clearly, $f_0(x) = f(x)$ and $N_0 = N^2$, where N is the number of unknowns. If we use the intervallic wavelets of Section 10.4 in solving (10.1), then number of unknowns (N) in (10.23), interestingly, does not depend on s_0. This number N is

$$N = 2^{s_u+1} + m - 1. \tag{10.64}$$

Table 10.1 gives an idea of the relative magnitudes of the largest and smallest elements in the matrix for conventional and wavelet MoM. As expected, because of their higher vanishing moment property, cubic spline wavelets give the higher A_{max}/A_{min} ratio.

With the assumption that the $\left[A_{\phi,\phi}\right]$ part of the matrix is unaffected by the thresholding operation, a fairly reasonable assumption, it can be shown that

$$S_\delta \leq \left[1 - \frac{1}{N} - \frac{(2^{s_0} + m - 1)(2^{s_0} + m - 2)}{N^2}\right] \times 100\,, \tag{10.65}$$

where N is given by (10.64).

As mentioned before [see (10.64)], the total number of unknowns is independent of s_0, the lowest level of discretization. However, it is clear from (10.65) that the upper limit of S_δ increases with the decreasing values of s_0. Therefore, it is better to choose $s_0 = \lceil \log_2(2m-1) \rceil$, where $\lceil x \rceil$ represents the smallest integer that is greater than or equal to x.

10.6 NUMERICAL EXAMPLES

In this section we present some numerical examples for the scattering problems described in Section 10.1. Numerical results for strip and wire problems can be found in [14]. For more applications of wavelets to electromagnetic problems, readers may refer to [22].

The matrix equation (10.23) is solved for a circular cylindrical surface [23]. Figures 10.6 and 10.7 show the surface current distribution using linear and cubic

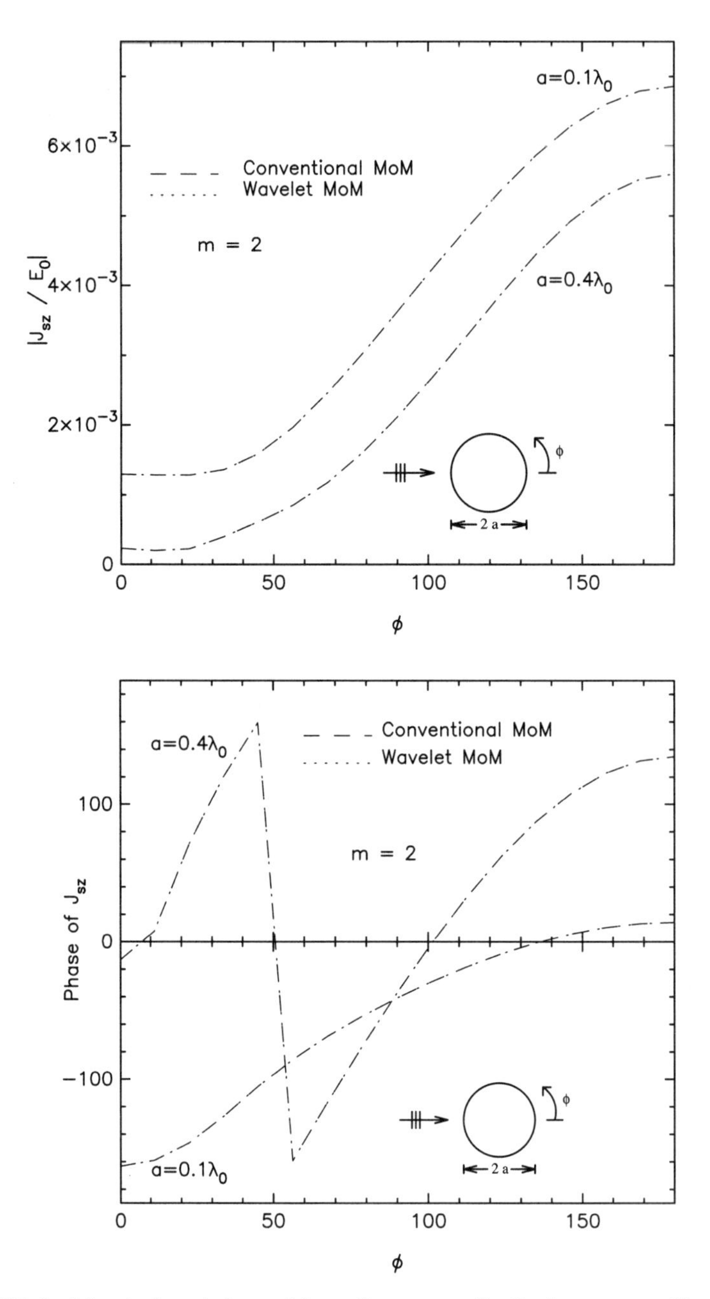

FIGURE 10.6 Magnitude and phase of the surface current distribution on a metallic cylinder using linear spline wavelet MoM and conventional MoM. (Reprinted with permission from [23], copyright © 1995 by IEEE.)

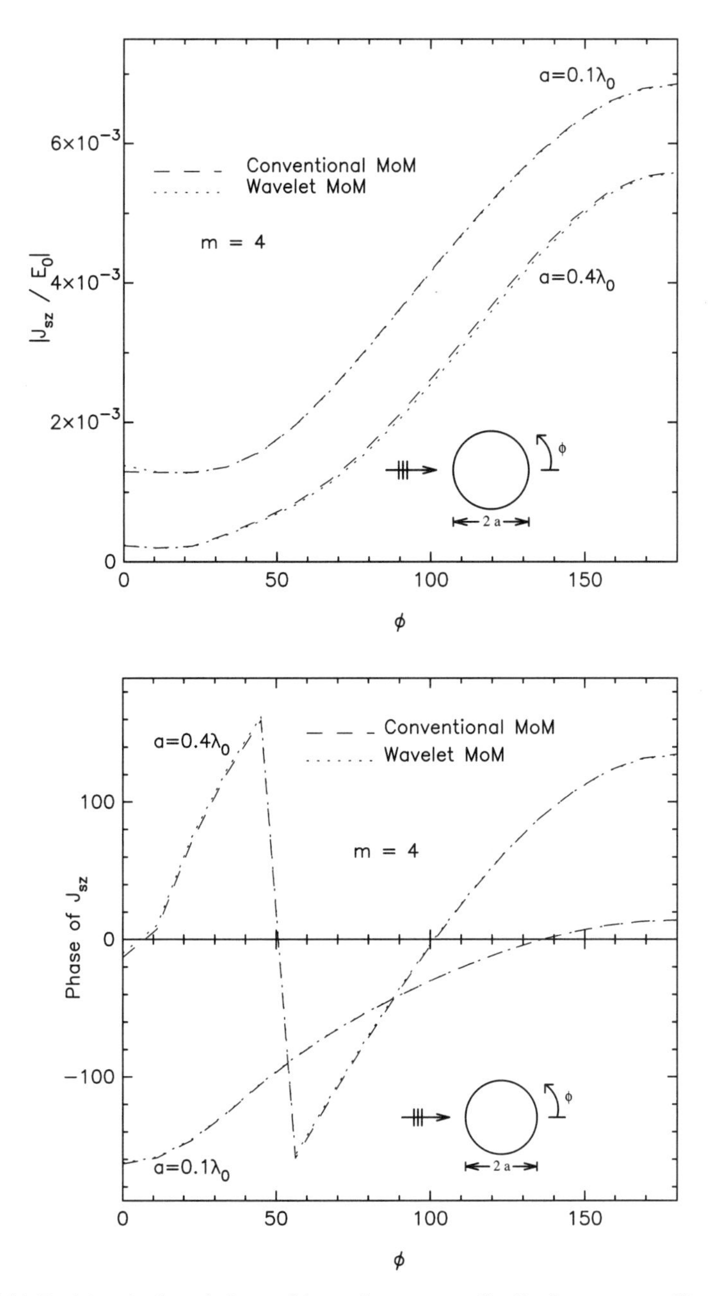

FIGURE 10.7 Magnitude and phase of the surface current distribution on a metallic cylinder using cubic spline wavelet MoM and conventional MoM.

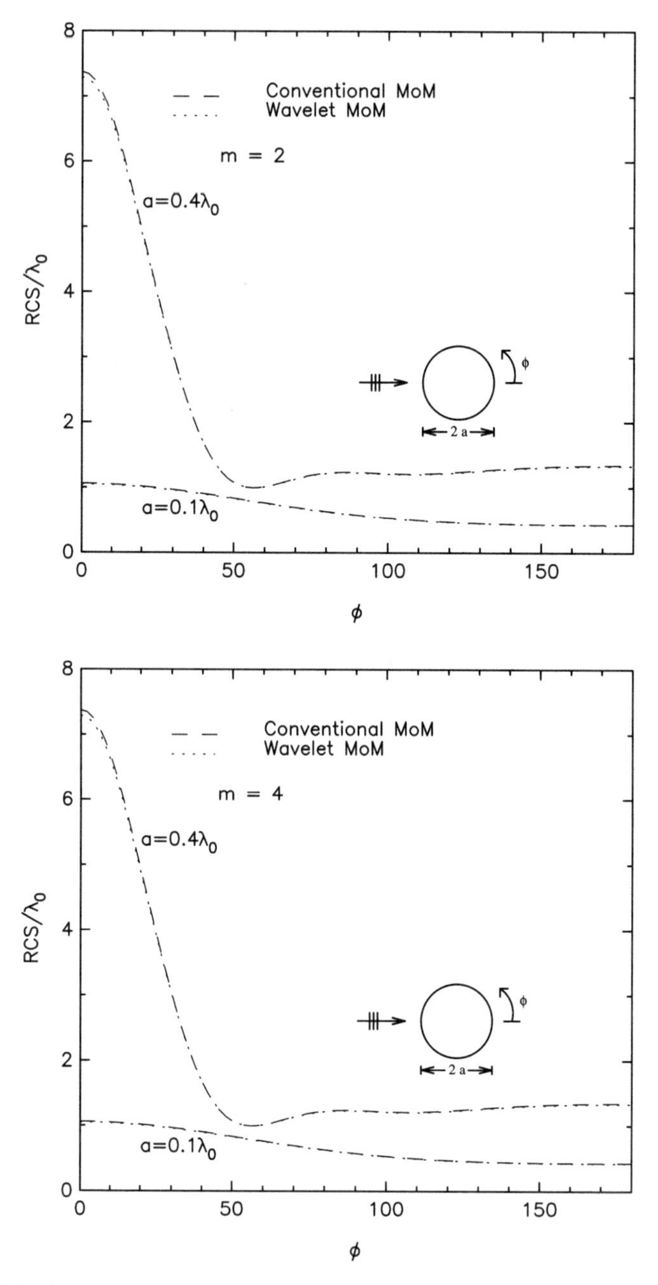

FIGURE 10.8 Radar cross section of a metallic cylinder computed using linear ($m = 2$) and cubic ($m = 4$) spline wavelet MoM and conventional MoM.

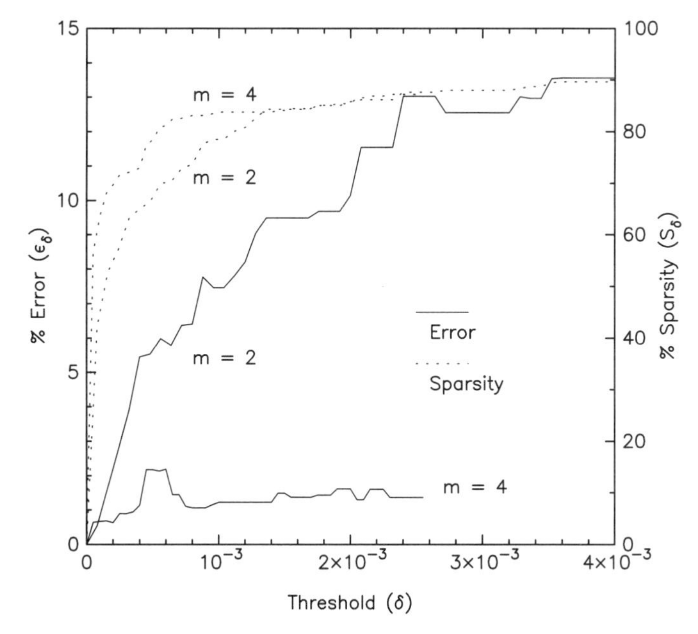

FIGURE 10.9 Error in the solution of the surface current distribution as a function of the threshold parameter δ. (Reprinted with permission from [23], copyright © 1995 by IEEE.)

splines, respectively, for different sizes of the cylinder. The wavelet MoM results are compared with the conventional MoM results. To obtain the conventional MoM results, we have used triangular functions for both expanding the unknown current distribution and testing the resultant equation. The conventional MoM results have been verified with the series solution [32]. Figure 10.8 gives the radar cross section for linear and cubic spline cases. The results of the conventional and wavelet MoM agree very well.

The effects of δ on the error in the solution and the sparsity of the matrix are shown in Figure 10.9. The magnitude of error increases rapidly for the linear spline case. Figure 10.10 shows a typical matrix obtained by applying the conventional MoM. A darker color on an element indicates a larger magnitude. The matrix elements with $\delta = 0.0002$ for the linear spline case are shown in Figure 10.11. In Figure 10.12 we present the thresholded matrix ($\delta = 0.0025$) for the cubic spline case. The $\left[A_{\psi,\psi}\right]$ part of the matrix is almost diagonalized. Figure 10.13 gives an idea of the point-wise error in the solution for linear and cubic spline cases.

It is worth pointing out here that regardless of the size of the matrix, only 5×5 in the case of the linear spline and 11×11 in the case of the cubic splines (see the top-left corners of Figures 10.11 and 10.12) will remain unaffected by thresholding; a significant number of the remaining elements can be set to zero without causing much error in the solution.

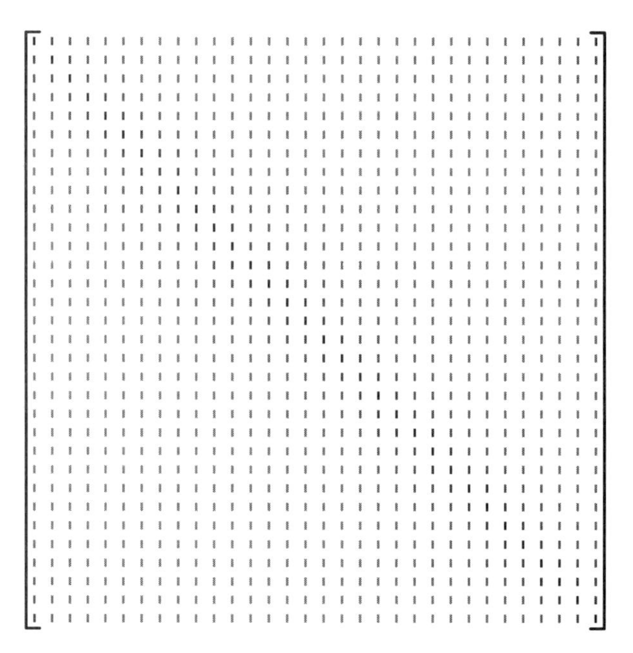

FIGURE 10.10 Typical gray-scale plot of the matrix elements obtained using conventional MoM. The darker color represents larger magnitude.

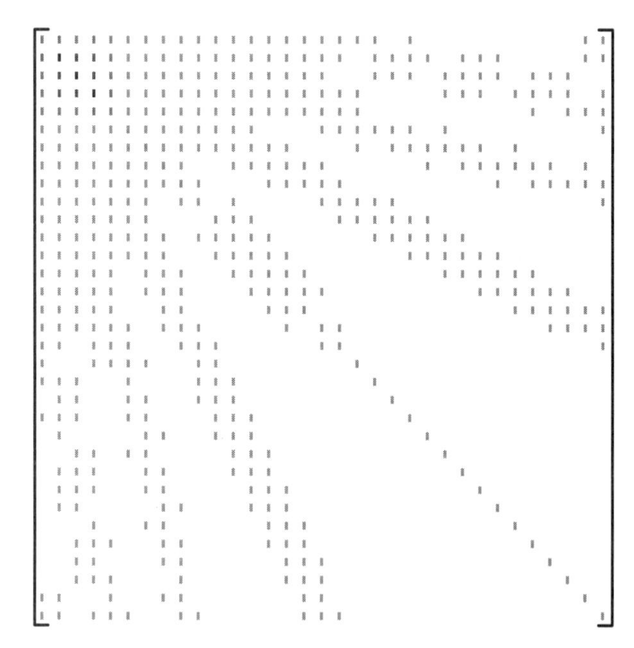

FIGURE 10.11 Typical gray-scale plot of the matrix elements obtained using linear wavelet MoM. The darker color represents larger magnitude. (Reprinted with permission from [23], copyright © 1995 by IEEE.)

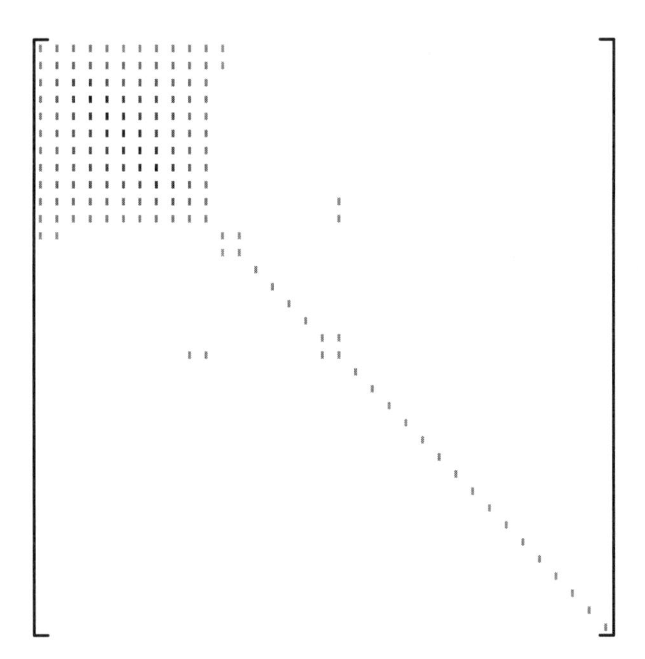

FIGURE 10.12 Typical gray-scale plot of the matrix elements obtained using cubic wavelet MoM. The darker color represents larger magnitude. (Reprinted with permission from [23], copyright © 1995 by IEEE.)

10.7 SEMIORTHOGONAL VERSUS ORTHOGONAL WAVELETS

Both semiorthogonal and orthogonal wavelets have been used for solving integral equations. A comparative study of their advantages and disadvantages has been reported in [14]. The orthonormal wavelet transformation, because of its unitary similar property, preserves the condition number (κ) of the original impedance matrix A; semiorthogonal wavelets do not. Consequently, the transformed matrix equation may require more iterations to converge to the desired solution. Some preliminary results comparing the condition number of matrices for different cases are given in Table 10.2 [17].

In applying wavelets directly to solve integral equations, one of the most attractive features of semiorthogonal wavelets is that closed-form expressions are available for such wavelets. Most of the continuous orthonormal wavelets cannot be written in closed form. One thing to keep in mind is that, unlike signal processing applications where one usually deals with a discretized signal and decomposition and reconstruction sequences, in the boundary value problem we often have to compute the wavelet and scaling function values at any given point. For a strip and thin wire case, a comparison of the computation time and sparsity is summarized in Tables 10.3 and 10.4 [14].

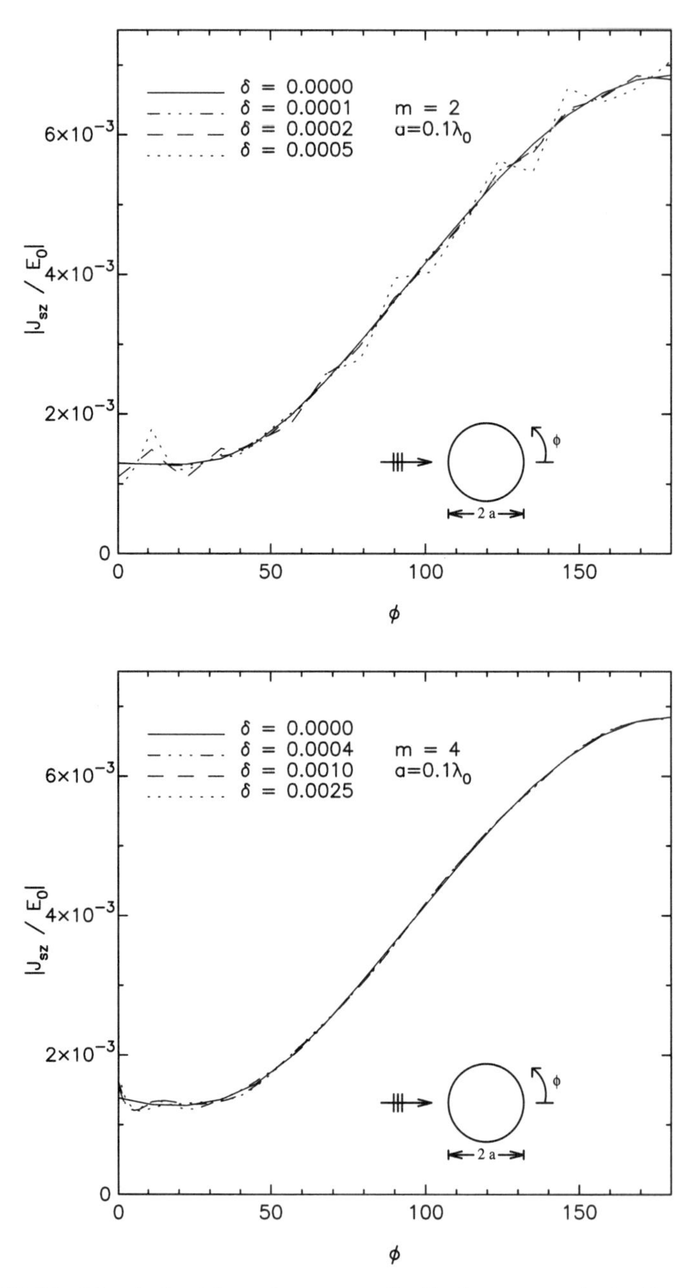

FIGURE 10.13 Magnitude of the surface current distribution computed using linear ($m = 2$) and cubic ($m = 4$) spline wavelet MoM for different values of the threshold parameter δ. (Reprinted with permission from [23], copyright © 1995 by IEEE.)

TABLE 10.2 Effect of Wavelet Transform Using Semiorthogonal and Orthonormal Wavelets on the Condition Number of the Impedance Matrix. Original Impedance Matrix Is Generated Using Pulse Basis Functions

Basis and Transform	No. of Unknowns	Octave Level	δ	S_δ	ϵ_δ	Condition No. κ Before Threshold	Condition No. κ After Threshold
Pulse and none	64	NA	NA	0.0	2.6×10^{-5}	14.7	—
Pulse and semiorthogonal	64	1	7.2×10^{-2}	46.8	0.70	16.7	16.4
Pulse and orthonormal	64	1	7.5×10^{-3}	59.7	0.87	14.7	14.5

TABLE 10.3 Comparison of CPU Time per Matrix Element for Spline, Semiorthogonal, and Orthonormal Basis Function

	Wire	Plate
Spline	0.12s	0.25×10^{-3}s
Semiorthogonal wavelet	0.49s	0.19s
Orthonormal wavelet	4.79s	4.19s

Source: Reprinted with permission from [14], copyright © 1997 by IEEE.

As discussed in previous chapters, semiorthogonal wavelets are symmetric, hence have generalized linear phase, an important factor in function reconstruction. It is well known [26, Sec. 8.1] that symmetric or antisymmetric, real-valued, continuous, and compactly supported orthonormal scaling functions and wavelets do not exist. Finally, as discussed previously, in using wavelets to solve spectral domain problems,

TABLE 10.4 Comparison of Percentage Sparsity (S_δ) and Percentage Relative Error (ϵ_δ) for Semiorthogonal and Orthonormal Wavelet Impedance Matrices as a Function of Threshold Parameter (δ)

Scatterer/ Octave Levels	Number of Unknowns: Semi-orthogonal	Number of Unknowns: Ortho-normal	Threshold δ	Sparsity S_δ: Semi-orthogonal	Sparsity S_δ: Ortho-normal	Relative Error ϵ_δ: Semi-orthogonal	Relative Error ϵ_δ: Ortho-normal
Wire /	29	33	1×10^{-6}	34.5	24.4	3.4×10^{-3}	4.3×10^{-3}
$j = 4$			5×10^{-6}	48.1	34.3	3.9	1.3×10^{-3}
			1×10^{-5}	51.1	36.5	16.5	5.5×10^{-2}
Plate /	33	33	1×10^{-4}	51.6	28.1	1×10^{-4}	0.7
$j = 2, 3, 4$			5×10^{-4}	69.7	45.9	4.7	5.2
			1×10^{-3}	82.4	50.9	5.8	10.0

Source: Reprinted with permission from [14], copyright © 1997 by IEEE.

we need to look at the time–frequency window product of the basis. Semiorthogonal wavelets approach the optimal value of the time–frequency product, which at 0.5 is very fast. For instance, this value for the cubic spline wavelet is 0.505. It has been shown [33] that this product approaches to ∞ with the increase in smoothness of orthonormal wavelets.

10.8 DIFFERENTIAL EQUATIONS

An ordinary differential equation (ODE) can be represented as

$$Lf(x) = g(x), \ x \in [0, 1] \tag{10.66}$$

with

$$L = \sum_{j=0}^{m} a_j(x) \frac{d^j}{dx^j} \tag{10.67}$$

and some appropriate boundary conditions. If the coefficients $\{a_j\}$ are independent of x, then the solution can be obtained easily via a Fourier method. However in ODE case, with nonconstant coefficients, and in PDEs, we generally use finite element or finite difference type methods.

In the traditional finite element method (FEM), local bases are used to represent the unknown function and the solution is obtained by Galerkin's method, similar to the approach described in previous sections. For the differential operator, we get sparse and banded stiffness matrices that are generally solved using iterative techniques, the Jacobi method for instance.

One of the disadvantages of conventional FEM is that the condition number of the stiffness matrix grows as $O(h^{-2})$, where h is the discretization step. As a result, the convergence of the iterative technique becomes slow and the solution becomes sensitive to small perturbations in the matrix elements. If we study how the error decreases with iteration in iterative techniques, such as the Jacobi method, we find that the error decreases rapidly for the first few iterations. After that, the rate at which the error decreases slows down [34, pp. 18–21]. Such methods are also called *high-frequency methods* since these iterative procedures have a smoothing effect on the high-frequency portion of the error. Once the high-frequency portion of the error is eliminated, convergence becomes quite slow. After the first few iterations, if we could re-discretize the domain with coarser grids and thereby go to lower frequency, the convergence rate would be accelerated. This leads us to a multigrid-type method.

Multigrid or hierarchical methods have been proposed to overcome the difficulties associated with the conventional methods [34–50]. In this technique one performs a few iterations of the smoothing method (Jacobi-type), and the intermediate solution and the operator are projected to a coarse grid. The problem is solved at the coarse grid and by interpolation one goes back to the finer grids. By going back and forth between finer and coarse grids, the convergence can be accelerated. It has been shown for elliptic PDEs, that for wavelet-based multilevel methods, the condition number is independent of the discretization step, that is, $\kappa = O(1)$ [45]. The multigrid method is too involved to be discussed in this chapter. Readers are encouraged to look at the references provided at the end of the chapter.

Multiresolution aspects of wavelets have also been applied in evolution equations [49, 50]. In evolution problems, the space and time discretization are interrelated to gain a stable numerical scheme. The time-step must be determined from the smallest space discretization. This makes the computation quite complex. A space-time adaptive method has been introduced in [50] where wavelets have been used to adjust the space-time discretization steps locally.

10.9 EXPRESSIONS FOR SPLINES AND WAVELETS

We provide formulas for the scaling functions $B_{m,0,k}(x)$ and the wavelets $\psi_{m,k,0}(x)$ for $k = -m+1, \ldots, 0$ and $m = 2, 4$. Formulas at the scale s_0 can be obtained by replacing x by $2^{s_0}x$ and scaling the intervals accordingly.

$$B_{m,-m+1,0}(x) = \begin{cases} (1-x)^{m-1}, & x \in [0, 1) \\ 0, & \text{otherwise} \end{cases} \tag{10.68}$$

$$B_{2,0,0}(x) = \begin{cases} x, & x \in [0, 1) \\ 2-x, & x \in [1, 2) \\ 0, & \text{otherwise.} \end{cases} \tag{10.69}$$

Formulas for cubic spline scaling functions except for $B_{4,-3,0}$ are given in Table 10.5. Formulas for $B_{2,-1,0}$ and $B_{4,-3,0}$ can be obtained from (10.68). Tables 10.6 to 10.8 contain the formulas for the wavelets. Functions are zero outside the intervals given in the tables. An empty entry indicates that the function is zero in the interval.

It should be pointed out that the scaling functions and wavelets described in this book can also be computed from their Bernstein polynomial representations [51, Sec. 4.4; 52, pp. 11–13]; however, the formulas presented here are direct and easy to implement.

TABLE 10.5 Cubic Spline Scaling Functions $B_{4,k,0}$ for Different Values of k [$6 \times B_{4,k,0}(x) = \sum_{i=0}^{3} a_i x^i$]: a_0, a_1, a_2, a_3 for Various Intervals

Interval	$k=-2$	$k=-1$	$k=0$
[0, 1)	0, 18, −27, 21/2	0, 0, 9, −11/2	0, 0, 0, 1
[1, 2)	12, −18, 9, −3/2	−9, 27, −18, 7/2	4, −12, 12, −3
[2, 3)		27, −27, 9, −1	−44, 60, −24, 3
[3, 4)			64, −48, 12, −1

Source: Reprinted with permission from [23], copyright © 1995 by IEEE.

TABLE 10.6 Linear Spline Wavelet $\psi_{2,k,0}$ for Different Values of k [$6 \times \psi_{2,k,0}(x) = a_0 + a_1 x$]: a_0, a_1 for Different Intervals

Interval	$k = -1$	$k = 0$
[0.0, 0.5)	−6, 23	0, 1
[0.5, 1.0)	14, −17	4, −7
[1.0, 1.5)	−10, 7	−19, 16
[1.5, 2.0)	2, −1	29, −16
[2.0, 2.5)		−17, 7
[2.5, 3.0)		3, −1

Source: Reprinted with permission from [23], copyright © 1995 by IEEE.

TABLE 10.7 Cubic Spline Wavelet $\psi_{4,k,0}$ for $k = -3, -2$ [$5040 \times \psi_{4,k,0}(x) = \sum_{i=0}^{3} a_i x^i$]: a_0, a_1, a_2, a_3 for Various Intervals

Interval	$k = -3$	$k = -2$
[0.0, 0.5)	−5,097.9058, 75,122.08345, −230,324.8918, 191,927.6771	1,529.24008, −17,404.65853, 39,663.39526, −24,328.27397
[0.5, 1.0)	25,795.06384, −110,235.7345, 140,390.7438, −55,216.07994	96.3035852, −8,807.039551, 22,468.15735, −12,864.78201
[1.0, 1.5)	−53,062.53069, 126,337.0492, −96,182.03978, 23,641.5146	−37,655.11514, 104,447.2167, −90,786.09884, 24,886.63674
[1.5, 2.0)	56,268.26703, −92,324.54624, 49,592.35723, −8,752.795836	132,907.7898, −236,678.5931, 136,631.1078, −25,650.52030
[2.0, 2.5)	−31,922.33501, 39,961.3568, −16,550.59433, 2,271.029421	−212,369.3156, 281,237.0648, −122,326.7213, 17,509.11789
[2.5, 3.0)	8,912.77397, −9,040.773971, 3,050.25799, −342.4175544	184,514.4305, −195,023.4306, 68,177.47685, −7,891.441873
[3.0, 3.5)	−904, 776, −222, 127/6	−88,440.5, 77,931.5, −22,807.5, 2218
[3.5, 4.0)	32/3, −8, 2, −1/6	21,319.5, −16,148.5, 4,072.5, −342
[4.0, 4.5)		−11,539/6, 1,283.5, −285.5, 127/6
[4.5, 5.0)		125/6, −12.5, 2.5, −1/6

Source: Reprinted with permission from [23], copyright © 1995 by IEEE.

TABLE 10.8 Cubic Spline Wavelet $\psi_{4,k,0}$ for $k = -1, 0$. $[5040 \times \psi_{4,k,0}(x) = \sum_{i=0}^{3} a_i x^i]$: a_0, a_1, a_2, a_3 for Various Intervals

Interval	$k = -1$	$k = 0$
[0.0, 0.5)	−11.2618185, 68.79311672, −242.2663844, 499.28435	0, 0, 0, 1/6
[0.5, 1.0)	330.8868107, −1,984.098658, 3,863.517164, −2,237.904686	8/3, −16, 32, −127/6
[1.0, 1.5)	−9,802.095725, 28,414.84895, −26,535.43044, 7,895.077856	−360.5, 1,073.5, −1,057.5, 342
[1.5, 2.0)	75,963.58449, −143,116.5114, 87,818.80985, −17,516.97555	8,279.5, −16,206.5, 10,462.5, −2,218
[2.0, 2.5)	−270,337.7867, 376,335.5451, −171,907.2184, 25,770.69585	−72,596.5, 105,107.5, −50,194.5, 7,891.5
[2.5, 3.0)	534,996.0062, −590,065.0062, 214,653.0021, −25,770.66691	324,403.5, −371,292.5, 140,365.5, −17,516.5
[3.0, 3.5)	−633,757.5, 578,688.5, −174,931.5, 17,516.5	−844,350, 797,461, −249,219, 77,312/3
[3.5, 4.0)	455,610.5, −355,055.5, 91,852.5, −7,891.5	4,096,454/3, −1,096,683, 291,965, −77,312/3
[4.0, 4.5)	−191,397.5, 130,200.5, −29,461.5, 2,218	−1,404,894, 981,101, −227,481, 17,516.5
[4.5, 5.0)	41,882.5, −25,319.5, 5,098.5, −342	910,410, −562,435, 115,527, −7,891.5
[5.0, 5.5)	−10,540/3, 1,918, −349, 127/6	−353,277.5, 195,777.5, −36,115.5, 2218
[5.5, 6.0)	36, −18, 3, −1/6	72,642.5, −36,542.5, 6,124.5, −342
[6.0, 6.5)		−5,801.5, 2,679.5, −412.5, 127/6
[6.5, 7.0)		343/6, −24.5, 3.5, −1/6

Source: Reprinted with permission from [23], copyright © 1995 by IEEE.

REFERENCES

1. G. B. Arfken and H. J. Weber, *Mathematical Methods for Physicsts.* San Diego: Academic Press, 1995.
2. G. M. Wing, *A Primer on Integral Equations of the First Kind.* Philadelphia: SIAM, 1991.
3. B. K. Alpert, G. Beylkin, R. Coifman, and V. Rokhlin, "Wavelet-like bases for the fast solution of second-kind integral equations," *SIAM J. Sci. Comput.* **14**, pp. 159–184, 1993.

4. J. Mandel, "On multi-level iterative methods for integral equations of the second kind and related problems," *Numer. Math.*, **46**, pp. 147–157, 1985.
5. J. C. Goswami, "An application of wavelet bases in the spectral domain analysis of transmission line discontinuities," *Int. J. Numer. Model.*, **11**, pp. 41–54, 1998.
6. J. H. Richmond, "Digital solutions of the rigorous equations for scattering problems," *Proc. IEEE*, **53**, pp. 796–804, August 1965.
7. R. F. Harrington, *Field Computation by Moment Methods.* New York: IEEE Press, 1992.
8. G. Beylkin, R. Coifman, and V. Rokhlin, "Fast wavelet transform and numerical algorithms I," *Commun. Pure Appl. Math.*, **44**, pp. 141–183, 1991.
9. R. L. Wagner, P. Otto, and W. C. Chew, "Fast waveguide mode computation using wavelet-like basis functions," *IEEE Microwave Guided Wave Lett.*, **3**, pp. 208–210, 1993.
10. H. Kim and H. Ling, "On the application of fast wavelet transform to the integral-equation of electromagnetic scattering problems," *Microwave Opt. Technol. Lett.*, **6**, pp. 168–173, March 1993.
11. B. Z. Steinberg and Y. Leviatan, "On the use of wavelet expansions in method of moments," *IEEE Trans. Antennas Propag.*, **41**, pp. 610–619, 1993.
12. K. Sabetfakhri and L. P. B. Katehi, "Analysis of integrated millimeter-wave and submillimeter-wave waveguides using orthonormal wavelet expansions," *IEEE Trans. Microwave Theory Tech.*, **42**, pp. 2412–2422, 1994.
13. B. Z. Steinberg, "A multiresolution theory of scattering and diffraction," *Wave Motion*, **19**, pp. 213–232, 1994.
14. R. D. Nevels, J. C. Goswami, and H. Tehrani, "Semi-orthogonal versus orthogonal wavelet basis sets for solving integral equations," *IEEE Trans. Antennas Propag.*, **45**, pp. 1332–1339, 1997.
15. Z. Xiang and Y. Lu, "An effective wavelet matrix transform approach for efficient solutions of electromagnetic integral equations," *IEEE Trans. Antennas Propag.*, **45**, pp. 1332–1339, 1997.
16. Z. Baharav and Y. Leviatan, "Impedance matrix compression (IMC) using iteratively selected wavelet basis," *IEEE Trans. Antennas Propag.*, **46**, pp. 226–233, 1997.
17. J. C. Goswami, R. E. Miller, and R. D. Nevels, "Wavelet methods for solving integral and differential equations," in *Encyclopedia of Electrical and Electronics Engineering*, J. H. Webster (ed.). New York: John Wiley & Sons, to appear in February 1999.
18. B. Z. Steinberg and Y. Leviatan, "Periodic wavelet expansions for analysis of scattering from metallic cylinders," *IEEE Antennas Propag. Soc. Symp.*, 20–23, June 1994.
19. G. W. Pan and X. Zhu, "The application of fast adaptive wavelet expansion method in the computation of parameter matrices of multiple lossy transmission lines," *IEEE Antennas Propagat. Soc. Symp.*, 29–32, June 1994.
20. G. Wang, G. W. Pan, and B. K. Gilbert, "A hybrid wavelet expansion and boundary element analysis for multiconductor transmission lines in multilayered dielectric media," *IEEE Trans. Microwave Theory Tech.*, **43**, pp. 664–675, March 1995.
21. W. L. Golik, "Wavelet packets for fast solution of electromagnetic integral equations," *IEEE Trans. Antennas Propag.*, **46**, pp. 618–624, 1998.
22. Special Issue on Wavelets in Electromagnetics, *Int. J. Numer. Model. Electron. Networks, Devices Fields*, **11**, 1998.

23. J. C. Goswami, A. K. Chan, and C. K. Chui, "On solving first-kind integral equations using wavelets on a bounded interval," *IEEE Trans. Antennas Propag.*, **43**, pp. 614–622, 1995.

24. J. C. Goswami, A. K. Chan, and C. K. Chui, "Spectral domain analysis of single and couple microstrip open discontinuities with anisotropic substrates," *IEEE Trans. Microwave Theory Tech.*, **44**, pp. 1174–1178, 1996.

25. J. C. Goswami and R. Mittra, "Application of FDTD in studying the end effects of slot line and coplanar waveguide with anisotropic substrate," *IEEE Trans. Microwave Theory Tech.*, **45**, pp. 1653-1657, 1997.

26. I. Daubechies, *Ten Lectures on Wavelets*. CBMS-NSF Ser. Appl. Math. 61, Philadelphia: SIAM, 1992.

27. C. K. Chui and E. Quak, "Wavelets on a bounded interval," in *Numerical Mathematics Approximation Theory*, Vol. 9, D. Braess and L. L. Schumaker (eds.). Basel: Birkhäuser Verlag, 1992, pp. 53–75,

28. E. Quak and N. Weyrich, "Decomposition and reconstruction algorithms for spline wavelets on a bounded interval," *Appl. Compu. Harmon. Anal.*, **1**, (3), pp. 217–231, June 1994.

29. A. Cohen, I. Daubechies, and P. Vial, "Wavelets on the interval and fast wavelet transform," *Appl. Comput. Harmon. Anal.*, **1**, 1, pp. 54–81, December 1993.

30. P. Auscher, "Wavelets with boundary conditions on the interval," in *Wavelets: A Tutorial in Theory and Applications*, C. K. Chui (ed.). San Diego, Calif.: Academic Press, 1992, pp. 217–236.

31. C. de Boor, *A Practical Guide to Splines*. New York: Springer-Verlag, 1978.

32. R. F. Harrington, *Time-Harmonic Electromagnetic Fields*. New York: McGraw-Hill Bosh Company, 1961.

33. C. K. Chui and J. Z. Wang, "High-order orthonormal scaling functions and wavelets give poor time-frequency localization," *Fourier Anal. Appl.*, pp. 415–426, 1996.

34. W. Hackbusch, *Multigrid Methods and Applications*. New York: Springer-Verlag, 1985.

35. A. Brandt, "Multi-level adaptive solutions to boundary value problems," *Math. Comput.*, **31**, pp. 330–390, 1977.

36. W. L. Briggs, *A Multigrid Tutorial*. Philadelphia: SIAM, 1987.

37. S. Dahlke and I. Weinreich, 'Wavelet-Galerkin methods: An adapted biorthogonal wavelet basis,' *Constr. Approx.*, **9**, pp. 237–262, 1993.

38. W. Dahmen, A. J. Kurdila, and P. Oswald (eds.), *Multiscale Wavelet Methods for Partial Differential Equations*. San Diego, Calif.: Academic Press, 1997.

39. H. Yserentant, "On the multi-level splitting of finite element spaces," *Numer. Math.*, **49**, pp. 379–412, 1986.

40. J. Liandrat and Ph. Tchamitchian, "Resolution of the 1D regularized Burgers equation using a spatial wavelet approximation," *NASA Report, ICASE*, No. 90–83, December 1990.

41. R. Glowinski, W. M. Lawton, M. Ravachol, and E. Tenenbaum, "Wavelet solution of linear and nonlinear elliptic, parabolic, and hyperbolic problems in one space dimension," in *Computing Methods in Applied Sciences and Engineering*, R. Glowinski and A. Lichnewsky (eds.), Philadelphia: SIAM, 1990, pp. 55–120.

42. P. Oswald, "On a hierarchical basis multilevel method with nonconforming P1 elements," *Numer. Math.*, **62**, pp. 189–212, 1992.

43. W. Dahmen and A. Kunoth, “Multilevel preconditioning,” *Numer. Math.*, **63**, pp. 315–344, 1992.
44. P. W. Hemker and H. Schippers, “Multiple grid methods for the solution of Fredholm integral equations of the second kind,” *Math. Comput.*, **36**, January 1981.
45. S. Jaffard, “Wavelet methods for fast resolution of elliptic problems,” *SIAM J. Numer. Anal.* **29**, pp. 965–986, 1992.
46. S. Jaffard and Ph. Laurençot, “Orhonomal wavelets, anlysis of operators, and applications to numerical analysis,” in *Wavelets: A Tutorial in Theory and Applications*, C. K. Chui (ed.). Boston: Academic Press, 1992, pp. 543–601.
47. J. Xu and W. Shann, ‘Galerkin-wavelet methods for two-point boundary value problems,’ *Numer. Math.* **63**, pp. 123–144, 1992.
48. C. Guerrini and M. Piraccini, “Parallel wavelet-Galerkin methods using adapted wavelet packet bases,” in *Wavelets and Multilevel Approximation*, C. K. Chui and L. L. Schumaker (eds.) New Jersey: World Scientific, 1995, pp. 133–142.
49. M. Krumpholz and L. P. B. Katehi, “MRTD: new time-domain schemes based on multiresolution analysis,” *IEEE Trans. Microwave Theory Tech.*, **44**, pp. 555–571, 1996.
50. E. Bacry, S. Mallat, and G. Papanicolaou, “A wavelet based space-time adaptive numerical method for partial differential equations,” *Math. Model. Numer. Anal.*, **26**, pp. 793–834, 1992.
51. C. K. Chui, *An Introduction to Wavelets.* San Diego, Calif.: Academic Press, 1992.
52. C. K. Chui, *Multivariate Splines.* CBMS-NSF Ser. Appl. Math. 54. Philadelphia: SIAM, 1988.

Index

Acoustic signals, 221
Adaptive resonance theory, 248
Admissibility condition, 69
Algorithms:
  CAD, 244
  computer programs, 184, 258, 263
  decomposition, 149
  FFT-based, 200
  filter bank, 141
  marching, 257
  pyramid decomposition, 210
  reconstruction, 153
  three-dimensional wavelet, 255
  two-dimensional wavelet, 233
  wavelet packet, 210, 212, 234
  zero-tree, 236
Aliasing, 143
Ambiguity function, 83
Analyzing wavelet, 94
Antialiasing condition, 174, 184
Approximation subspace, 148
Arithmetic code, 235
ART, 248
Autocorrelation, 95, 135, 149
Averaging filter, 25

Basis functions, 9
  biorthogonal, 11
  dual, 11
  entire-domain, 272
  global, 272
  Haar, 13
  local, 13, 272
  orthonormal, 11
  Riesz, 13
  semiorthogonal, 11
  stable, 15
  subdomain, 272
Battle–Lemarié wavelet, 123
Bernstein polynomial, 295
Best basis selection, 220
Binary phase-shift key, 252
Binary scales, 187
Binomial theorem, 136
Biorthogonal basis, 11
Biorthogonal decomposition, 93
Biorthogonal filter bank, 174
Biorthogonality condition, 93, 179
Biorthogonal wavelets, 95, 108, 129
  Cohen, Daubechies, and Feaveau, 130
Bit plane coding, 235
Boundary condition, 269
Boundary value problems, 267
Boundary wavelets, 281
Bounded interval, 280
BPSK, 252
B-spline, 45
BVP, 267

CAD algorithm, 244
Cardinal *B*-splines, 112
Causal filter, 176
Causality, 176
Centered integral wavelet transform, 196

Change of bases, 21, 154, 157, 160
Chirp signal, 68, 204
Chui–Wang wavelet, 112
CIWT, 196
Cluster detection, 244
Coding:
  arithmetic, 235
  bit plane, 235
  entropy, 238
  EZW, 238
  Huffman, 235
  image, 235
  lossy, 235
  predictive, 235
  runlength, 235
  wavelet tree, 236
Cohen, Daubechies, and Feaveau wavelet, 130
Coiflets, 137
Comb function, 33
Computer programs:
  B-spline, 106
  1-D algorithm, 184
  2-D algorithm, 258
  iterative method, 139
  short-time Fourier transform, 85
  wavelet, 138
  wavelet packet, 263
  Wigner–Ville distribution, 86
Condition number, 291
Continuous wavelet transform, 67
Conventional bases, 273
Conventional MoM, 268
Convolution, 24, 39, 147
Corner smoothing functions, 120
Correlation function, 81
Crack detection, 202
Critical sampling, 150
Cubic spline, 159, 162
Cut-off frequency, 205
CWT, 67, 150

Danielson and Lanczos, 55
Daubechies, 211
Daubechies' orthogonal filters, 176
Daubechies wavelet, 125
DCT, 245
Decimation, 142, 147
Decomposition relation, 97, 151
Design origin, 183
Determinant, 19
DFT, 55
Differential equations, 294
Differential operator, 272
Differentiating filter, 25
Digital signal processing, 210
Dilation (refinement) equation, 91
Dilation parameters, 69
Direct sum, 91
Direction decomposition, 210
Direct-sum decomposition, 91
Dirichlet kernel, 55
Discontinuity problem, 276
Discrete Fourier basis, 50
Discrete Fourier transform, 54
Discrete Gabor representation, 65
Discrete short-time Fourier transform, 64
Discrete-time Fourier transform, 52
Discrete wavelet transform, 67, 72
Dispersive nature, 205
Doppler, 83
Doppler frequencies, 165
DTFT, 52
Dual basis, 11
Duality, 93
Duality principle, 155
Dual scaling function, 157
Dual spline, 154
Dual wavelet, 93
DWT, 67, 182
Dyadic points, 156, 187

Edge effect, 281
Eigenmatrix, 19
Eigensystem, 19
Eigenvalue, 19
Eigenvalue method, 134
Eigenvectors, 20
Electromagnetic scattering, 268
Elliptic PDEs, 295
Embedded zerotree wavelet, 238
Entire-domain basis, 272
Entropy coder, 238
Error considerations, 282
Euclidean vector space, 7
Euler Frobenius Laurent polynomial, 108, 114, 157
Evolution problems, 295
EZW, 238

False positives, 249
Fast integral transform, 187
Faulty bearing, 221
Feature extraction:
  spectral, 227
  wavelet packets, 227
  wavelets, 226
FEM, 294
FFT, 55, 226

FFT-based algorithm, 200
Filter bank, 141, 180
Filter bank algorithms, 141
Filters:
  averaging, 25
  biorthogonal, 174
  causal, 176
  differentiating, 25
  filter bank, 141, 180
  FIR, 114, 156, 172
  half-band, 172
  HBF, 172
  highpass, 25
  IIR, 114, 172
  lowpass, 25, 102
  noncausal symmetric, 172
  perfect reconstruction, 177
  PR condition, 167, 184
  PR filter bank, 166, 182
  PR requirements, 177
  QMF, 171
  quadrature mirror, 171
  smoothing, 25
  two-channel biorthogonal filter bank, 179
Finer scale resolution, 190
Finer time resolution, 188
Finite element method, 294
Finite-energy function, 90
FIR filters, 114, 156, 172
First-kind integral equation, 268
FIWT, 196
Fourier method, 31
Fourier series, 31
Fourier series kernel, 34
Fourier transform, 35
FPs, 249
Frequency scaling, 38
Frequency shifting, 38, 63
Functional, 5

Gabor transform, 61
Galerkin method, 272, 274, 279
Gaussian function, 42, 103
Generalized linear phase, 171
Generalized self-similarity tree, 244
Gibb's phenomenon, 50
Global basis, 272
Graphical display, 132
GST, 237

Haar, 96
Haar basis, 13
Haar wavelet, 105
Half-band filter, 172
Hard thresholding, 216
Hat function, 84
HBF, 172
Hermitian matrix, 22
Hierarchical methods, 294
Hierarchical wavelet decomposition, 211
High frequency methods, 294
Highpass filter, 25
Histogram thresholding, 248
Huffman code, 235

Identity matrix, 19
IIR filters, 114, 172
Image coding, 235
Image compression, 210, 235
Image visualization, 210
Impedance matrix, 271
Inner product, 8
Integral equations, 268
Integral operator, 272
Integral wavelet transform, 67
Interference suppression, 210, 219
Interoctave approximation subspaces, 194
Interoctave parameter, 190
Interoctave scales, 190
Interpolation, 144, 147
Interpolator, 145
Interpolatory representation, 103
Interpretations, 74
Intervallic wavelets, 281
Inverse, 19
Inverse wavelet transform, 69
Inversion formula, 61
Iteration method, 132
IWT, 67

Knot sequence, 281

Linear function, 197
Linear operator, 272
Linear phase, 171
Linear spaces, 5
Linear spline, 159, 162
Linear transformation, 18
Linearity, 37, 62
Local basis, 13, 272
Logarithmic division, 220
Lossy coding, 235
Lowpass filter, 25, 102

Mammogram, 244
Marching algorithm, 257
Marginal properties, 80

Matrix:
  determinant, 19
  eigenmatrix, 19
  eigensystem, 19
  eigenvalue, 19
  eigenvectors, 20
  Hermitian, 22
  identity, 19
  inverse, 19
  minor, 19
  singular, 19
  sparse, 273
  transpose, 18
  unitary, 23
MCCS, 250
Mean square error, 16
Medical image visualization, 252
Method of moments, 268, 272
Meyer Wavelet, 119
  corner smoothing functions, 120
Microcalcifications, 244
Minor, 19
MoM, 268
Moment property, 200
Moments, 38
Morlet wavelet, 71
Moving average, 24
MRA, 89, 151
MRA space, 103
Multicarrier communication systems, 210, 249
Multivoice per octave, 187
Multigrid, 294
Multiresolution analysis, 89
Multiresolution spaces, 89
Music signal, 204
MVPO, 187

NEG, 238
Neural network, 223
Noise reduction, 210
Noncausal symmetric filter, 172
Nondyadic points, 187
Nonlinear contrast enhancement, 245
Nonlinear time–frequency distributions, 219
Norm, 6
Normalization factor, 155
Normed linear space, 6
Nyquist rate, 47

Octave levels, 191
ODE, 294
OFDM, 250
Order of approximation, 102
Orthogonal, 92, 291
Orthogonal decomposition, 10, 92
Orthogonal frequency division multiplexing, 250
Orthogonal projection, 16, 148
Orthonormal basis, 11
Orthonormalization, 14
Orthonormal scaling function, 118, 212
Orthonormal wavelets, 108, 114

Parseval identity, 157
Parseval's theorem, 40
Partial sum, 35, 49
Partition of unity, 45, 102, 281
Pattern recognition, 221
PDE, 267
Percentage thresholding, 218
Perfect reconstruction, 177
Periodic wavelets, 281
Photonegative x-ray, 244
Piecewise sinusoids, 276
Plane wave, 269
Point matching, 272
Poisson's sum, 35, 43
Polynomial reproducibility, 103
Polyphase domain, 180
Polyphase representation, 180
POS, 238
PR condition, 167, 184
Predictive coding, 235
PR filter bank, 166, 182
Processing domain, 182
Processing goal, 183
Product filter, 175
Projection, 8
PR requirements, 177
PWS, 276
Pyramid decomposition, 210

QMF, 171
Quadratic chirp, 82
Quadratic superposition principle, 81
Quadrature mirror filter, 171

Radar cross section, 270
Radar imaging, 82
RCS, 270
Reconstruction algorithm, 153
Reconstruction process, 164
Rectangular time-scale grid, 193
Region of convergence, 26
Rendering techniques, 256
Riesz basis, 13
Riesz bounds, 14, 114
Riesz lemma, 127

RMS bandwidths, 191
Runlength code, 235

Sampling, 23
Sampling theorem, 35, 46
Scalar product, 8
Scaling function, 91
Scattered electric fields, 269
Second-kind integral equations, 273
Semiorthogonal, 92, 108, 291
Semiorthogonal basis, 11
Semiorthogonal decomposition, 95
Semiorthogonal spline wavelets, 112
Semiorthogonal subspaces, 154, 156
Shannon function, 13
Shannon wavelet, 118
Shift-invariant systems, 24
Short-time Fourier transform, 60
Signal representation, 148
Signature identification, 210, 221
Singular matrix, 19
Slots, 276
Smoothing filter, 25
Smoothing function, 120
Soft thresholding, 217
SOT, 237
Space-time adaptive method, 295
Sparse matrices, 273
Sparsity, 282
Spatial oriented tree, 242
Spectral-domain analysis, 168
Spectral domain methods, 275
Spectral feature extraction, 227
Spectral method, 132
Spline functions, 98
Splines, 45
  computer programs, 106
  cubic, 159, 162
  linear, 159, 162
  properties, 102
Spline wavelets, 160
Splitting method, 211
Stable basis, 15
Stable numerical scheme, 295
Standard wavelet decomposition, 149, 189
STFT, 58, 67
Strang–Fix condition, 102
Strips, 276
Subdomain basis, 272
Surface current distribution, 269
Synthesis transform, 153
Synthesis wavelet, 94

Target detection, 210
Target identification, 210
Testing functions, 272
Thin strip, 270
Thin wire, 270
Three-dimensional wavelets, 255
Thresholding, 205, 214, 275, 282
  hard, 216
  histogram, 248
  percentage, 218
  soft, 217
Time-domain analysis, 176
Time-domain biothogonal condition, 179
Time–frequency analysis, 57
  continuous wavelet transform, 67
  discrete Gabor representation, 65
  discrete short-time Fourier transform, 64
  discrete wavelet transform, 67, 72
  Gabor transform, 61
  integral wavelet transform, 67
  short-time Fourier transform, 60
  time–frequency window, 62
  time–frequency window, 70
  time–frequency window product, 60, 279
  uncertainty principle, 60
  wavelet series, 73
Time–frequency window, 62, 70
Time–frequency window product, 60, 279
Time-scale analysis, 69
Time-scale grid, 187
Time-scale map, 150
Time scaling, 38
Time shifting, 38, 63
Tonal equalizer, 165
Total positivity, 102, 275
TP, 249
Translation, 69
Transmission line, 276
Transmission line discontinuity, 276
Transpose, 18
Transverse magnetic, 269
True positives, 249
Two-channel biorthogonal filter bank, 179
Two-channel perfect reconstruction, 165
Two-dimensional wavelet algorithm, 233
Two-dimensional wavelet packets, 231
Two-dimensional wavelets, 228
Two-scale relations, 96, 153

Uncertainty principle, 60
Unitary matrix, 23
Unitary similar, 273

Vanishing moments, 69, 275, 285
Vector spaces, 7

Waveguide, 205
Wavelet coefficient, 150, 246
Wavelet decomposition, 149, 245
Wavelet feature extraction, 226
Wavelet MoM, 268
Wavelet packet algorithms, 210, 212, 234
Wavelet packet based cross-term deleted
    representation, 220
Wavelet packet feature extraction, 227
Wavelet packets, 84, 152, 211, 226, 252, 280
Wavelet packet tree decomposition, 210
Wavelets:
  Battle–Lemarié, 123
  biorthogonal, 129
  boundary, 281
  Chui–Wang, 112
  Cohen, Daubechies, and Feaveau, 130
  coiflets, 137
  computer programs, 138, 139
  Daubechies, 125
  graphical display, 132
  Haar wavelet, 105
  intervallic, 281
  Meyer, 119
  Morlet, 71
  orthonormal, 108, 114
  periodic, 281
  Shannon, 118
  spline, 112
  three-dimensional, 255
  two-dimensional, 228
Wavelet series, 73
Wavelet subspaces, 154
Wavelet techniques, 273
Wavelet tree coder, 236
Weighting functions, 272
Wideband correlation processing, 187
Wigner–Ville distribution, 76, 83, 219
  auto term, 80, 220
  computer programs, 86
  cross term, 80, 220
  properties, 80
  wavelet packet based cross-term deleted
      representation, 220
  WPCDR, 220
Window function, 58
  center, 59
  radius, 59
  time–frequency window, 62
  time–frequency window product, 60
  uncertainty principle, 60
  width, 59
WPCDR, 220
WS, 73

Zero-tree algorithm, 236
ZTR, 238
z-transform, 25
  inverse z-transform, 28
  region of convergence, 26

WILEY SERIES IN MICROWAVE AND OPTICAL ENGINEERING

KAI CHANG, Editor
Texas A&M University

FIBER-OPTIC COMMUNICATION SYSTEMS, Second Edition • *Govind P. Agrawal*

COHERENT OPTICAL COMMUNICATIONS SYSTEMS • *Silvello Betti, Giancarlo De Marchis and Eugenio Iannone*

HIGH-FREQUENCY ELECTROMAGNETIC TECHNIQUES: RECENT ADVANCES AND APPLICATIONS • *Asoke K. Bhattacharyya*

COMPUTATIONAL METHODS FOR ELECTROMAGNETICS AND MICROWAVES • *Richard C. Booton, Jr.*

MICROWAVE RING CIRCUITS AND ANTENNAS • *Kai Chang*

MICROWAVE SOLID-STATE CIRCUITS AND APPLICATIONS • *Kai Chang*

DIODE LASERS AND PHOTONIC INTEGRATED CIRCUITS • *Larry Coldren and Scott Corzine*

MULTICONDUCTOR TRANSMISSION-LINE STRUCTURES: MODAL ANALYSIS TECHNIQUES • *J. A. Brandão Faria*

PHASED ARRAY-BASED SYSTEMS AND APPLICATIONS • *Nick Fourikis*

FUNDAMENTALS OF MICROWAVE TRANSMISSION LINES • *Jon C. Freeman*

MICROSTRIP CIRCUITS • *Fred Gardiol*

HIGH-SPEED VLSI INTERCONNECTIONS: MODELING, ANALYSIS, AND SIMULATION • *A. K. Goel*

FUNDAMENTALS OF WAVELETS: THEORY, ALGORITHMS, AND APPLICATIONS • *Jaideva C. Goswami and Andrew K. Chan*

PHASED ARRAY ANTENNAS • *R. C. Hansen*

HIGH-FREQUENCY ANALOG INTEGRATED CIRCUIT DESIGN • *Ravender Goyal (ed.)*

MICROWAVE APPROACH TO HIGHLY IRREGULAR FIBER OPTICS • *Huang Hung-Chia*

NONLINEAR OPTICAL COMMUNICATION NETWORKS • *Eugenio Iannone, Francesco Matera, Antonio Mecozzi, and Marina Settembre*

FINITE ELEMENT SOFTWARE FOR MICROWAVE ENGINEERING • *Tatsuo Itoh, Giuseppe Pelosi and Peter P. Silvester (eds.)*

SUPERCONDUCTOR TECHNOLOGY: APPLICATIONS TO MICROWAVE, ELECTRO-OPTICS, ELECTRICAL MACHINES, AND PROPULSION SYSTEMS • *A. R. Jha*

OPTICAL COMPUTING: AN INTRODUCTION • *M. A. Karim and A. S. S. Awwal*

INTRODUCTION TO ELECTROMAGNETIC AND MICROWAVE ENGINEERING • *Paul R. Karmel, Gabriel D. Colef, and Raymond L. Camisa*

MILLIMETER WAVE OPTICAL DIELECTRIC INTEGRATED GUIDES AND CIRCUITS • *Shiban K. Koul*

MICROWAVE DEVICES, CIRCUITS AND THEIR INTERACTION • *Charles A. Lee and G. Conrad Dalman*

ADVANCES IN MICROSTRIP AND PRINTED ANTENNAS • *Kai-Fong Lee and Wei Chen (eds.)*

OPTICAL FILTER DESIGN AND ANALYSIS: A SIGNAL PROCESSING APPROACH • *C. K. Madsen and J. H. Zhao*

OPTOELECTRONIC PACKAGING • *A. R. Mickelson, N. R. Basavanhally, and Y. C. Lee (eds.)*

ANTENNAS FOR RADAR AND COMMUNICATIONS: A POLARIMETRIC APPROACH • *Harold Mott*